Wakefield Press

Exploring Wild Law

The Philosophy of Earth Jurisprudence

Exploring Wild Law

The Philosophy of Earth Jurisprudence

Edited by
Peter Burdon

Wakefield
Press

Wakefield Press
16 Rose Street
Mile End
South Australia 5031
www.wakefieldpress.com.au

First published 2011

Cover design by Stacey Zass, page 12
Typeset by Wakefield Press

National Library of Australia Cataloguing-in-Publication entry

Title:	Exploring wild law: the philosophy of earth jurisprudence / edited by Peter Burdon.
ISBN:	978 1 86254 946 3 (paperback).
Subjects:	Environmental law.
	Natural law.
Other Authors/ Contributors:	Burdon, Peter.
Dewey Number:	341.4

the
Gaia
foundation
www.gaiafoundation.org

Dedicated to the life and memory of

Fr Thomas Berry (1914–2009)

and

Ng'ang'a Thiong'o (1958–2010).

'Wild law is a gathering force beginning to sweep around the world. This excellent compendium is a must for all those who wish to practise it.'

– James Thornton, CEO ClientEarth

'These essays collectively advocate a new legal paradigm of "Earth Jurisprudence", based on recognition of the connections and continuity between our legal systems and the Earth system. The various, expert contributors offer a rich, informed and transformational perspective that builds on past jurisprudence in the form of natural law and embodies more recent conceptions such as deep ecology and bioregionalism. The fundamental and compelling message is, to quote the editor, Peter Burdon, that "we must question the values and legitimacy of any law that surpasses the ecological limits of the environment to satisfy the needs of one species".

This is a most timely and compelling volume, given the urgency of the threats posed to humanity by climate change, loss of biodiversity and ongoing economic and population growth. Its message is of universal importance and will hopefully be received and absorbed by policy-makers around the globe.'

– Professor Rob Fowler, Chair, IUCN Academy of Environmental Law

'This work, *Exploring Wild Law*, breaks new ground in the field of Earth Jurisprudence. Such a perspective is sorely needed for responding to multiple environ-mental issues from a legal perspective. We are immensely grateful to Peter Burdon for his important efforts in compiling this timely book.'

– Mary Evelyn Tucker, Yale University, Forum on Religion and Ecology

'This book is a wonderfully diverse bouquet of perspectives on wild law that will delight anyone who is interested in creating communities and societies that flourish in harmony with Nature. By drawing the varied insights and perspectives of leading thinkers and activists from around the world together in a single volume, Peter Burdon has created a multi-dimensional and nuanced understanding of wild law and Earth jurisprudence. This is a book that Earth-loving people will be reading and re-reading for many years to come.'

– Cormac Cullinan, Enact International and author of
Wild Law: A Manifesto for Earth Justice

'*Exploring Wild Law* is multi-dimensional and multi-disciplinary, both practical and wildly innovative. This complex, multi-vocal work contributes elegantly to a growing chorus of scholars, activists and artists who are singing up visionary and pragmatic forms of more-than-human conviviality here on planet Earth.'

– Professor Deborah Bird Rose, author of
Reports from a Wild Country: Ethics for Decolonisation

Contents

Preface

In his final book, *The Great Work*, Fr Thomas Berry (1914–2009) called on human society to enter a new covenant with nature. He writes, 'history is governed by those overarching movements that give shape and meaning to life by relating the human venture to the larger destinies of the universe. Creating such a movement might be called the Great Work of a people'. In the context of our present environmental crisis, he notes that the Great Work is 'to carry out a transition from a period of human devastation of Earth to a time when humans would be present to the planet in a mutually beneficial manner'. This book is intended to be one step in the fulfilment of this work and while it is focused on the discipline of law, those engaged in science, philosophy, religion and cultural studies will find sections engaging. Certainly the eclectic nature of the book is unique and represents a deliberate effort to connect distinct fields and provide vital context for understanding Earth Jurisprudence.

This book represents the effort of countless people, working across four countries. I would like to extend particular thanks to the staff and editors at Wakefield Press, in particular Michael Bollen and Jessica Marshallsay for their vision and hard work publishing this book. I would also like to thank the University of Adelaide School of Law for providing such a stimulating environment in which to lecture, study and think up new projects. Finally, special thanks to Friends of the Earth Adelaide for support, friendship and handwork in bringing this project together.

Thank you also to *Resurgence Magazine* for permission to re-print Thomas Berry's article, 'The Rights of Nature'; Cormac Cullinan for permission to reprint 'If Nature Had Rights, What Would We Need to Give Up?', Stephan Harding and Imprint Academic Publishers for permission to reprint parts of 'Gaia and Earth Jurisprudence' and *Penn State Environmental Law Review* for permission to reprint sections of Judith E. Koons' 'What is Earth Jurisprudence? Key Principles to Transform Law for the Health of the Planet', vol. 18, 2009, p. 47.

I would also like to individually thank Dr Paul Babie, Associate Professor Alexander Reilly, Professor Pamela Lyon, Joel Catchlove, Sophie Green, and my beautiful wife Shani and daughter Freya for their love, support and inspiration during the editing process.

This book is dedicated to the life and memory and of Fr Thomas Berry. May his work continue to inspire and his spirit infuse us all. Shortly before the publication of this book, barefoot lawyer and inspirational orator of Earth Jurisprudence, Ng'ang'a Thiong'o, suddenly passed. This book is also dedicated to his memory and the hope that his words find a place in all of our hearts.

Peter Burdon
12 March 2010,
Tandanya Bioregion, South Australia

Part One

What is Earth Jurisprudence?

Dedication to Thomas Berry

Jules Cashford

Anyone who ever heard Thomas Berry laugh would know why this book on Earth Jurisprudence could only be dedicated to him! It was the most resoundingly generous, compassionate, all-embracing laugh, resonating with joy and wonder at the beauty of the universe. It would follow a devastating critique of the current state of western civilisation as easily as it would a chanting of his favourite poem to a circle of friends outside on the grass. It was everything he wanted us to be with Nature: spontaneous, intimate, wild, reverential, authentic, passionate, imaginative, mutually enhancing – gloriously expressive of the manifold dimensions of the sacred.

The sacred was at the heart of his work: why we had lost it and how we could recover it, not just for ourselves but, more crucially, for the celebration of the universe in the new cosmology:

> We will recover our sense of wonder and our sense of the sacred only if we experience the universe beyond ourselves as a revelatory experience of that numinous presence whence all things come into being. Indeed, the universe is the primary sacred reality. We become sacred by our participation in this more sublime dimension of the world about us.[1]

We could believe him because he entirely embodied what he said in who he was. In his deep gravelly tones, in talks and lectures across many continents, he would tell again and again the story of the universe, with humour, gravity and passion, in order to move us beyond our small human-centred preoccupations into the sublime wonder of the vision of the whole. The story of the universe, he would say, 'is the new sacred story,' the only story that will help us to renew the world.[2] If we do not start with the universe, how can we understand ourselves in any context but our own? We cannot see that all our human institutions – governments, religions, universities, corporations, and especially the legal systems underpinning them – all rest upon 'a mode of consciousness that has established a radical discontinuity between the human and other modes of being and the bestowal of all rights on the humans'.[3]

A perspective centred exclusively on the human finds other modes of being inferior; they become 'a collection of objects not a communion of subjects'; an 'it', not – as in earlier times and still today by indigenous people – a profoundly respectful 'Thou'. This results, inevitably, in unlimited plunder and exploitation of other life forms: they are given no intrinsic value of their own: 'They have reality and value only through their use by the human',[4] no inherent right to their own specific life.

The 'Great Work,' as Thomas called it, is then to carry out the transition from a period of human devastation of the Earth to a period when humans would be present to the planet in a mutually beneficent manner'.[5] The task has to begin by moving to an Earth-centred norm of reality and value, which recognises that the universe is the enduring reality, and the source of value. This is a work not chosen by us; indeed, it was chosen *for* us, by the fact of our being born into this time of crisis when the very structure of the Earth is threatened and the extinction of species continues unimpeded. It requires of us that we align our own work to the Great Work of the universe, for the 'small self of the individual reaches its completion in the Great Self of the universe'.[6] To put it the other way, 'The story of the universe is the story of each individual being in the universe',[7] and so the journey of the universe – forever evolving, continually emerging – 'is the journey of each individual being in the universe'.[8] But that means everyone, *all* the children of the universe, to whom he gave voice in one of his own favourite poems:

> To all the children
> To the children who swim beneath
> The waves of the sea, to those who live in
> The soils of the Earth, to the children of the flowers
> In the meadows and the trees of the forest,
> To all those children who roam over the land
> And the winged ones who fly with the winds,
> To the human children too, that all the children
> May go together into the future in the full
> Diversity of their regional communities.[9]

Thomas loved poetry. Many a serious discussion would end with Thomas striding out in the direction of trees and flowers and, his flock gathered around him, closing his eyes and raising his arms to the sky while the poems broke forth from him in sheer joy. 'Loss of Imagination and loss of Nature are the same thing. If you lose one you lose the other',[10] he said, elaborating this idea in everything he wrote:

Only if the human imagination is activated by the flight of the great soaring birds in the heavens, by the blossoming flowers of Earth, by the sight of the sea, by the lightning and thunder of the great storms that break through the heat of summer, only then will the deep inner experiences be evoked within the human soul.[11]

In his poetry he wrote that it is imagination which 'awakens the child to a world of beauty, emotions to a world of intimacy'. He was fond of quoting William Blake's

'Divine Imagination', followed by Shelley's conclusion to his *Defence of Poetry* that 'Poets are the unacknowledged legislators of the world'.[12]

Thomas became aware of his own lyrical Imagination when he was eleven. As he remembered it, in the telling of it, his face would light up with an extraordinary radiance. It was an early afternoon in late May when he saw 'the meadow across the creek':

> The field was covered with white lilies rising above the thick grass. A magic moment, this experience gave to my life something that seems to explain my thinking at a more profound level than almost any other experience I can remember. It was not only the lilies. It was the singing of the crickets and the woodlands in the distance and the clouds in a clear sky. It was not something conscious that happened just then. I went on about my life as any young person might do.[13]

This transforming experience became a touchstone for him – 'normative' throughout the whole range of his thinking: 'Whatever preserves and enhances this meadow in the natural cycles of its transformation is good; whatever opposes this meadow or negates it is not good. My life orientation is that simple. It is also that pervasive. It applies in economics and political orientation as well as in education and religion'.[14]

Thomas, letting the meadow express itself through him, yet again completely embodies what he taught; for here it is as though, in the particular human mode that was himself, the universe is indeed 'reflecting on and celebrating itself in a unique mode of conscious self-awareness'.[15]

The meadow taught him to read the Book of Nature. The universe is the primary text, he would insist, the source of order and law, the primary educator and the primary healer. But it could teach him only because he felt deeply and intimately for it and was profoundly moved by it. It taught him through wonder and awe and gratitude, and he kept his word to it and to all that it evoked in him. He looked at the human-centred world he saw around him through its eyes.

When we came to North America, he wrote in disbelief:

> we brought our sacred traditions with us in a book. We never thought that this continent, its mountains and rivers and deserts, its forests and wildlife, its birds and butterflies, had anything to teach us concerning the deeper meaning of our existence.[16]

In the Earth-centred world, the integral Earth community would learn by 'sensitising' the human to 'the profound communications made by the universe about us',[17] and this would apply to all human activities: economics, medicine, politics, spiritual reorientation and law. For this we need to learn again the language of Nature – a language that is poetic, musical, symbolic, subjective, a language of feeling and intuition, held from the beginning in the archetypal images of the psyche. We can no longer learn from the literal, objective language of the old science, which reduces the world to human categories in the illusion that it explains reality to the human mind.

Yet if the poet opens up the multivalent language of Earth, the philosopher is also necessary to perceive and reflect upon the creative power in the intelligible order we observe in the universe: 'The philosopher is controlled ultimately by the balance and harmony of things, by reasoning intelligence'. The artist 'revels in the ultimate disequilibrium of things,' from which wildness spontaneously breaks forth. 'Both are valid, both are needed',[18] for between them they manifest the expanding and contracting forces in the universe.

As well as being a poet and a philosopher, Thomas Berry was also a prophet. He diagnosed the sickness of the age and proposed the way to heal it, bearing witness to the divine in new form. 'We are in between stories,' he wrote in 1978, observing that the old story 'shaped our emotional attitudes, provided us with life purpose, energised action. It consecrated suffering, integrated knowledge, guided education'.[19] Who is to lead us from the old story to the new story but one who has been wholly rooted in the old and profoundly moved to the depths of his being by what was missing from the old story: the Earth as a sacred mode of being of the Universe?

It takes a man who became a monk at the age of 20 (there were two places where you could be alone to think, he would say with a grin, a prison and a monastery, and I chose the monastery), who belonged to the Passionist Order, and whose mentor was his namesake Thomas Aquinas, to be able to say that Christianity has failed the Earth: 'Nature gradually disappeared from Christian consciousness'.[20] He did not believe that the present situation could be explained simply as a consequence of post-Cartesian empirically based sciences, or eighteenth century Enlightenment, or the excesses of the industrial age. Neither did it arise out of a Buddhist, Hindu, Chinese or Islamic context. Rather, it 'arose out of a civilisation with a biblical religious and Greek humanist matrix'.[21] Thomas came to see one aspect of this inheritance as a tragedy:

> the tragedy of so emphasising the transcendence of the divine that the earth becomes desacralised, the tragedy also of so emphasising the spiritual dimension of the human that the earth becomes only a resource base for whatever humans choose to do with the earth.[22]

While, in the medieval period, Aquinas had embraced the universe as a whole, saying that 'the whole universe together participates in the divine goodness and represents it better than any single being whatsoever',[23] the church's later preoccupation with human redemption and salvation largely replaced discussions on creation. The increasing focus on a linear process of revelation through history diminished the earlier spatial mode of understanding and participating in the great cyclical dramas of the Earth's transformation that had taken place for thousands of years. These had always been occasions for the celebration of the 'great liturgy of the universe'[24]: in spring and summer, autumn and winter, in the dawn and dusk, in the ever-changing phases of the moon, and in the darkness of night which reveals the universe.

Further, the replacement of the cyclical with the temporal linear mode of revelation gave rise to an expectation of an 'infra-historical millennial period in which the human condition would be overcome'.[25] This gave a destructive mythic drive to the intrinsically secular doctrine of progress, which helps to explain the 'technological trance' in which humans knowingly cut off the source of their own, and everyone else's, life. Like all prophets, Thomas saw though the false forms of his age and gave them a name. He described the industrial age as:

> a period of technological entrancement, an altered state of consciousness, a mental fixation that alone can explain how we came to ruin our air and water and soil and to severely damage all our basic life systems under the illusion of 'progress'. Now the trance is passing, we need to 'reinvent the human' as integral with the whole of the Earth community.[26]

He alerted us to a 'deep cultural pathology'; he warned that the industrial mind had co-opted the language in which our values are expressed. It is the language of profit and loss – and what of 'the Earth deficit'? – of being 'productive,' of regarding the Earth as a 'resource' to be 'used,' a 'product' to be 'developed,' 'exploited,' 'dominated,' discarded when no longer useful. It perceives limitation as a 'demonic obstacle to be eliminated',[27] heroically conquering everything in its way as a manifestation of 'a deep inner rage' at the human condition.[28] Indeed, this 'degradation' of the Earth is seen as the condition for 'progress of humans,' which makes the Earth 'a kind of sacrificial offering'.[29]

If it would be the Poet who feels the sacrifice passionately as his own, and the philosopher who makes it intelligible as a deviation from the true path of cosmogenesis – the continual unfolding of the universe – then it may be the Prophet who finally refuses this sacrifice, setting up an opposing value in its place.

'This little meadow,' as he called the meadow of his childhood, was also the most eloquent advocate for 'Earth Jurisprudence'. For who has the right to deprive that meadow of its own life, to sacrifice it on the altar of human progress? On the contrary, 'what is good recognises the rights of this meadow and the creek and the woodlands beyond to exist and flourish in their ever-renewing seasonal expression even while larger processes shape the bioregion in its sequence of transformations'.[30]

As with all other forms of human culture, the anthropocentrism of human law for human beings had to become integral to the primary lawgiver, the universe. This is what Thomas called the 'Great Jurisprudence', the inherent order and lawfulness of the cosmos which structures and sustains all life within it. This lawfulness is written into the pattern of birth and death for all of creation, and into the rhythms of all of our lives. Consequently, the new Earth Jurisprudence begins with the universe, derives its validity from the universe, and upholds the sacred values of the new Universe Story.

Thomas was not entirely happy with the language of rights, but it was the best we had to be going on with. If, in 1886, corporations could, incredibly, be given the rights of individuals, then so should the Earth in all its gloriously diverse manifestations have the rights of an individual. 'We begin from where you are,' he said. For Law was necessary where morality had failed, and morality was necessary where love had failed. If you love the meadows and the woodlands and the rivers and all the creatures that live in them, then you would not want to harm them, you could not do it. Their wounds are your wounds: we are all mutually dependent, mutually reflecting and mutually enhancing. That is why we need to learn the language of mountains and rivers, trees and birds and all the animals and insects, 'as well as the languages of the stars in the heavens'.[31]

However, the language of rights answers the legal establishment in its own terms. As it is now, existing legal structures in all nation states, so closely allied with the industrial establishment, not only cannot protect the natural world but actually legitimise the destruction of the Earth inherent in the old story. But where do rights come from in the first place? Thomas insists they come from existence itself, not from other humans, and that means that rights cannot belong exclusively to humans, enclosed in their little worlds, cut off from the Earth they take for granted. Indeed, rights are not theirs to give away, to award or withhold from other beings on Earth. In an interdependent world, where every mode of being depends on every other mode of being, then every mode of being has rights derived from the universe which brought them into being and made them who they are. In this sense, every mode of being is equal: 'The well-being of each member of the Earth community is dependent on the well-being of the Earth itself'.[32]

Within this context, Thomas made the following ten proposals expressed in terms of rights which he believed should be recognised in national constitutions and courts of law:

Ten Principles of Jurisprudence
1. Rights originate where existence originates. That which determines existence determines rights.
2. Since it has no further context of existence in the phenomenal order, the universe is self-referent in its being and self-normative in its activities. It is also the primary referent in the being and the activities of all derivative modes of being.
3. The universe is composed of subjects to be communed with, not objects to be used. As a subject, each component of the universe is capable of having rights.
4. The natural world on the planet Earth gets its rights from the same source that humans get their rights: from the universe that brought them into being.

5. Every component of the Earth community has three rights: the right to be, the right to habitat, and the right to fulfil its role in the ever-renewing processes of the Earth community.

6. All rights are role-specific or species-specific, and limited. Rivers have river rights. Birds have bird rights. Insects have insect rights. Humans have human rights. Difference in rights is qualitative, not quantitative. The rights of an insect would be of no value to a tree or a fish.

7. Human rights do not cancel out the rights of other modes of being to exist in their natural state. Human property rights are not absolute. Property rights are simply a special relationship between a particular human 'owner' and a particular piece of 'property,' so that both might fulfil their roles in the great community of existence.

8. Since species exist only in the form of individuals, rights refer to individuals, not simply in a general way to species.

9. These rights as presented here are based on the intrinsic relations that the various components of Earth have to each other. The planet Earth is a single community bound together with interdependent relationships. No living being nourishes itself. Each component of the Earth community is immediately or mediately dependent on every other member of the community for the nourishment and assistance it needs for its own survival. This mutual nourishment, which includes the predator-prey relationship, is integral with the role that each component of the Earth has within the comprehensive community of existence.

10. In a special manner, humans have not only a need for but also a right of access to the natural world to provide for the physical needs of humans and the wonder needed by human intelligence, the beauty needed by human imagination, and the intimacy needed by human emotions for personal fulfilment.[33]

* * *

The 'mythic vision' has been set into place, he wrote, and he trusted the great powers of the universe to assist us in the realisation of the new story, to help us summon the 'vast psychic energy' required to 'realign our thinking' with the well-being of the entire planet and so to move into what he called the Ecozoic Era. And perhaps we should trust him in his trust of the universe. In his own journey of passionate dedication to the Great Work, as monk, scholar, cultural historian, philosopher, prophet, visionary, and always poet, he has taken us further into the great mystery of existence than we could have gone without him and, if anyone should know, it would be him:

If the dynamics of the universe from the beginning shaped the course of the heavens, lighted the sun, and formed the Earth, if this same dynamism brought forth the continents and seas and atmosphere, if it awakened life in the primordial cell and then brought into being the unnumbered variety of living beings, and finally brought us into being and guided us safely through the turbulent centuries, there is reason to believe that this same guiding process is precisely what has awakened in us our present understanding of ourselves and our relation to this stupendous process. Sensitised to such guidance from the very structure and functioning of the universe, we can have confidence in the future that awaits the human venture.[34]

> **It Takes a Universe**
> The child awakens to a universe.
> The mind of the child to a world of wonder.
> Imagination to a world of beauty.
> Emotions to a world of intimacy.
> It takes a universe to make a child
> both in outer form and inner spirit.
> It takes a universe to educate a child
> It takes a universe to fulfil a child.
> And the first obligation of each generation
> To the next generation is to bring these two together
> So that the child is fulfilled in the universe
> And the universe is fulfilled in the child
> While the stars ring out in the heavens.[35]

Notes

1   Thomas Berry, *The Great Work: Our Way into the Future,* Bell Tower, New York, 1999, p. 48.

2   Thomas Berry, *Evening Thoughts: Reflecting on Earth as Sacred Community,* ed. Mary Evelyn Tucker, Sierra Club Books, San Francisco, 2006, p. 57.

3   Berry, *The Great Work,* p. 4.

4   Ibid.

5   Ibid., p. 3.

6   Ibid., p. 190.

7   Ibid., p. 27.

8   Ibid., p. 164.

9   Ibid., dedication.

10   Phone call from Thomas.

11   Berry, *The Great Work,* p. 55.

12   Percy Bysshe Shelley, 'The Defence of Poetry', in *Poems and Prose,* Everyman, J.M. Dent, London, and Charles E. Tuttle, Vermont, 1995, p. 279.

13 Berry, *The Great Work*, p. 12.

14 Ibid.

15 Ibid., p. 56.

16 Thomas Berry, Introduction to Tom Hayden *The Lost Gospel of The Earth: A Call for Renewing Nature, Spirit & Politics*, Sierra Club Books, San Francisco, 1996, p. xiii.

17 Berry, *The Great Work*, p. 64.

18 Ibid., p. 53.

19 Quoted from *The New Story* in *Evening Thoughts*, Appendix 3, p. 166.

20 *Thomas Berry and the New Cosmology*, ed. Anne Lonergan and Caroline Richards, Twenty-Third Publications, Mystic, Connecticut, 1987, p. 16.

21 Berry, *The Lost Gospel*, p. xii.

22 Ibid., p. ix.

23 Thomas Aguinas, *Summa Theologica (Part I, Question 47, article 1)*, quoted by Berry in *Thomas Berry and the New Cosmology*, p. 30, and *The Great Work*, p. 77.

24 Berry, *The Great Work*, p. 17.

25 Lonergan, *New Cosmology*, p. 16.

26 Ibid., p. 19.

27 Berry, *The Great Work*, p. 67.

28 Ibid., p. 165.

29 Ibid,. p. 62.

30 Ibid., p. 13.

31 Brian Swimme & Thomas Berry, *The Universe Story: From the Primordial Flaring Forth to the Ecozoic Era*, HarperSanFrancisco, San Francisco, 1992, p. 258.

32 Berry, *Evening Thoughts*, p. 110.

33 Appendix 2 in *Evening Thoughts*, pp. 149–50. A few words have been omitted to coincide with the List of Rights originally presented by Thomas at the conference on Earth Jurisprudence hosted by the Gaia Foundation at Airlee House, Washington in 2001.

34 From *The New Story*, quoted by Mary Evelyn Tucker, *Evening Thoughts*, Appendix 3, p. 169.

35 Thomas Berry, *Every Being Has Rights*, Twenty-Third Annual E.F. Schumacher Lectures, Stockbrige, Massachusetts, October 2003, E.F. Schumacher Society, 2004, pp. 7–8.

A History of Wild Law

Cormac Cullinan

Attempting to map the history and development of wild law is a bit like tracking water droplets on a windscreen on a misty morning. At first, tiny micro-droplets form at some distance from each other. As time passes more and more tiny particles of moisture precipitate out of the atmosphere and each one swells into a shining globule, tremulous under its increasing weight. First one drop, and then another, and another overcomes the forces holding it to the windscreen and snakes suddenly down the glass absorbing others into its quickening stream.

From my perspective, what we now refer to as Earth Jurisprudence and wild law started as tiny insights and inklings of ideas. As I began to make connections with other people and their thoughts on the subject, the ideas rapidly swelled and cohered until they acquired a life of their own and became a stream of thought moving away like quicksilver from the original droplet. From the perspective of each drop, it is the source of the trickle that becomes the rivulet, the stream, the river, and the sea. At first, my understanding of the emergence of wild law was the subjective experience of a single droplet. Now with the benefit of time and a broadening perspective, I see it as only one of the streams of a new consciousness and worldview which is precipitating out of the atmosphere of these times. As with the droplets, receptive minds enable a multitude of thoughts to form from the formlessness of the mist, and these are now gathering together to form an increasingly coherent and fast-moving flow impelled by our inherent desires for connection and fulfilment.

The wild law story is then both the subjective, individual stories of how each of us came to these ideas and connected with other minds, and is simultaneously part of a wider cultural story. That cultural story is about replacing the dream of dominating, controlling and using Earth for the benefit of a few humans, with a worldview in which the role of human beings is to celebrate our participation in the awe-inspiring community of life we call Earth. This new cultural dream is closely aligned with many ancient wisdom traditions and cosmologies of indigenous peoples, but also incorporates contemporary scientific understandings of how each aspect of our reality is present to, and interconnected with, every other part of the Earth community which manifested us.

Earth Jurisprudence

It is now clear that *Wild Law* and the ideas in it are part of the almost simultaneous emergence on several continents of initiatives to bring about a fundamental change in human governance systems. These initiatives all share the belief that one of the primary causes of environmental destruction is the fact that our governance systems are designed to perpetuate human domination of Nature, instead of fostering mutually beneficial relationships between humans and the other members of the Earth community. They all advocate an approach to law and governance which is referred to in *Wild Law* as 'Earth Jurisprudence' (see Box). Earth Jurisprudence is a philosophy of law and human governance that is based on the idea that humans are only one part of a wider community of beings and that the welfare of each member of that community is dependent on the welfare of the Earth as a whole. From this perspective, human societies will only be viable and flourish if they regulate themselves as part of this wider Earth community and do so in a way that is consistent with the fundamental laws or principles that govern how the Universe functions (the 'Great Jurisprudence').

This approach requires looking at law from the perspective of the whole Earth community and balancing all rights against one another (as we do between humans) so that fundamental rights like the right to life take precedence over less important ones such as rights to conduct business. Currently the rights of humans, and particularly corporations, automatically trump the rights of all others. It also means that while a fox eating a rabbit

Principles of Earth Jurisprudence

- The Universe is the primary law-giver, not human legal systems.
- The Earth community and all the beings that constitute it have fundamental 'rights', including the right to exist, to habitat or a place to be, and to participate in the evolution of the Earth community.
- The rights of each being are limited by the rights of other beings to the extent necessary to maintain the integrity, balance and health of the communities within which it exists.
- Human acts or laws that infringe these fundamental rights violate the fundamental relationships and principles that constitute the Earth community ('the Great Jurisprudence') and are consequently illegitimate and 'unlawful'.
- Humans must adapt their legal, political, economic and social systems to be consistent with the Great Jurisprudence and to guide humans to live in accordance with it, which means that human governances systems at all times take account of the interests of the whole Earth community and must:
- determine the lawfulness of human conduct by whether or not it strengthens or weakens the relationships that constitute the Earth community;
- maintain a dynamic balance between the rights of humans and those of other members of the Earth community on the basis of what is best for Earth as a whole;
- promote restorative justice (which focuses on restoring damaged relationships) rather than punishment (retribution);
- recognise all members of the Earth community as subjects before the law, with the right to the protection of the law and to an effective remedy for human acts that violate their fundamental rights.

could be seen as a violation of the rabbit's right to life, it does not violate the Great Jurisprudence because the maintenance of predator-prey relationships (as distinct from killing for sport) is fundamental to preserving the integrity of the whole community (i.e. the greater good).

Thomas Berry

My story of the conception, birth and early life of wild law has roots extending deep into my personal history that have been nourished by many great thinkers and writers, most notably Thomas Berry. Here I can do no more than describe a few of the milestones along the road I have taken.

One of the first significant events in my journey occurred when I met Liz Hosken of the Gaia Foundation of London at a workshop in Cape Town in 2000. At the coffee break, I mentioned that in the course of drafting wildlife legislation for Namibia I was encountering conceptual difficulties that could not be resolved by simply amending legislation because they had their roots in the underlying legal philosophies (jurisprudence). In response, she told me about the work of the eminent American cultural historian, religious scholar and philosopher Thomas Berry and his call for the development of a 'new jurisprudence'. Thomas argued that:

> [W]e need a jurisprudence that would provide for the legal rights of geological and biological as well as human components of the Earth community. A legal system exclusively for humans is not realistic. Habitat of all species, for instance, must be given legal status as sacred and inviolable.[1]

I was fascinated and became more so when I read *The Universe Story* by Thomas Berry and Brian Swimme and subsequently Thomas Berry's *The Great Work*. In fact, I was so engrossed in reading *The Universe Story* on the train back from work that I completely missed my stop. As I retraced my journey I realised that these ideas were going to be particularly significant for me.

One of the most important milestones on my journey occurred in April 2001 when I met Thomas Berry at the Airlie Centre in Virginia.[2] Liz Hosken and Ed Posey of the Gaia Foundation sponsored Bruce Dell (an experienced wilderness guide and director of the Wilderness Leadership School) and me to make the long journey from South Africa to attend a weekend workshop which they co-convened in response to Thomas's call for the development of a new jurisprudence that took account of the fact that we are members of the Earth Community. In the months leading up to the meeting I had begun to refer rather tentatively to this new jurisprudence as 'Earth Jurisprudence', but it was only when I walked into the dining room at the Airlie Centre and saw cardboard signs reserving tables for 'Earth Jurisprudence' that I realised that Thomas's vision of a 'new jurisprudence' had already come into being.

Meeting Thomas made a deep impression on me. As with everyone who met him, I was soon in awe of the depth and breadth of his scholarship. However, what made the deepest impression on me then was the sense of urgency and deep purpose which he communicated. He spoke repeatedly of the 'ultimacy' of the situation in which we find ourselves and was clearly intent on passing the baton to younger people who would take 'the Great Work' forward into the future. The authenticity of his being, his humour, and appreciation of the beauty of the spring time, were for me further validation of the wisdom and passion I had encountered in his writings.

The Earth Justice Network

At that time, I had no thought of writing a book and my primary objective was to help catalyse the emergence of a worldwide movement of people and organisations who would commit to working collaboratively to bring about a new form of human presence on Earth and usher in the Ecozoic age which Thomas Berry spoke of. My experiences as an activist in the United Democratic Front during the struggle to liberate South Africa from apartheid had convinced me that environmental and social justice organisations throughout the world had arrived at the point where they needed to join forces (without losing their individual identities or sense of purpose) and become conscious of themselves as being part of a coherent global movement. I believed that this was needed to enable us to move beyond reactive protest politics and to begin setting the agenda and building the society which we wanted to see.

I had discussed these ideas with Liz Hosken and Ed Posey while walking and breakfasting on Hampstead Heath in London, and they decided that the Gaia Foundation should catalyse this process. In December 2000 they circulated a draft document entitled 'A Proposal to Launch the Movement for the Regeneration of the Earth Community' or 'Earth Movement' for discussion among a group of friends and associates, who generally responded positively. I continued to develop these ideas in informal discussion documents during late 2000 and early 2001[3] and in March 2001 described the proposed 'Whole Earth Movement' as follows:

> We are people and organisations that have come together in a global social movement to co-create a positive present and a future for all the communities of life on our planet. We believe that by working in concert we can have a profound effect on bringing about an inclusive and fairer world society based on a recognition of the laws of the Universe that govern us all and respect for all life.

Gaia soon arranged a series of meetings and gatherings of activists from many different countries, which lead to the formation of what was finally named the 'Earth Justice Network' or simply 'EJN'. At that time the discussion of Earth Jurisprudence within this emerging community of activists was very much part of developing a common understanding and sense of purpose which could be used to unify a diverse group of activists from around the world.

Going public

We then decided that we needed to test how these ideas would be received by the public and decided to start with what we believed would be a relatively sympathetic audience at the 7th World Wilderness Congress being held in November 2001 in Port Elizabeth, South Africa. The organisers of the conference agreed to let us run breakaway sessions on three consecutive afternoons and in preparation for leading those discussions I prepared a document of about 30 pages setting out the essential ideas of Earth Jurisprudence, which was later to form the basis for *Wild Law*. The 7th World Wilderness Congress passed a resolution proposed by Professor Wangari Maathai (who was later awarded the Nobel Peace Prize) resolving that delegates should: 'Initiate programmes and provide resources to develop a jurisprudence that recognises humans as inseparable from the planetary ecosystem and that for it to function properly, human societies must regulate their behaviour in a way that supports rather than undermines the integrity and health of the community of life on earth'.[4]

Encouraged by the positive response at the World Wilderness Congress the EJN turned its attention to the World Summit on Sustainable Development (WSSD) which was being convened in September 2002 in Johannesburg, South Africa, ten years after the historic 'Earth Summit' in Rio de Janeiro. As it became apparent that the focus of the WSSD would be on the interests of developing countries and that the protection of the environment was to be relegated to secondary status, the EJN and its allies decided to convene a 'Peoples Summit' to occur simultaneously in Johannesburg.

We saw the WSSD as an ideal opportunity both to popularise the concept of the EJN among a global community of activists, and also to spread Earth Jurisprudence. However I was very aware of how long it takes to communicate a very different worldview, and particularly to make a good case for rethinking our entire approach to law and governance. At the World Wilderness Congress we had been given enough time to develop our arguments but in future we were going to need a way of communicating these ideas much more swiftly. I expressed my concerns to friends at dinner one evening and mentioned that it would be helpful to have a book that explained the ideas more fully. One of my friends, Simon, asked why I didn't write that book. I reluctantly conceded that I could probably write it but pointed out that the WSSD was now a few months away and that it would not be possible to get it done in time. Simon is a legal publisher and he immediately seized a serviette and pen and began calculating the date the manuscript would have to be completed in order for a book to be published in time for the WSSD. Finally, he announced that if I could write it in four to six weeks he would publish it in time for the WSSD. With some trepidation (and with hindsight, much naivety) I accepted the challenge and *Wild Law* was written, edited, and published over an intense period of a few months, finally emerging in August 2002, the week before the WSSD.

The period around the WSSD was one of intense activity. My firm, EnAct International, and the Gaia Foundation co-hosted the first Earth Jurisprudence Colloquium in the Ithala game reserve in South Africa. Time was short so we picked up the delegates from the airport in Johannesburg, presented them with copies of *Wild Law* hot off the press and flew them in chartered DC-10 directly to a landing strip in the reserve. After a weekend of intense and productive discussions about the ideas and how to spread them, I went on to speak at two international environmental law conferences in South Africa and at the Peoples World Summit. *Wild Law* and the ideas in it were now in the public domain.

Connecting with traditional wisdom in Africa

The African Biodiversity Network has played a leading role in spreading Earth Jurisprudence through its member organisations across Africa. Experiential courses on African customary law and culture in Botswana run at a bush lodge in Botswana by Colin and Niall Campbell (supported by the Gaia Foundation) have proved to be particularly powerful in developing a core of people with a deep understanding of, and commitment to, the Earth Jurisprudence approach. The Gaia Foundation has also been instrumental in forging links between traditional communities in Africa and Columbia, which has given rise to the emergence of a similar approach known as 'community ecological governance' that emphasises the importance of drawing on the wisdom of community elders.

The idea that African customary law and traditional practices contained important wisdom that we could draw on to address contemporary social and environmental issues has had particular resonance in Africa. Professor Melesse Damtie, who teaches law at the Civil Service College of Ethiopia, has inspired government officials to return to their regions and rediscover the ancient and almost forgotten wisdom of African customary law systems with a view to inspiring the development of more effective governance systems that are consistent with both people and place.

In Kenya, the NGO Porini has been inspired by reconnecting with ancient African legal lineages to recover indigenous practices that teach people to respect and care for the Earth community. This has lead to renewed interest in the wisdom of tribal elders and to a successful application to court to restore to a community the right to conserve and manage sacred sites. I vividly remember the excitement and sincerity with which the late Kenyan human rights lawyer and co-founder of Porini, Ng'ang'a Thiong'o said to me: 'You know, I have practised law my whole life but it has always been the law of the colonisers. For the first time in my life I feel that I can now practise an authentic law of my people and of Africa'.

The United Kingdom

John Elford of Green Books published another edition of *Wild Law* in the United Kingdom in November 2003 and I made several trips to London to speak at events organised by the Gaia Foundation. In April 2004 my former colleague Donald

Reid organised the first 'workshop' in the UK to discuss and develop the principles of Earth Jurisprudence. Entitled 'Wild Law Wilderness Workshop: A Walking Workshop on Earth Jurisprudence' the workshop was in fact a memorable hike through the beautiful, remote and wild Knoydart Peninsula interspersed with animate debates in the 'bothy' where we took shelter from the inclement weather.

Elizabeth Rivers, an environmental mediator and former solicitor who heard me speak at the Gaia Foundation in London, was later approached by Simon Boyle of the UK Environmental Law Association (UKELA) to speak to UKELA members about environmental mediation. Liz suggested that she speak about wild law and Earth Jurisprudence instead. Simon was intrigued, read *Wild Law* and became a committed advocate of Earth Jurisprudence. This led to the convening of the first UKELA Wild Law conference held at the University of Brighton in November 2005. Every year since, UKELA, in collaboration with the Gaia Foundation and the Environmental Law Foundation, have hosted Wild Law conferences or 'Wild Law weekends'. In 2009 Peter Burdon took the initiative of convening the first Wild Law conference in Australia and Liz Rivers was one of the speakers.

Earth democracy in India

In India, the celebrated environmental activist Dr Vandana Shiva coined the phrase 'Earth democracy' to describe a world-view and political movement promoted by Navdanya (an organisation which she founded). As Dr Shiva explains:

> Earth democracy is both an ancient worldview and an emergent political movement for peace, justice and sustainability ... It incorporates what in India we refer to as *vasudhaiva kutumbkam* (the earth family) – the community of all beings supported by the earth ... Earth democracy is not just a concept, it is shaped by the multiple and diverse practices of people reclaiming their commons, their resources, their livelihoods, their freedoms, their dignity, their identities, and their peace.[5]

One of the reasons for the success of the Earth democracy movements is because it has reconnected people with pre-consumerist cultural understandings of the sacred dimensions of seeds, food, water and land. Navdanya has also drawn on traditions of resistance to colonial authority, for example, by employing the strategies of Mahatma Gandhi's salt *satyagraha* to resist legislation that allowed the patenting of seeds and other life forms.[6] As in Africa, the ideas of Earth Jurisprudence are being carried forward by new organisations and approaches that reconnect people with an ancient sense of identity and fulfilment within a sacred community of life.

EJ in the USA

The story of wild law in the USA so far revolves around two initiatives – the founding of the Center for Earth Jurisprudence (CEJ) in Florida by the Catholic universities of Barry and St Thomas, and the campaigns for local democracy undertaken by the Community Environmental Legal Defense Fund ('CELDF'). However many American thinkers had already planted the seeds of a non-anthropocentric approach to law and governance. In *A Sand County Almanac* (1949) Aldo Leopold articulated a 'land ethic', stating that: 'a thing is right when it tends to preserve the integrity, stability, and beauty of the biotic community. It is wrong if it tends otherwise'. Leopold pointed out that the land ethic 'changes the role of Homo sapiens from conqueror of the land-community to plain member and citizen of it. It implies respect for his fellow members, and also respect for the community as such'.

In 1972 Professor Christopher Stone of the University of Northern California published a seminal article, entitled 'Should Trees Have Standing? Towards Rights for Natural Objects'.[7] In *Trees*, Professor Stone pointed out that the widening of society's 'circle of concern' had lead to the recognition of more extensive legal rights for women, children, native Americans and African Americans. There was no good reason, he argued, why increasing public concern for the protection of Nature could not lead to the recognition of rights for Nature. This would enable legal suits to be instituted on behalf of trees and other 'natural objects' and damages to be recovered and applied for their benefit.

The Center for Earth Jurisprudence

In early 2006 I received an email from Sister Patricia Siemens, who wrote to tell me that she and her colleagues had received some funding from the Catholic Church to initiate a Center for Earth Jurisprudence (CEJ) inspired by the work of Thomas Berry. I enthusiastically offered to help if I could and Sister Pat subsequently invited me to speak at the first Earth Jurisprudence conference in the USA which was convened by the CEJ in Miami, Florida in 2007 and at a subsequent conference held in Orlando, Florida in 2008.

The Community Environmental Legal Defense Fund

Prior to the first Earth Jurisprudence Conference hosted by the CEJ, *Orion* magazine asked me to write a retrospective article on the impact of Christopher Stone's seminal 1972 article (*Trees*). The features editor suggested that I speak to Thomas Linzey of the Community Environmental Legal Defence Fund (CELDF). I emailed Thomas and he wrote back immediately saying that, serendipitously, the first local Ordinance recognising the rights of natural communities had been passed the night before by the Tamaqua Borough Council. I asked for a copy and was astounded to see an example of legislation that explicitly implemented key aspects of Earth Jurisprudence even though the members of the CELDF had never heard the term or seen *Wild Law*. I sent Thomas a copy of *Wild Law* and we were mutually delighted to discover that we had arrived at very similar conclusions via different routes, and in different continents.

The CELDF had been successful for many years in protecting communities from a range of environmentally destructive activities (including the disposal of sewage sludge on land, the establishment of massive pig farms and mining) by using legal processes to have the authorisations set aside on the basis of deficiencies in the authorisation processes. However, despite their initial successes, Linzey soon realised that the victories were short-lived because the corporations simply repeated the process in a manner that complied with all legal requirements and eventually triumphed.

The CELDF and the communities that they worked with realised that local communities could not secure their own well-being without protecting the integrity and functioning of the ecological communities within which they lived. They concluded that in order for a local community to protect itself from being harmed by corporations it would have to assert its democratic right to self-governance and recognise and enforce rights for nature. Consequently the CELDF began assisting communities to draft local ordinances that: (a) re-asserted their right to prohibit activities harmful to their wellbeing; (b) recognised rights for natural communities; (c) enabled local governments and individuals to sue for damages to be used for the restoration of any damage to ecological communities; and (d) stripped corporations who contravened the ordinances of legal personality (and hence their right to benefit from the civil rights in the Constitution of the United States).

Pat Siemens had also invited Thomas Linzey and Richard Grossman of CELDF to speak at the first Earth Jurisprudence conference held by the CEJ, and Thomas then suggested that he and I go on a 'Rights for Nature' speaking tour to US law schools. I agreed and immediately after the CEJ conference, Thomas and I embarked on a gruelling coast-to-coast speaking tour that encompassed eleven law schools. Along the way we got to know each other and discussed ideas, strategies and tactics from Florida to Seattle and back to Pennsylvania. This laid the basis for ongoing cooperation between us, and after the second CEJ conference in Orlando in 2008, I met with Thomas and his colleagues at Green Gulch, the beautiful Buddhist retreat centre outside San Francisco. The meeting was arranged by Linda Sheehan of California Waterkeepers. Thomas Linzey and Mari Margill briefed us about their recent trip to Ecuador at the invitation of the Pachamama Alliance to discuss the possibility of including rights for nature in the draft Ecuadorian Constitution, and we wrestled with how to simultaneously work from the ground up (as CELDF was doing with local communities in the USA) and frame the debate at the national and international levels so that Earth Jurisprudence did not become pigeonholed as a subcategory of environmental law. Unbeknown to us, our concerns about how to frame the international debate were soon to be answered by a series of dramatic developments in Latin America.

The rights of Pachamama[8]
As early as 2002 a Chilean lawyer, Godofredo Stutzin, had argued in an article entitled 'Nature's Rights' that:

The development of law has reached a crucial moment: the idea and the ideal of justice have to acquire a new universality which comprises the entire biosphere, adding not only new objects, but also new subjects to the legal establishment.[9]

He argued that recognising Nature's rights 'constitutes an act of justice by which the law, advancing in its process of development, confirms the distinctive values inherent in the natural world, leaving behind the indefensible anthropocentric vision of earth according to which the planet and all that exists upon it are but the environment of humankind, having no other value than their usefulness for the human species'. As Stutzin points out, one of the practical advantages of recognising rights for Nature is that anyone seeking to alter or destroy any aspect of Nature would have to put forward reasons to justify why this should be permitted, instead of those who wish to prevent destruction having to prove why Nature should be conserved.

Ecuador
An enormously significant milestone in the wider story of Earth Jurisprudence occurred in September 2008 when the people of Ecuador adopted a constitution by popular referendum that commits the state and citizens to seeking well-being in a manner that is harmonious with Nature and that recognises the rights of Nature. These remarkable provisions in the constitution of Ecuador came into being as a result of the collaboration between indigenous people's organisations and environmental organisations in Ecuador, the CELDF and certain key individuals in the Constitutional Assembly charged with drafting the new constitution. Earth Jurisprudence was no longer 'unthinkable' or merely a philosophical ideal, it had become a constitutional reality. Ecuador changed the debate from whether or not it was possible to recognise rights for Nature to whether or not doing so would be effective.

The Bolivian initiative
On 22 April 2009, the United Nations General Assembly adopted a resolution proposed by Bolivia proclaiming 22 April as 'International Mother Earth Day'. In his speech to the General Assembly on that day, Bolivian President Evo Morales Ayma expressing the hope that, as the twentieth century had been called 'the century of human rights', the twenty-first century would be known as the 'century of the rights of Mother Earth'. He called upon the member states to begin developing a 'Declaration on the Rights of Mother Earth' that, among other rights, would enshrine the right to life for all living things; the right for Mother Earth to live free of contamination and pollution; and the right to harmony and balance among and between all things.

This was followed on 17 October 2009 by a declaration of the nine countries of the Bolivarian Alliance for the Peoples of Our America (ALBA) supporting the call for the adoption of a Universal Declaration of Mother Earth Rights. The

Declaration expresses the fundamental principles of Earth Jurisprudence with great clarity, stating:

> In the 21st Century it is impossible to achieve full human rights protection if at the same time we do not recognise and defend the rights of the planet earth and nature. Only by guaranteeing the rights of Mother Earth can we guarantee the protection of human rights. The planet earth can exist without human life, but humans cannot exist without planet earth.[10]

After the failure of the December 2009 Copenhagen meeting of the Conference of the Parties of the United Nations Framework Convention on Climate Change to agree on an international legal instrument to tackle climate change, the Bolivian President announced that Bolivia would host a Peoples World Conference on Climate Change and Mother Earth's Rights in April 2010. The Conference is intended to provide a forum to discuss a range of issues, including a Universal Declaration of the Rights of Mother Earth, and an International Tribunal on Climate Justice. The extent to which existing organisations and networks support this initiative is likely to be a crucial factor in determining the impact of the initiative launched by the Bolivian President and the speed at which Earth Jurisprudence ideas spread.

The path ahead

Earth Jurisprudence has deep roots in the cosmologies of many indigenous peoples, the customary practices of rural people in Africa, India and elsewhere, as well as in the writings of visionaries such as Aldo Leopold and Thomas Berry. Thomas Berry articulated most clearly the need for the jurisprudence of the cultures that dominate the contemporary world to be rethought as part of a wider cultural movement towards achieving a benign human presence within the Earth community. Berry also articulated fundamental principles of this new jurisprudence, which has been articulated in *Wild Law* as 'Earth Jurisprudence'.

One of the most significant aspects of the emergence of Earth Jurisprudence is that it has resonated with a wide range of diverse people and communities throughout the world, many of whom had already reached similar conclusions from their own, widely different, experiences. In many cases the initial reaction of people who first read or hear of Earth Jurisprudence is one of recognition – as if they are hearing for the first time something that they already guessed. This means that despite the many different origins of these ideas, as soon as they make contact with one another, like water drops, they rapidly cohere and absorb one another.

The evolution of these ideas is now entering a new phase as they become visible as a distinctive stream of thought. In a period of just over a year since the adoption of the Ecuadorian Constitution in September 2008, the issue of rights for Nature or Mother Earth and the accompanying philosophy of Earth Jurisprudence have burst onto the international agenda. If the Peoples World Conference on Climate

Change and the Rights of Mother Earth in April 2010 adopts a draft Universal Declaration of the Rights of Mother Earth that is widely supported, it is likely that the flow of these ideas will intensify and more and more examples of wild laws will appear throughout the world.

Notes

1   T. Thomas, *The Great Work: Our Way Into the Future*, Bell Tower, New York, 1999, p. 161. Thomas Berry proposed a new jurisprudence based on the following core principles (among others):

'The universe is a communion of subjects, not a collection of objects. As subjects the component members of the universe are capable of having rights.

'The natural world on the planet Earth gets its rights from the same source that humans get their rights, from the universe that brought them into being.

'Every component of the Earth Community has three rights. The right to be, the right to habitat, and the right to fulfil its role in the ever-renewing processes of the Earth Community'.

2   See M. Bell, 'Thomas Berry and an Earth Jurisprudence: An Exploratory Essay', *The Trumpeter*, vol. 19, no. 1, 2003.

3   For example an ideas paper entitled 'Building the Earth Centric Movement' and a document entitled 'The Whole Earth Movement', which was intended as a rough draft of text for a website.

4   7th World Wilderness Congress Resolution 33 of 8 November 2001.

5   V. Shiva, *Earth Democracy: Justice, Sustainability, and Peace*, South End Press, Cambridge, Massachusetts, 2005, p. 1 and p. 5. For a synopsis see V. Shiva, 'Paradigm Shift: Earth Democracy. Rebuilding true security in an age of insecurity', *Resurgence Magazine*, no. 214, September/October, 2002.

6   Mahatma Gandhi encouraged Indians to make their own salt as an act of non-violent resistance (*satyagraha*) to the unjust laws of the British Empire which had created a monopoly in salt manufacture. See www.navdanya/earthdcracy/index.htm, accessed on 17 August 2009.

7   C.D. Stone, 'Should Trees Have Standing? Towards Legal Rights for Natural Objects', *California Law Review*, vol. 45S, 1972, p. 450. Stone's article motivated the famous dissenting judgment by Justice Douglas in the case of *Sierra Club v Morton*, 405 US 727, 741–42 (1972) in which he stated that: 'The critical question of 'standing' would be simplified and also put neatly into focus if we fashioned a federal rule that allowed environmental issues to be litigated before federal agencies or federal courts in the name of the inanimate object about to be despoiled, defaced or invaded by roads and bulldozers and where injury is the subject of public outrage. Contemporary public concerns for protecting nature's ecological equilibrium should lead to the conferral of standing upon environmental objects to sue for their own preservation'.

8   Pachamama is a goddess revered by the indigenous people of the Andes. Pachamama is usually translated as mother earth or mother world.

9   G. Stutzin, 'Nature's Rights', *Resurgence Magazine,* no. 210, January-February, 2002, pp. 24–26.

10   VII ALBA-TCP Summit: Special Declaration for a Universal Declaration of Mother Earth Rights, 2009, <http://motherearthrights.org/2009/10/17/vii-alba-tcp-summit-special-declaration-for-a-universal-declaration-of-mother-earth-rights/> accessed 1 December 2009.

Reflections on an Inter-cultural Journey into Earth Jurisprudence

Liz Hosken

As the rather stooped elder peered over the lectern with his piercing eyes, his voice boomed across the room: 'The industrial process is now in its terminal phase. This is the inevitable consequence of civilisations that destroy their life support system. The difference this time is that the dominant civilisation has colonised the farthest reaches of the Earth ...'

Twenty-four hours later, Jules Cashford, a mythologist from Ireland, Ed Posey, co-founder of the Gaia Foundation, Satish Kumar, editor of Resurgence Magazine and myself, found ourselves on a train with Thomas Berry heading towards Schumacher College in Devon, UK. This was in 1996 when Father Thomas, as he was fondly known, was thinking of his next book, *The Great Work*. As he said, in his usual understated way, 'I have been doing some scribbling on what I see as our major challenges today – the transformation of the Western industrial institutions of religion, education, governance and politics from an anthropocentric preoccupation, to an Earth-centred understanding of our role and responsibilities as humans, embedded in the larger Earth community'.

Father Thomas had a quality that is rare in a culture with a long history of domination – a capacity to understand the impact of the industrial process from the inside and from the outside. In other words, to see both the magnificent things it has achieved and the fatal cost to both those who have apparently benefited and those who have suffered. And in an equally compassionate yet dispassionate way, he could sweep from a radical political analysis to a profound mystic perspective of what this means for the individual and collective psyche – of humans and of Nature – which for him were clearly not separate.

To me, Thomas embodied the quality of humanity Einstein called for in his wonderful invocation:

> A human being is a part of a whole, called by us 'Universe', a part limited in time and space. He now experiences himself, his thoughts and feelings as something separated from the rest ... a kind of optical delusion of his consciousness. This delusion is a kind of prison for us, restricting us to our personal desires and to affection for a few persons nearest to us. Our task must be to free ourselves from this prison by widening our circle of understanding and compassion to embrace all living creatures and the whole of Nature in her beauty.

On many an occasion I would hear Thomas's booming voice as the voice of our Earth speaking through him, and resonating with the deep knowing that must surely lie in the human heart. After all, this prison of consciousness to which we have reduced ourselves, is only a few hundred years old in its acute form. Yet we have been co-evolving with other species for millennia. And the greater period of our memory is understanding ourselves as part of a living Universe. Our indigenous history, including for those of us from Europe, is older and longer than our modern conception of who we are as humans.

Thus it was in 1999, when *The Great Work* was published, that a few of us from the Gaia Foundation went to visit Thomas in his character-full home on a lake in Greensboro, USA. He wanted to talk about what he felt was our most pressing challenge – the transformation of the industrial governance system and its legal underpinnings. Of all the industrial institutions, this was the most pernicious, he said, as it legitimised the destruction of the planet and the last remaining traditions which had escaped its domination.

We talked for days about the need for a new jurisprudence – legal philosophy – that recognises that the Earth, indeed the Universe, is lawful and ordered. Human beings are inextricably webbed into this order. For most of humanity's short history on this planet, we have understood that we had to comply with this greater order if we were to maintain our own well-being. The life force that runs through us as humans runs through the whole of life. This order is a dynamic living process, which maintains a dynamic equilibrium. Humans have, over centuries, developed sophisticated ways of governing themselves to maintain this order. They have derived their ways of regulating themselves from the greater order of Nature, the ecosystem in which they are embedded. Those who have destroyed their ecosystem have disappeared. This is now what we face, but this time it is the planetary ecosystem that the dominant culture has destabilised, and in the process created enormous social and ecological injustices.

Thomas reflected on how our loss of intimacy with the living world of Nature, which the industrial process has fostered, has resulted in our loss of perspective of who we are and what truly sustains us. We have become obsessed with ourselves and with the material, rational world. This has diminished our human potential and our memory and enabled us to become manipulated by commercial interests.

The two major sources of inspiration in this quest for a new jurisprudence are the Earth herself and indigenous traditions – living traditions today and the ancestral traditions of Europe, where the industrial process began. The Earth is the source of law and order, and hence law has to be read from the book of Nature. This requires us to re-learn the language of Nature, and to redevelop those senses, which have been atrophied by our obsession with the rational mind and the material world. As a priority, our intimacy with Nature needs to be rekindled.

Indigenous traditions remind us that humans are capable of what Einstein calls for: a more expansive and generous consciousness which 'embraces all living creatures and the whole of Nature in her beauty'; a reverence for the ancestry out of which humans have evolved, including other species; a deep sense of connectivity with our evolutionary journey, which connects us to our present reality; and our inextricable relationship with the larger Earth community.

This made complete sense to us in the Gaia Foundation, as our journey over the decade of the nineties had led us into an intimate relationship with the indigenous people in the Amazon. We realised that what Thomas had identified was the legal principle which is universal to all indigenous governance systems. The Earth is the primary source of law. Law exists. We are born into a lawful and ordered Universe. Humans cannot make law but must become aware of it. To do this, we need to comply with the *a priori* laws of life. The governance systems of indigenous communities do not come from the individual interests of their leaders. They themselves are accountable to a higher law, beyond human interest. Indigenous leaders are also accountable to past, present and future generations. This is what helps to contain the excesses of human ego. A worldview that holds humans within a larger world of interspecies and intergenerational accountability is a vital antidote to human self-centredness and power, and nurtures a more expansive and generous human consciousness.

We left Thomas's book-filled home, feeling ourselves transported into a more expanded consciousness. We could see how Earth Jurisprudence (EJ) recognised and named the universal principle for human governance, exemplified by indigenous peoples, and also gives those within the industrial process a beacon to reach for, which indigenous peoples can help to inspire. For ultimately the common ground which unites, inspires and orientates us all is the living world of Nature – out of which we are born, nourished and in which we are decomposed – whether we are conscious of this or not!

The Journey Unfolds

In April 2001, Father Thomas, Andrew Kimbrell, an advocate from USA who had known Father Thomas for many years, and Gaia, invited a few people to meet at Airlie House, outside Washington, to explore the idea of Earth Jurisprudence. Some of those whom Gaia invited were Jules Cashford, for her scholarship in the stories of origin of Goddess cultures, Martín von Hildebrand who brought his extensive experience with the Amazonian cultures of Colombia, and Cormac Cullinan, a lawyer who had shown a great interest in the idea of Earth Jurisprudence. Already at this first meeting some core principles began to emerge. Those who believed Earth Jurisprudence challenged us with a profound shift in thinking, rallied to meet outside, so that we could dislocate our habits of mind and be informed by the bounty of Nature's spring celebration, expressed in the awe-inspiring light shining through the pink apple blossom and translucent green leaves. Others were distinctly uncomfortable, both with the idea that Nature might have rights, saying these ideas would not stand up in a court of law, and that they needed the comfort and familiarity of tables and flip-charts to think through this romantic idea. In response, Thomas Berry drew up the Airlie Principles, to encourage us to go back to first principles and reach beyond our comfort zones.

As a follow-up to this meeting, we worked intensely in Africa to prepare for the Earth Summit, to be held in South Africa in 2002 (10 years after the first Rio Earth Summit). The vision was to support a strong African presence, calling for 'another way' to the industrial development model: Cormac committed to write a book on Earth Jurisprudence from a legal perspective, called *Wild Law*; Gaia and Cormac's consultancy, Enact International, agreed to co-host a colloquium on Earth Governance; and Gaia hosted a Peoples Earth Summit with other colleagues from Africa, to raise the peoples' voice for the Earth.

Prof. Jacqueline McGlade, now Director of the European Environment Agency, Dr Vandana Shiva, environmental activist from India, Colin Campbell, a traditional doctor from South Africa, Ng'ang'a Thiong'o, Prof. Wangari Maathai's lawyer from Kenya, and Don Pinnock, writer and editor, were some of the participants in the Colloquium. Principles emerging out of these various adventures were:

1. Reading the Law in the Land

By recognising that Nature is the source of law, humanity's primary responsibility is to learn how to read her law, her language. This is the role of shamans and elders in traditional societies, who are trained for years in this skill. The last few centuries of the industrial process however, have radically disabled its followers from being eco-literate. The faculties we require as humans to understand the symbolic language of Nature – intuition, feeling, sensing – have atrophied as the rational mind and human hubris have been exalted. Bill Clinton graphically exemplified this when, commenting on the Human Genome Project, he said, 'Now we know what God knew when he made Man'.

2. Different Ways of Knowing

In order to free ourselves from our prison of consciousness and develop our other ways of knowing (feeling, sensing, intuition) we need to create experiential learning processes, where more than the rational mind is engaged. Cultures across the world, including pre-industrial European traditions, educate their societies through experience-based rites of passage and apprenticeships. As indigenous traditions say, the law of Nature is written in the heart. A society's responsibility is to affirm what people innately know because they are part of Nature. This is a self-discipline that begins from childhood, where children are taught to feel the consequences of breaking this law, for themselves, for Nature and for future generations. This comes from experiencing one's embeddedness in a larger Universe, beyond the self.

As many people have observed over the years, we have enough information to tell us that the consequences of violating the Earth's laws is leading to dire consequences for ourselves, the Earth and future generations. Yet our behaviour does not change. Information and rational understanding is not enough. Change of behaviour and following the laws of Nature comes from an inner moral imperative, when you simply cannot cut the forest or pollute a river or eat food riddled with pesticides or enslave animals and humans in industrial factories – because you feel the consequences of this in the core of your being.

The challenge is how can enough of us get to this point to trigger a cultural shift on a large enough scale? This has been a recurring question over the years.

Taking Stock

By 2003, we felt the momentum had grown amongst our partners, and it was time to gather again and to reflect on next steps. Gaia hosted what became known as the 'May Summit' in London, in 2003, for its core associates to meet Thomas Berry and take stock. Amongst those present were Prof. Wangari Maathai from the Green Belt Movement, Gathuru Mburu (Kenya) and Million Belay (Ethiopia) from the African Biodiversity Network (ABN), Lara Lutzenberger, daughter of José Lutzenberger (Brazil), Brian Goodwin and Satish Kumar from Schumacher College, Ian Mason, UK barrister, and a number of those who had been to earlier gatherings.

We all agreed that Earth Jurisprudence was the underpinning philosophy of our work, although we may have used other terms to describe it previously. Together we provided a range of experiential learning opportunities within our various cultural contexts. We agreed that this work needed to continue, to grow and to interconnect more.

We had been exploring more accessible terms for Earth Jurisprudence, our worry being that Jurisprudence was both not a very well used term and made it sound as if we were talking about legislation rather than the philosophy and ethics underpinning an inherently lawful and just world. However, Thomas was adamant that Earth Jurisprudence was the correct term and our job was to make the idea more accessible!

The Flowering of Things

During 2004 many initiatives flowered. Together with Colin Campbell and his brother Niall, Gaia and the newly formed African Biodiversity Network developed what has become an annual learning experience in Botswana. This entailed living for one week in a traditional Tswana Lodge, following the customs of this Southern African tradition, which is rooted in the principles of Earth Jurisprudence. This process has enabled African leaders and others to explore the EJ underpinnings of African traditions, which had highly sophisticated ways of understanding and teaching Nature's laws and how this fosters social and ecological justice. This training has become the foundation from which young leaders in Africa have regained respect and confidence in their ancient heritage and its richness.

Also in 2004, Gaia arranged for two African lawyers, Ng'anga' Thiong'o and Mellese Damtie, Dean of the Civil Service College in Ethiopia, and two shamans, Colin and Niall Campbell, to visit the Colombian Amazon through Gaia Amazonas. This exploratory visit resulted in a five-year process that, together with what became known as the Botswana experience, resulted in the emergence of many EJ-related initiatives. We learned through this how a few potent experiences can catalyse widespread change, as the transformed become the transformers.

The indigenous people in the Colombian Amazon have been through an intensive process of reviving and adapting their community cohesion and their EJ governance system, with the guidance of the shamans and elders. They now administer an area of rainforest larger than the UK, receiving funds which they negotiate from the state each year. At the centre of their governance system are the sacred sites, places of ecological and spiritual potency, such as sources of rivers, animal breeding grounds, wetlands, and mountains. These places of potency are the vital coordinates of the territory, like acupuncture points in the body of the Earth. They exist across the whole planet and for all indigenous people these places must not be violated, as they maintain the vitality and energy flow of the planet. Governance for them is about maintaining a healthy relationship between the human community, the larger Earth community and the spiritual or energetic dimension of life. The shamans play an essential role in guiding the communities to maintain this order. This experience of witnessing how it is possible to revive EJ practices on such a large scale, and interact with the state on the basis of these principles, has been an enormous inspiration for the ABN.

It was during the first Botswana process in 2004 that two organisations were formed – Porini in Kenya and MELCA-Mahiber in Ethiopia – dedicated to pioneering a path for EJ, rooted in reviving African traditional ecological governance systems. Out of this emerged an initiative of EJ discussion groups in Kenya, Ethiopia, Ghana, UK and South Africa. These created the conditions through which those interested in EJ could reflect and develop initiatives. In Ethiopia, ABN and Gaia supported Mellese Damtie to teach EJ in the Civil Service College. For three years he selected 10 students to go back to their communities to study the EJ principles which were embedded in their customs. After the first year Mellese wrote a gleeful message to the network: 'EJ is alive in Ethiopia! We can now work towards a similar vision here as that of the Colombian Amazon – to recognise indigenous territorial governance based on EJ principles!'

Out of this, MELCA's work grew in the Sheka forest in southwest Ethiopia, the last remaining indigenous forests where traditional communities practice EJ governance. They have worked with the communities to reinforce their traditional structures and gained recognition from the local government for the communities to govern their territories. Development projects have been stopped from destroying more forest. But the challenge of strengthening ecosystem and community resilience continues, in the face of the growing scramble for land and 'resources' in Africa.

In Kenya, the community elders have read Thomas Berry's books and Cormac Cullinan's *Wild Law*, and have been enormously encouraged to restore their EJ traditions, which they recognised in these texts.

Porini accompanied the communities to revive their sacred forests and EJ-based customs and practices, and they have prevented further destruction. The local government now recognises the community's ecological governance system. Other local organisations have been formed, like the Institute for Culture and Ecology (ICE), which does similar work with communities in Kenya to regenerate and protect their sacred sites, territories and traditional livelihoods.

A Global Alliance is Formed

In 2005, Gaia supported another gathering of the growing EJ practitioners. The Community Ecological Governance (CEG) Global Alliance was formed at a meeting in South Africa with participants from Peru (Jorge Ishizawa), Sweden (Anders Tivell), Colombia (Silvia Gomez of Gaia Amazonas) as well as South Africa, Kenya, Ethiopia and Ghana. The focus was primarily on reviving and enhancing EJ traditions, which we call Community Ecological Governance (CEG), and on communities regaining legal recognition for their governance systems. The term CEG emerged to describe ecologically compliant, socially just and culturally diverse values, customs and practices which embody EJ principles.

The group then visited the Venda community in South Africa, which has since become a strong advocate for promoting the EJ principles underpinning their traditions. In fact community leaders have requested that some of Thomas Berry's writings be translated into Venda. His thinking not only resonates with theirs, but also helps them in their advocacy work to show that their campaign to protect their territory and strengthen their EJ practices is part of a growing trend to secure a viable future for generations to come.

By 2005, Bakari Nyari from the ABN was teaching EJ in the University of Development Studies in Northern Ghana and sharing lessons with Mellese Damtie, who had introduced a number of courses on EJ in the Civil Service College in Ethiopia.

In 2005 there was a further breakthrough in the EJ process. Anders Tivell, based at the Swedish Agricultural University, invited Gaia to visit one of the communities he was working with in Kalix, on the Baltic Sea, to explain Gaia's work and the EJ principles. To our surprise, this community completely resonated with these ideas – even the idea that some elders in their community perform the role of shamans, who read the law in the land. They are the ones who advise the community when to fish and to hunt, and in this are far more accurate than the government and policy makers. The community has an ongoing battle with the local authorities who impose laws on governance of the land from Stockholm or Brussels, and completely disregard the community's lived knowledge and experience. This is why they were keen to be part of the CEG Global Alliance as their struggle in Sweden was similar to elsewhere in the Southern Hemisphere. The breakthrough was seeing how the EJ philosophy and practice could inspire European communities, where the industrial process is dominant.

1. Intimacy as a principle

Herein lies another EJ principle. In EJ the law is read from the land through establishing an intimate and respectful relationship with the land. The shamanistic role of mediating between the land, the community and the spiritual or energetic domain of the ecosystem, requires ongoing daily awareness and 'reading'. This is a living law and a living governance system. Laws and policies based on long 'objective' studies have their place, but as we see in the recent climate change science 'scandal', can also be wrong or misinterpreted, and vested interests can take advantage of this. A wise and mature society will consider all sources of knowledge and information but, in the end, discernment of what is the right action is best taken by those living in the place and committed to maintaining a mutually enhancing relationship with their place. EJ is a subjective relationship within a community of subjects, not an objective relationship with a collection of objects – one of Thomas Berry's core EJ principles born out in community governance.

2. Territory is more than the sum of its parts

This relates to another central indigenous EJ tradition. Asserting the concept of 'territory' has been a hard-won battle, especially strong in Latin America. For indigenous people, land is not only soil and biodiversity and water – it is also the place from which your cultural role and identity as part of a larger Earth Community is defined. The law which governs your life and your community, and which has done so for generations, is born with you, from the land. Language and cultural identity are derived from the land. Your ancestors and the spiritual dimensions of the ecosystem and other species all reside in the territory. The land nurtures the psyche. All this together is the territory from which you are shaped.

This understanding of law and territory has been deeply eroded by the transient globalised society. The loss of identity with land, with territory, is one of the root causes of our alienation from the Earth and her laws, other species, and thereby ourselves. The social movements such as Transition Towns in the industrialised countries are the beginning of the recognition of our need to reconnect with place in order to find identity, well-being and to learn once again how to live with ecological integrity, in compliance with the laws which inherently govern our lives. These laws exist, whether we are conscious of them or recognise them or not. The localisation movements arising as an antidote to the globalisation of the industrial process, and the unaccountable powers that go with it, can be seen as a response from the psyche, the *anima mundi*, to find its way back home, at a time of multiple crises and extreme alienation from who we are as human beings, inextricably born of the Earth.

In the UK, the EJ discussion group, initiated by Ian Mason and Gaia, helped to nurture a collaborative initiative with the UK Environmental Law Association (UKELA) and the Environmental Law Foundation (ELF) to run workshops on EJ in the UK, beginning tentatively in 2004. These developed into annual events and by 2009 they had spawned a number of different initiatives led by different groups. The EJ process in the UK was not so much about reviving and affirming EJ traditions, although this is beginning to happen. This process is perhaps more about affirming people's instincts, intuition, sensing and feeling aspects of the psyche, which know something has to change. This work, in the heart of the industrial society, at the centre of its origin, is vital in the transition process. The memory of our true identity has been smothered for many generations. It is for this reason that Gaia has been encouraging inter-cultural exchanges to help stimulate the indigenous memory both within the southern and in the northern hemisphere. As indigenous people living today remind us, our cultural identity with the land never dies, but it needs a lot of watering if we are to remember our roots, when they have been buried for so long.

A Turning Point

In our experience over the last decade, 2007 stands out as a turning point, the point at which EJ shifted to another level of acceptance across the planet. It was the year that the impact of climate change and its recognition also shifted, and perhaps the two are connected.

The Center for Earth Jurisprudence (CEJ), initiated by Pat Siemen, was formed in the USA, hosted Barry and St Thomas universities, in Florida. It held its inaugurating gathering early in 2007. The Community Environmental Legal Defense Fund (CELDF) had established a number of ordinances recognising community and ecosystem rights, and this news spread around the network as a great affirmation that the recognition of Nature's rights was an idea whose time has come. Ecuador's new Constitution recognising the Rights of Nature is another landmark in 2008. In 2009 the Gaia Foundation and the UK Environmental Law Association (UKELA) published a research paper exploring whether Earth Jurisprudence exists in law and practice, revealing that there are some wild laws. And in Australia the first EJ conference was held.

Here, perhaps, we see the emergence in the human psyche of a search to restore our relationship with Nature. This, as Father Thomas reminds us, is the Great Work of our time. Climate change is the ultimate indicator of the extent to which we have violated the laws which govern life – for the first time since our presence on Earth, we humans have destabilised the equilibrium of the whole planet. In essence, this is because of the breakdown in our relationship with our source of life. As indigenous shamans say, when we violate Mother Earth, we ourselves become sick and dehumanised. Ecological and social injustices grow. And we end up where we are now – with a myriad of interconnected ecological and social crises. We can see from the climate negotiations in Copenhagen in December 2009, that the global governance structures are completely dysfunctional. They bring out the worst in human nature – a fanatical belief in economic growth and commercial opportunism where governments are held to ransom by corporate power – as if this can carry on while the fabric of life is unravelling beneath our feet.

Another way is more critical than ever. As one of the great elders of the ecological movement, Teddy Goldsmith said, in times of great crisis nothing less than a U-turn is required of us:

> The ecology we need is not the ecology that involves viewing the ecosphere on which we depend for our survival at a distance and with scientific detachment. We will not save our planet by means of a conscious, rational and unemotional decision, a sort of ecological contract based on a cost-benefit analysis.
>
> A moral and emotional commitment is required. Indeed, one of the key tasks of ecology must be to redirect our emotions so that they may fulfil the role they were designed for: to commit us to what should be the overriding human enterprise of maintaining the critical order of the ecosphere.[1]

A Legacy to be Upheld

In 2009, Thomas Berry passed on, leaving a great legacy for us to uphold. Gaia held an EJ Retreat for our partner practitioners to reflect on the decade, the legacy of Father Thomas and plan for the next decade. Out of this came the recognition that each continent needs to develop its own discourse on EJ, as the context is different. Now that the core principles are clear and the idea is recognised, further refinements are required. There was also a lot to share such as content for EJ courses being run in various countries, including Malaysia and UK. We recognised the enormous challenge ahead, to continue to reach for the breadth and depth of thinking that Father Thomas embodied, as the crises intensify.

Once again, as a core part of the EJ process, the gathering recognised that the hardest challenge for those shaped by the industrial process is to learn to create the conditions for our other ways of knowing to grow, to overcome our powerful desire to control, analyse and rationalise our reality. Especially now as we face these multiple crises, enabling planet Gaia's self-healing processes to take the lead is incredibly difficult. Yet often the most powerful and potent way for the planet to heal herself is for us to stop interfering. Learning to listen and to wait, not to do, and to allow right action to emerge, is the most potent healing for our over-active, over-productive psyche. But this is the hardest thing for us to do. Our atrophied senses work on a different timing. They require the ego's mind to wait – to 'suspend disbelief', in order for something new to emerge.

In the end then, EJ is about practising to live according to a different rhythm – where Nature and our own inner nature have time to reconnect as a reflection of each other. It is about taming the rational mind to stop grasping after reason and knowing the answers, to contain the excesses of our over-inflated ego. It requires self-discipline, to allow the law of the Earth, written in the heart, to emerge. It is about reconnecting with that inner knowing, the inner ethic – where at the deepest level we know that what is right for our own well-being and what is right for the planet's well-being are one. This understanding gives us the moral strength to hold unaccountable power to account, because it violates the good of the whole. That is the true EJ law, inherent in life, connecting us with the web of life. To find this is an inner journey as well as an outer one, where we learn humility from the awesome Universe and the great Mother Earth, Gaia, who gives life and law to all. It is where we all meet. This is the living law of our Earth Community. Our indigenous heritage is evidence that human societies are capable of creating conditions to nurture the better aspects in human nature and to live with ecological integrity. This is the Great Work, the Great U-Turn that is required of us now. The future generations of all species, including our own, depends on it.

Notes

1   Edward Goldsmith, *The Way: An Ecological World-view*, Shambhala Publications, Boston, Massachusetts, 1993, p. 77.

Bibliography

Berry, T, *Evening Thoughts: Reflecting on Earth as Sacred Community,* Sierra Club Books, San Francisco, 2006.
Berry, T, *The Great Work: Our Way into the Future*, Three Rivers Press, New York, 1999.
Cullinan, C, *Wild Law: Protecting Biological and Cultural Diversity*, Green Books, Devon, 2003.
Filgueira B, & Mason, I, *Wild Law: Is there any Evidence of Earth Jurisprudence in existing Law and Practice?* 2009. Available at: http://www.earthjurisprudence.org/documents/WildLaw_Report.pdf

One In All:
Principles and Characteristics of
Earth Jurisprudence

Ian Mason

It was dark and quiet when I woke up. I slipped from under the mosquito net, silently opened the flimsy door and stepped onto the porch outside. Nothing stirred. There was a wooden seat from which I could survey the bush: shadows and mystery against a bright, starry darkness dimly reflected on the still surface of a nearby lake. Everything was silence: no traffic, no generators, no distant aircraft, no electronic hums, no voices, no movement, no lights; perfect silence. Mind and heart filled with awe, reverence, wonder, quiet. There was nothing to do but join in, rest, enjoy: all senses alive but making no barrier between inner stillness of being and the outer stillness of the pre-dawn bush.

Time passed timelessly and it was lighter than before. Silence deepened, nothing moving, still no sound, not even lapping water on the lakeshore. Now there was a horizon, dim, wild, jagged, African, against the faintest light. Stars faded, shadows hardened, shapes defined. Suddenly, momentarily, a harsh cry filled the silence and was gone. A bird? A baboon? Nothing else stirred. Then another call, another single pulse, different, lighter, higher pitched, there and gone – and then another, deep and throaty, and another and more as the wild, disordered, joyous, dawn orchestra awakened.

As the light strengthened, a group of large dark boulders took shape on the far side of the lake. Surely they were not there yesterday – did one of them move? A long wait and another movement: hippos, a whole family. A snort and a rustle in some nearby bushes, a disturbance in the still water and then a steady wake behind an odd irregular shape as a submerged hippo swam the lake. And suddenly the sun was in the sky, the ever-present silence was filled with the vibrant sounds of Nature and a new day began as day after day has begun here for thousands of years.

But there are far fewer hippos than there were, even on this sequestered game reserve, because there are too few wardens to combat poachers. A few antelope and baboons remain, but the lions, elephants, rhinos and even zebras and giraffes that once shared the bush have gone, along with the simple people who once shared it with them: hunted or driven from their habitat to meet the demands of distant markets. The people, like the animals, knew the bush and how to live here. They knew there was a different law for the bush and for the homestead and that their survival depended on observing the law and knowing the difference. They understood the rhythms of days and seasons, the migrations and cycles of their wild companions, the qualities and messages of plants and flowers that they depended on for continued existence; and they understood the nature of their dependence – that for every taking there is a giving, that survival depends on mutual respect and that Nature provides for all Earth's creatures. Earth Jurisprudence is a philosophy of law that applies this intimate understanding of Nature to modern human law and law-making.

Earth Jurisprudence was conceived in the wake of Thomas Berry's call in his book, *The Great Work*, for a new jurisprudence to re-shape the human relationship with Earth. He saw that the combination of Enlightenment philosophy and reductionist science had transformed the human relationship with the rest of the natural world into a tyranny of exploitation where nature is simply regarded as a resource for the satisfaction of human ends. He saw this relationship as the cause of contemporary human depredations – species loss; deforestation; pollution of air, water, land; climate change – and saw the need to rediscover a relationship with nature that is mutually enhancing, benefiting Nature and its human component together. Such a relationship would be based on human respect for Nature's intrinsic value as the life force in everything.

His book describes a new jurisprudence, which came to be called Earth Jurisprudence, as an essential part of a cultural transformation involving theology/philosophy, education, economics and law in which the key will be a sense of the unity and interdependence of the natural world including its human component. This view takes the universe as its primary reference, relating all fields of knowledge to a universal whole and trying always to see component parts in the context of the whole to which they belong, understanding as a universal principle that the whole really is greater than the sum of the parts. Such a view postulates hierarchies of 'holons', or unities within unities. The context of an atom is a molecule; of a molecule, a cell; of a cell, an organ; of an organ, a body; of a body, its habitat, or eco-system; of habitats, elements – earth, air, sunshine, water; and of elements the universal governing principle that unites them into a coherent living whole. Each of these forms is a holon with a relationship to all of the others. Although each is itself a recognisable and defined entity, none of them can exist or survive except in the context of the complete complex of relationships with all the other holons in the system. Each is utterly dependent on all of them.

This principle of wholeness is the key principle running through the entire philosophy of Earth Jurisprudence. It applies as much to human laws and institutions as to anything else. In essence, no individual exists apart from a family; no family apart from community; no community apart from its tribe or nation; no tribe or nation apart from humanity; no humanity apart from Earth; and no Earth apart from its parent universe. Each of these units of human organisation is a holon which is related both to its component holons and also to the greater holons in which it is held and upon which it depends.

In this view Nature ceases to be a collection of objects for human use and is regarded instead as a communion of subjects that includes the human race: subjects because everything is subject to the laws of its own creation, maintenance and dissolution and cannot avoid them, and also because the laws governing all creatures are intimately connected. In this sense we live in an entirely lawful universe in which the laws are already given. All that human beings can do is to discover them and live by them.

For example, all living things depend on some combination of earth, air, water and sunshine for survival. It follows that any interference with any of these essential life support elements affects everything that depends on them, ultimately, everything that lives. This can be observed everywhere. The canopy of a rainforest, for example, affects every creature on the forest floor by limiting the sunshine that reaches them. Similarly, its ability to seed rain clouds affects global climate conditions, which in turn affects the ability of all living things to survive. All this is entirely lawful in the sense that it is based on clear and discoverable principles of cause and effect that do not change. This principle of lawfulness also runs through the whole philosophy of Earth Jurisprudence, which postulates that law is discovered, not made, and that the health of each holon depends on the health of the whole.

This gives rise to a third principle which particularly affects the human realm. This is the principle of care. Simply put, the principles of wholeness and lawfulness require that human energy and consciousness are directed to leaving Earth in as good or better condition than we find her. The principle of care is necessary because human beings have immense creative power which they can use to modify their surroundings. Used without regard to the laws of those surroundings, that power can equally be immensely destructive. The principle of care directs human action to benefit the wider Earth community in the interests of the whole. Earth Jurisprudence assumes that there is a proper place for the human in the Earth community and that human powers and abilities to modify nature are subject to a duty of care for the health or well-being of the whole.

Philosophic Foundations
The philosophic foundations for such respect for nature are not new. They can be found in what Leibnitz and Huxley described as the perennial philosophy – that is to say, the undercurrent of philosophical principles that can be discovered at the heart of most of the world's greatest religions and philosophies. In their different ways all the major religions and many philosophies from ancient Egypt to modern Christianity and from indigenous America to Taoist China have pointed to some kind of hidden Reality, Being or Great Spirit which at the same time transcends ordinary human experience, and is immanent throughout the material world of

Earth. Understanding Earth as the outer expression of that inner reality is the philosophic or spiritual foundation of Earth Jurisprudence.

The idea is that the inner reality is unified in the way that the life in all that lives is the same life appearing in countless different forms. All creatures have this in common, that they are animated by the same conscious power or consciousness which runs through and unites them. They all share in existence by the fact that they exist. This is the real wholeness that is the first principle of Earth Jurisprudence. It is the foundation of the Christian commandment of love, of Buddhist compassion, and of Hindu concepts of harmlessness: 'The man who can see all creatures in himself, himself in all creatures, knows no sorrow' [*Eesha Upanishad*]; 'The Self dwelling in all beings, all beings dwelling in the Self, sees he whose mind has been made steadfast by yoga, who everywhere sees the same' [*Bhagavad Gita*].

The same idea can be found on the songs of indigenous people:

> I am the shine of sun;
> I am the shine of moon;
> I give shine to stars;
> Because they are all within me.
>
> I am the shine of sea;
> I live in the wind;
> I live in the forest;
> Because it's within me.
> *Santo Daime people, South America*

And in English poetry:

> If the wild bowler thinks he bowls ...
> Or if the batsman thinks he's bowled,
> They know not, poor misguided souls,
> They too shall perish, unconsoled.
>
> I am the batsman, and the bat,
> I am the bowler and the ball,
> The umpire, the pavilion cat,
> The roller, pitch and stumps and all.
> *Andrew Lang*

The message is that no-one has an existence or life separate from everything else and that consequently everything is connected to everything else in a relationship of interdependence. It follows that human beings, who have some freedom of choice and freedom of action, have a responsibility to act from understanding that their actions affect everything else in Nature. To cut a tree, to dam or divert a river, to blast the summit from a mountain or strip the vegetation from a plain, or to exterminate a species, are not isolated acts. They are acts which disrupt whole ecosystems and everything that depends on them, from microbes and insects through animal and human communities to temperature and precipitation. If human actions are destructive and careless, we end up destroying the very things that we depend on to sustain us through life and to sustain the life in us. This principle of care completes the core triumvirate of principles of Earth Jurisprudence: Wholeness, Lawfulness and Care are the key guides to human action.

A Materialist View

For those who doubt the existence of the inner reality of the perennial philosophy, Earth can still be seen as alive with the same life that animates the human form. Modern, scientifically attested, Gaia theory and systems theory confirm the interconnectedness of life, nature and humanity. We remain dependent on our environment and its ecosystems; the world around us is governed by relationships of cause and effect; our creative and destructive tendencies must still be tempered by recognising the need to maintain the systems that support our continued existence. Nature still demands our respect because of our utter dependence on her (or 'it', for the truly materialist). Recognising this common sharing and celebration of life and the respect it engenders is the basis of the mutually enhancing relationship at the heart of Earth Jurisprudence.

Mutual Enhancement

Thomas Berry's observation was that the human–Earth relationship enhances human life and experience in a flourishing world of nature. As he put it, 'Every component of the Earth community is immediately or mediately dependent on every other member of the community for the nourishment and assistance it needs for its own survival. This mutual nourishment, which includes the predator-prey relationship, is integral with the role that each component of the Earth has within the comprehensive community of existence'.

But although it includes physical nourishment, human dependence on nature is not purely material because of the power of the natural world to enrich the human mind and spirit. Human mind and emotion are also nourished by contact with and experience of nature; nourished by the beauty, variety, fecundity, mystery and power of nature as well as by the companionship she offers and the wonder she evokes. The human spirit too is enriched when nature is seen and understood as the immediate clothing of the immanent and ever-present reality that pervades everything at all times. In this perception care for nature becomes a sacred trust.

Learning by Intimacy

Because of the closeness of the human–Earth relationship, Earth Jurisprudence draws on human faculties of intuition, empathy, wonder, humility and intelligence to rediscover and rebuild the intimate relationship with nature that enriches the human psyche and enlivens nature herself. Here, it is important to understand that Earth Jurisprudence is not some new invention of the human mind, because it is based on the principle that law is discovered, not made. Mike Bell, who has worked for many years with the indigenous Inuit of the Canadian Arctic, tells the story of the Inuit carver who, asked how he managed to put such beauty, life and spirit into the polar bear he was carving from stone, replied, 'I don't put the spirit of the bear into the stone. It's already in the stone. I just chip away everything that doesn't look like a bear'. This is the real spirit of Earth Jurisprudence. It is only a matter of chipping away the accretions of human-centredness to rediscover the naturalness of relationships with the wider Earth community. The real law of Earth Jurisprudence is that intimate relationship governing all human responses to Nature.

For this reason, great attention is paid to the customs and practices of indigenous peoples when looking for the practicalities of Earth Jurisprudence. Such people have a very different approach to law when they live, as they traditionally have, in close communion with Nature and do not see themselves as separate from it. Indigenous peoples do not rely on statutes and ordinances, or on courts or parliaments for their laws. They find them instead in their own intimate relationships with their own particular surroundings. The relationships themselves are the laws, just as the relationship of mother and child determines the conduct of each towards the other without assistance from externally imposed laws. The law works from within the participants in a relationship where each understands that they are not just here for themselves, but are each also here for the other – which is not really other at all – and also that the health of the relationship depends on them both.

Intimate Law

For that reason, Earth Jurisprudence works on two levels. The first is the intimate, personal level in which each individual finds and establishes their own relationship with Nature. This is the real origin of the sense of lawfulness to the extent that, where it is well established and understood, no other law is necessary. This is the law of the bush, the real law of the jungle which, far from being competitive, destructive or exploitative, in fact allows for mutual coexistence and mutual nourishment. Even the predator/prey relationship is held in this law, tempered by a principle of necessity which requires that the only acceptable taking, the only acceptable killing, is taking or killing that is necessary for survival and that every taking of life is tempered by an expression of gratitude for the life that has been forfeit so that life may continue.

When human beings live in such intimate communion with nature it becomes impossible to act in ways that are actively harmful to nature, just as it is impossible for loving parents to act deliberately in ways that are actively harmful to their children. The relationship itself makes harmfulness impossible.

Wild Law

It is when this intimate communion with nature is lost or forgotten that the second, political aspect of Earth Jurisprudence becomes necessary. This political jurisprudence is the use of laws and governance systems of the dominant contemporary kind to reflect the more intimate jurisprudence of close communion with nature. This has given rise to the concept of 'wild law', which is lawyers' law drawn from and reflecting the intimate human–Earth relationship and recognising the limitations of formal law. The reality is that formal law in the conventional sense cannot *command* respect and intimate relationship with Nature – all it can do is expect and enforce conduct which is consistent with such relationship and prohibit conduct which is not consistent with it. Wild law is formal law founded on principles of Earth Jurisprudence.

Earth-Centred Law

Such law has a number of characteristics. The first of these is 'Earth centredness'. What this means is that, instead of formulating laws from a purely human perspective and for purely human purposes, laws are formulated from the perspective of the whole of nature epitomised by Earth. The distinction can be illustrated by the US Endangered Species Act 1973 which is a major US federal law directed to protecting species and the ecosystems to which they belong, in accordance with international agreements. In order to benefit from the protection of the Act, a species has to be listed as either 'endangered' or 'threatened', and currently around 1000 species are so listed. The Act is enlightened and effective legislation in the context of an anthropocentric legal environment in which species do become endangered by human interruptions of delicately balanced ecosystems. Earth Jurisprudence, however, would change that context.

The real difference from an Earth-centred point of view, is that all species would be considered as protected by the common duty of care. Far from making special provisions for endangered species, special provision would have to be made to allow humans to use or affect species in any way which might jeopardise their survival.

The important point here is that Earth, epitomising all its components, is recognised by the law as having a value in its own right regardless of any particular value it may have to the human species. The attitude is already expressed in the United Nations World Charter for Nature:

> Every form of life is unique, warranting respect regardless of its worth to man, and, to accord other organisms such recognition, man must be guided by a moral code of action …

And more recently in the 2009 Constitution of the Republic of Ecuador:

> Art 71 – Nature or Pachamama, from which life reproduces and enfolds itself, has the right to the integral respect for its existence and the maintenance and regeneration of its vital cycles, structures, functions and evolutionary processes.

Such provisions reflect the idea of Earth as a communion of subjects and there-fore enjoying a sort of equality before the law. Thomas Berry saw this in terms of three key rights of all beings, or at least of all species: 'the right to be, the right to habitat, and the right to fulfil its role in the ever-renewing process of the Earth community'. No human action would be permitted to compromise those funda-mental rights of species, imposing an effective law of restraint on human activity whose effect would be to ensure that species did not become so endangered as to need the special protection of an Endangered Species Act. In Earth Jurisprudence, environmental protection becomes a central principle and underlying presumption of all law, not a special provision for particular circumstances.

Commons

A second characteristic of wild law is that the whole natural world is regarded as a commons which cannot be reduced to private ownership. This attitude inevitably has significant implications for property law because it calls into question the nature and extent of human rights over the natural world and therefore over land. When dealing with land, or real property, most contemporary jurisdictions, fuse a number of elements which are in fact quite distinct. These include (a) the solid part of the Earth's surface; (b) human rights over or pertaining to the Earth's surface (such as liberties, easements, purchase and sale, etc.); (c) human additions or adap-tations of the Earth's surface such as permanent fixtures, buildings, paths, planta-tions, gardens etc.; and (d) the soil and everything above and below it including water, water courses and anything that grows or lives in or on the land.

Earth Jurisprudence makes a key distinction between the naturally occurring element of land and the human impact on it. Land, in Earth Jurisprudence, means the whole natural environment including the Earth's solid surface, all natural vegetation and wildlife, the soil and naturally occurring organisms within it, rivers, standing water, watercourses etc. What we call human rights over land such as ownership, easements and rights of way are seen by Earth Jurisprudence not as rights but as special privileges which carry obligations. Such privileges do not form part of the definition of land, although they will invariably be attached to the use and occupation of land. The same applies to human adaptations or additions to land. They are quite distinct from the land itself and need and deserve separate treatment. A wild law definition of land might therefore look something like this: *'land means the whole of nature external to the human form including the Earth's solid surface, anything existing naturally beneath it, all naturally occurring vegetation and animal life, the soils and naturally occurring organisms within it, rivers, standing water and water courses'.* Any human privileges in relation to land occupation and use would necessarily be subject to the three key Earth rights, the right to be, the right to habitat and the right to fulfil its role.

Another aspect of Earth Jurisprudence-based property law is that it would recognise sacred places as a special category of land of particular importance both to human communities and to nature. These are places which have traditionally been used by people, often without disturbing their wild character, to relate to the natural world, the universe and the hidden powers and forces within them. Sacred places would be subject to special protection so that there would be no question of their being used for any other purpose.

Restorative justice

A third characteristic of wild law is that its system of remedies is based on restorative, rather than retributive, justice. This too is based on the principle of wholeness, the idea being to restore the damaged integrity of both human and Earth communities and make it possible for the offender or wrongdoer to make appropriate amends and rejoin the community as quickly and efficiently as possible in the interests of the well-being of the whole. Mike Bell describes the rationale from his experience with the Inuit people:

> the Inuit had one overwhelming concern – survival. Because they lived in what is probably the most severe environment on the planet, every person had to contribute to that survival. They had laws and lived according to a code of conduct and a jurisprudence that was designed to ensure that survival ... When someone committed an offense, the offense was not seen as a breaking of the law, but, rather, it was seen as the severing of a relationship with other members of the community. If the offense was not dealt with, the offender could jeopardise the survival of the group. So every effort was made to get the individual to admit his failings, change behaviour, and restore his relationship with the group ... This process usually involved: mediation, especially through the intervention of elders or camp leaders; healing – the internal acknowledgment of the offense and an apology to the group and, in particular, to the person offended; and restitution – beginning with a change of behaviour and often ritualised by some public effort or compensation to the victim or the group to 'heal the relationship.

Community Ecological Governance

A fourth characteristic of wild law is community ecological governance. This idea comes from observing the customs and practices of indigenous people who have retained their close intimacy with the natural environment in a way which enables them to live simply and well without causing harm. It is observable that such peoples have found their own solutions to the problems of living closely with nature and in doing so have developed a profound wisdom and ecological knowledge. This is almost invariably combined with a sense of community cohesion and ecological integrity in which sustainable food and livelihood security are developed and maintained.

Community ecological governance recognises that community knowledge holders, or elders, are custodians of a long heritage of wisdom and understanding which urgently needs to be preserved and passed on to further generations. It also recognises the significance and importance of sacred places as living libraries of natural biodiversity and ecosystems, and a living link with the long chain of evolution. Such people and places require careful preservation and attention as centres of community life and knowledge and as a vital reference for us today.

Although they would not have used the term themselves, such communities exemplify the Earth Jurisprudence way. Their governance starts locally and is integrated with the Earth community. Their culture derives from that intimacy. The aim of Earth Jurisprudence in the extended holons of regional, national and international governance is to develop an enabling and supportive legislative framework that encourages the survival, health and well-being of such communities and also encourages hitherto less sensitive communities to learn from them and adapt accordingly.

Community ecological governance works from the principle that human health and well-being are dependent upon the well-being of the whole planet Earth. It springs from the observation that a holistic and culturally sensitive approach that recognises the basic principles of wholeness, lawfulness and care, can enable communities to reduce their vulnerability to poverty and ecosystem collapse while enhancing the well-being of the entire Earth community.

Conclusion

Many of the ideas presented here will seem far-fetched, idealistic and improbable to many people. However, as Professor Philipe Sands Q.C. remarked when commenting on Earth Jurisprudence: '… experience teaches us that what may seem as over-reaching at one time soon becomes conventional wisdom'. With the world facing an urgent need to forge a new and healthier relationship between the human race and the planet that sustains us, Earth Jurisprudence offers an intellectual and jurisprudential framework for doing so.

Bibliography

Bell, M., 'Thomas Berry and an Earth Jurisprudence', unpublished essay, 2001

Berry, T., *The Great Work*, Bell Tower, New York, 1999.

Cullinan, C., *Wild Law: A Manifesto For Earth Justice*, Green Books, Devon, 2003.

Harding, S., *Animate Earth*, Green Books, Devon, 2006.

Huxley, A., *The Perennial Philosophy*, Triad Grafton Books, London, 1985.

Schumacher, E.F., *A Guide for the Perplexed*, Abacus, London, 1978.

Shastri, T.R. (ed)., *Bhagavad Gita* Samata Books, Madras, 1985.

Stone, C., *Should Trees Have Standing: And Other Essays on Law, Morals and the Environment*, Oceana Publications, New York, 1996.

Wilber, K., *A Brief History of Everything*, Gateway, New York, 2000.

Yeats, W.B. & Purohit, S. (eds)., Eesha Upanishad, Faber & Faber.

Key Principles to Transform Law for the Health of the Planet

Judith E. Koons

> But we have only begun to love the earth.
> We have only begun to imagine the fullness of life.
> How could we tire of hope? – so much is in the bud.[1]

Earth Jurisprudence is an emerging legal theory based on the premise that rethinking law and governance is necessary for the well-being of Earth and all of its inhabitants. Earth Jurisprudence is an inclusive and systems-based theoretical perspective that supports robust environmental regulation and recognises a kinship with the field of environmental ethics.[2] In addition, Earth Jurisprudence embraces the connection between Earth justice and social justice.

Earth Jurisprudence brings an innovative jurisprudential dimension to the environmental movement. At the heart of this dimension lies the premise of a necessary shift – proposed by ecological philosopher Thomas Berry and others – from a human-centred to an Earth-centred system of law and governance, recognising humanity as part of the greater Earth community. Without such a jurisprudential shift, Earth and humanity remain at peril.

The purpose of this essay is to contribute to the development of the field of Earth Jurisprudence by suggesting some key principles and their applications to law and governance. Drawing from the functioning of the universe, this essay will explore a vision of Earth Jurisprudence through three principles: the intrinsic value of Earth; the relational responsibility of humanity toward Earth; and the democratic governance of the Earth community. These jurisprudential principles will be illustrated through a legal framework of rights, responsibilities and duties, and through the representative legal doctrines of standing, the public trust doctrine and intergenerational equity. To begin the creative enterprise envisioned in this essay, the next section invites the redesign of our systems of law and governance.

A New Vision of Jurisprudence

> Two things are needed to guide our judgment and sustain our psychic energies for the challenges ahead: a certain alarm at what is happening at present and a fascination with the future available to us if only we respond creatively to the urgencies of the present.[3]

If citizens' commissions were convened to rethink our systems of law and governance for the twenty-first century, where might the members begin? They might start with current snapshots of what is happening in the world. The first snapshot would depict global warming, with images of melting glaciers, rising oceans, cataclysmic weather events, and perishing species.[4] A second snapshot would show humanity at war with itself over efforts to amass power and resources as well as over ethnic and religious differences. A third snapshot would illustrate the disparity in wealth across the world. Although other snapshots would depict areas of peace and cooperation among people, the new visions of law and governance should be equipped to address overarching problems in the world while also preserving its successes.

The environmental, social, and economic distress depicted in the snapshots should provide great impetus to consider ways to design human institutions to preserve ecological and human health. A threshold step would be to conceive of Earth at the centre of law and governance, shifting away from purely human-focused systems and appreciating the role of humankind as a part of the broader community of being. An aligned step would be to refocus on ways that law and governance could support ecosystems and the complex interactions among animate and inanimate entities upon which life depends.

Some citizens might recognise that Earth-centredness as a guiding philosophy is not new. This theme has long animated environmental reformers, social justice activists, indigenous rights movements, and grassroots campaigns for sustainability. Once independent of each other, these movements are coalescing to effect a wider change in consciousness that is necessary to bring about peace, social justice, and environmental health. Into this momentous niche of time and onto the foundation laid by a multitude of environmental groups and workers, Earth Jurisprudence is stepping forward to formalise and systematise Earth-oriented concepts in the field of law.

In considering a philosophical framework for Earth-centred systems of jurisprudence, the citizens might focus on principles that govern the workings of Earth and the universe. With such a focus, human systems of governance would reflect the attributes of the natural systems in which they are embedded. According to Thomas Berry, the universe is organised according to three main themes – subjectivity, communion, and differentiation. As precepts that arise out of scientific theory and philosophy, these themes could serve as a platform for rethinking law and governance.[5]

The first theme is subjectivity. Through subjectivity (autopoiesis), Thomas Berry and Brian Swimme see the universe as self-organising, with self-manifesting power. Stars regulate hydrogen and helium to produce light and chemical elements. Earth is a self-regulating system; the balance of chemicals in the atmosphere, oceans, and soil is continually renewed and adjusted. Every atom of the universe is a self-organising system, 'a storm of ordered activity'.[6]

The second theme is communion. Through communion (interdependence),

Berry and Swimme see the universe as a 'web of relationships' that form a unity that is comprehensive. From the first moment of existence, when the first particles exploded into being, each particle has been related to every other particle in the universe. Scientists, particularly in the twentieth century, have noted the full extent of relatedness of the universe. Isaac Newton brought forth our understanding of gravitational attraction; Darwin offered evidence of genetic connections in the web of beings; Einstein and quantum theorists presented new understandings of relatedness in the universe at sub-atomic levels.[7] 'To be', according to the universe, 'is to be related'.[8]

The third theme is differentiation. Through differentiation (complexity), Berry and Swimme see the universe as a reality of 'unending diversity'. The originating explosion expressed a creativity that formed galaxies 'of highly individuated starry oceans of fire'. That creativity is ongoing. Multiplicity governs the structures of galaxies, stars, and planets. On Earth, life is reflected in an abundance of diversity. We humans manifest ourselves in an astonishing array of modes of being. '[T]o be', according to the universe, 'is to be different'.[9] Throughout the universe, tiny particles and enormous spiralling nebulae are expressing, 'I am fresh'.[10]

How could the philosophical and scientific themes of subjectivity, communion, and differentiation translate into principles for systems of jurisprudence and into working legal standards? A jurisprudential reflection of subjectivity may lie in the principle that all beings, systems, and entities in Nature have intrinsic value, to be expressed in law and governance. The theme of communion may be translated into jurisprudence as the responsibility of humanity to appreciate our relationship with Earth as a sacred trust. Finally, differentiation may be reflected in the notion of an Earth democracy that supports, at all levels of governance, legal recognition of all components of our Earth community, both present and future.

The next three sections will elaborate on the jurisprudential principles of intrinsic worth, relational responsibility, and Earth democracy. To illustrate how these principles may be reflected in legal doctrine, this essay will highlight standing, the public trust doctrine, and intergenerational equity.

The Principle of Subjectivity: Intrinsic Value of Earth

Each individual thing in the universe is ineffable.[11]

In discussing subjectivity, Berry and Swimme affirm that all of existence – from atoms to galaxies and from colonies of ants to the sun – exhibits creative and self-organising dynamics. Despite the often careless way we interact with Earth and the entities of Earth, Nature is a subject and not a collection of objects.[12]

Recognising the subjectivity of Nature carries legal, philosophical, and moral significance. Western law, philosophy, and morality have long been structured around dualistic thinking. For example, law focuses on plaintiff and defendant, judge and jury, law and facts, theory and practice. In philosophy, key polarities are reason and passion, mind and body, community and autonomy, culture and nature. In morality, we think in terms of right and wrong, is and ought, good and evil, cognition and volition, liberty and constraint. Our everyday thinking is structured in terms of male and female, fast and slow, early and late, tall and short, loud and soft, thin and fat.

One of the chief dualisms underlying Western thought is subject and object.[13] Subjects (those like me) are assigned value and everything unlike me is an 'other'. The consequences of this dualism include justifying mistreatment of others based on perceived differences and interacting with the world through privilege, but being unaware of it. Some scholars propose that othering or objectification is the basis of violence because the belief is internalised that only beings identified as subjects are capable of suffering cognisable harm.[14]

For Berry and Swimme, the subjectivity of the universe is manifest everywhere. Within everything in the universe is an 'inner principle of being' that connotes a power to participate in the ongoing creation story. In this cosmology, sentience and potential sentience pervade the world.

While the subjectivity of 'higher life forms', such as mammals, may be granted, some may balk at considering natural entities or 'mere things' to possess subjectivity. Consider one of the most challenging cases – the subjectivity of rock formations. Russian biologist Vladimir I. Vernadsky defines life in terms of dispersal of rock, or rock that is rearranging itself. The crust of the Earth, to Vernadsky, has sufficient energy to transform the passive geological parts into living parts through metabolic action. In this way, living organisms may be understood as composed of inorganic minerals from the crust of Earth, which cycles living matter into inorganic minerals and then transforms those minerals back to living form. It was of some significance to Vernadsky that the same atoms alternate between animate and inanimate matter.

From the perspective of the Universe Story, Berry and Swimme imagine that Earth, once a fiery rock, now 'fills its air with the songs of birds'. Out of the dynamic activity of molten magma, the self-organising power of the universe brought forth new shapes – 'animals capable of being racked with terror or stunned by awe of the very universe out of which they emerged'.

Subjectivity may be translated into Earth Jurisprudence as the principle of the intrinsic worth of Nature. This claim stands on the premise that beings, systems, and entities in Nature warrant *moral consideration*. In 1978, the notion of 'moral considerableness' was first used in the environmental context by ethicist Kenneth Goodpaster.[15] To Goodpaster, moral considerableness means that 'something falls within the sphere of moral concern, that it is morally relevant, that it can be taken into account when moral decisions are made'.[16] Having moral considerableness is broader than holding moral rights and is 'like showing up on a moral radar screen – how strong the signal is or where it is located on the screen are separate questions'.[17]

The claim of the intrinsic value of Nature also stands on the premise that beings, systems, and entities in Nature warrant *legal consideration* and should be given legal recognition. Christopher D. Stone presented the argument of the legal considerateness of Nature in various writings beginning in 1972.[18] Stone argues that having legal consideration, like moral consideration, should not be confused with holding rights.[19] An entity that has legal recognition may or may not be a rights-bearing entity. Legal recognition may be given in a number of different ways. A jural person may be granted rights, be given duties and responsibilities, be the recipient of immunities and privileges, or be held liable – all of which are inter-mediate, operative notions that flow from the broader principle of legal considera-bleness. Having legal status means being enabled to participate in the legal system, although not necessarily as a rights-holder.

Legal doctrines routinely allow 'persons' that are not human beings to partici-pate in the legal system. Among the 'persons' permitted to sue are ships, trusts, municipalities, estates, joint ventures, universities, railroads, churches and states, not to mention business corporations.[20] Stone notes that lack of moral decision-making capacity does not undermine the recognition of moral and legal status, for example, of humans with mental disabilities. Furthermore, guardians and trustees regularly appear in our legal system to give voice to people and entities who are unable to speak. Federal and state agencies already serve as guardians and trustees of natural entities such as public lands, marine mammals, and 'natural resources' that have suffered damage. There are many ways of bringing natural entities into legal considerateness.

The legal status of natural entities may be understood in terms of a floor of commonalities as well as a ceiling of limitations. For commonalities, we all share this ground, this air, this water, and this history of Earth and the universe. To give effect to these commonalities, Berry asserts that each component of Earth embodies three rights: 'the right to be, the right to habitat or a place to be, and the right to fulfill its role in the ever-renewing processes of the Earth community'. For limitations, Stone reasons that a natural entity's legal status must be 'intelligible'. If a tree were to be granted rights, for example, it would not be to sit on a jury, but perhaps to be given voice through a guardian to be saved from a chainsaw. In similar fashion, Berry understands rights to be role- and species-specific: 'Difference in rights is qualitative, not quantitative. The rights of an insect would be of no value to a tree or a fish'. Rights may vary for different rights-holders, but also allow for participation in the legal system.

How the legal status of jural natural entities is to be recognised – via rights, duties, or responsibilities, for example – and how that legal status is to be consid-ered when in conflict with the rights, duties, and responsibilities of other jural persons and entities are matters for complex weighing. However, the imagined difficulties of adjudicating and legislating Earth's legal status does not alter the principle that Nature, having intrinsic value, is worthy of legal consideration.

Courts and legislators commonly sort through weighty conflicts in complex cases, including those raising negligence in mass disasters, competing rights and duties of terminally ill patients and caregivers, criminal liability of corporations, patentability of life forms, and responsibilities of nations for war crimes. In similar fashion, our legal system must be able to consider rights and obligations of other-than-human animals and ecological entities.

One way the principle of intrinsic value of Earth could be given legal expression is through the doctrine of standing. A number of scholars have called for rethinking the doctrine of standing, which at present denies other-than-humans and natural entities the right to sue in their own status. Instead, in efforts to protect other-than-human species and natural entities, human plaintiffs must allege injury to their own associational, recreational, aesthetic, scientific, and educational interests. The result is often strained, if not tortured and sad.[21]

A prototypical allegation to support standing appears in *Lujan v Defenders of Wildlife*.[22] The affiant, a member of an environmental organisation, alleged that she had travelled to Sri Lanka and 'observed the habitat' of the endangered Asian elephant and the leopard. Although the affiant was not able to see the endangered species, she was harmed because she 'intend[s] to return to Sri Lanka in the future and hope[s] to be more fortunate in spotting at least the endangered elephant and leopard'. Allegations such as these miss the appropriate focal point for judicial inquiry, which should be on the threatened injury to endangered species, not on fictionalised human injury to gain access to court. That the doctrinal requirements for entry into our legal system find their expression in sworn statements that are superficial and off-focus serves not only to diminish the dignity of human beings and our legal system, but also to ignore and jeopardise Nature. Focusing on real harm to valued participants in our legal system would bring nurturing depth and meaning to law and governance.[23]

The Principle of Communion: Relational Responsibility

> The global environment with its finite resources is a common concern of all peoples. The protection of Earth's vitality, diversity, and beauty is a sacred trust.[24]

The prior section discussed the notion of dualistic thinking, which sets up subject-object relationships that operate as a hierarchy. Linguists are careful to note that each binary is marked by a favoured pole, generally expressed first, followed by a disfavoured pole (for example, judge-jury, male-female, reason-passion, theory-practice).[25] This type of thinking, embedded in law and language, has been cited as the basis for the subject-object relationships that structure gender, race, and class injustice.[26] Ecofeminists extend this idea to argue that binary thinking also supports the exploitation and degradation of Nature, viewed as a 'resource' to be used by humans without compunction and as a wilderness to be tamed, as in Humanity versus Nature.[27] The consequences of a worldview based on dualistic thinking are tragically apparent in the separation of humanity from Earth and the grotesque overuse of the goods of Earth to support consumptive lifestyles. Dualistic thinking creates and reinforces humanity's disassociation from Nature.

However, the functioning of the universe is not reflected in hierarchy or separation, but in a circling dance of spheres, orbits, and rotations.[28] Life on Earth may be seen as a circle, with the cycle of seasons, the rhythm of birth and death, and the movement of water from clouds to rain to transpiration in plants and back again.[29] Cullinan advises, 'Western physicists confirm that the same atoms and sub-atomic particles may be part of the soil on Monday, a plant on Tuesday and us on Wednesday'. Even the predator and prey relationship bespeaks of a circle of intimacy.[30] Beings on Earth serve as food for others.

Berry refers to this tendency of the universe as communion. In the circle of interdependence, humankind is part of the whole. Our proper relationship with Earth is not one of separation and exploitation, but one of membership in the Earth community.[31] Healthy natural systems function according to 'whole-maintaining' characteristics so that each part of a system acts in a way that supports the well-being of the entire system.[32] Any aspect of the system that functions to undermine the whole will eventually stop operating, along with the full system.

To take a lesson from natural systems on Earth, humanity must begin to function as a component of a larger natural community. Much of humankind has been behaving in ways that are at odds with being 'part of a whole'. Failing to orient toward our relationship to the whole, the bulk of humanity has been acting as a whole within itself, as the centre of the universe. The current environmental distress serves as a witness to humanity's inattention to or rejection of our interdependence with Nature.

The theme of communion in the universe may be translated into jurisprudence as a principle of relational responsibility. Berry reminds us that humanity is endowed with special capacities of thought and consciousness that are a means for the universe to reflect on itself with gratitude and wonder. Nurtured by Earth, humanity has developed abilities to establish systems of law and governance that should reflect our role as guardians of the Earth that birthed, clothed, and fed us. Because we have the capacity to understand and appreciate Earth, as well as our place within the whole, we bear a unique responsibility for developing and using that knowledge to preserve the Earth community.[33]

In a more concrete way, humanity's relationship to Earth may be best expressed as a trust and our responsibility as a trustee.[34] The public trust doctrine gives legal effectiveness to the notions of communion and relational responsibility. With roots in the Magna Carta and Roman law, the ideas and values of the public trust doctrine have been traced by Charles F. Wilkinson to ancient societies in Europe, Africa, and East Asia, as well as to Native American and Muslim cultures.

As traditionally expressed, courts have employed the public trust doctrine to apply trust principles to watercourses, shorelines, and underwater lands as the inherent property of the public at large or as subject to inherent easements for certain public purposes. Wilkinson asserts that this tradition reflected a widespread appreciation for the public value of water and a deep reluctance to allow our waterways to be subject to extensive private acquisition.

In the United States, courts began using trust language to describe the relationship between states and waterways in 1842. However, the case that established the viability of the public trust doctrine in the United States was decided fifty years later, in 1892. In *Illinois Central Railroad v Illinois*, the United States Supreme Court affirmed that the state could revoke an absolute grant of more than one thousand acres of waterfront and submerged lands in Chicago on Lake Michigan.[35] According to the court, the state had received title to the harbour at statehood, but the title was impressed with a trust to maintain the waterways for public use.

A milestone in the evolution of the public trust doctrine in the United States took place in 1970 with the publication of *The Public Trust Doctrine in Natural Resources Law* by Professor Joseph L. Sax. Acknowledging the conventional boundaries of the public trust doctrine, Sax also 'unhooked it from its traditional moorings on or around water bodies and applied it to dry land as well'.[36] In widening the concept of the public trust, Sax was instrumental in shifting the focus of the doctrine to environmental protection. As a reflection of changing public values toward Earth, the public trust doctrine has addressed 'conservation, scenic resources, open space, generation of energy, and preservation of ecosystems and historical sites'.[37]

As applied to water, Wilkinson refers to the public trust doctrine as a 'set of modest beliefs', including a belief in the propriety of short-term private interests accommodating broader public values, an understanding of the necessity of property rights yielding to responsible regulation, a recognition that polluting rivers is wrong, as well as 'a belief that our rivers and canyons are more than commodities, that they have a trace of the sacred'. As applied to other aspects of Nature, Mary Christina Wood affirms the potential in the public trust doctrine to catalyse us into the next phase of our relationship with Earth, a phase in which human law and governance express our responsibility to safeguard the well-being of Earth as a trust. With this catalyst, what is changed is not only the law, but also human hearts and minds. Without a change in human consciousness to embrace our responsibilities as members of the Earth community, no set of legal doctrines will resolve the environmental crises of the twenty-first century.

The Principle of Differentiation: Earth Democracy

Nature abhors uniformity.[38]

To Berry and Swimme, creativity is at the heart of the workings of the universe. With ever-expanding complexity, the universe expresses an 'outrageous bias for the novel, for the unfurling of surprise in prodigious dimensions throughout the vast range of existence'. On Earth, Nature produces an unending demonstration of diversity, from species and structures to individuals and dynamics: 'No two days are the same, no two snowflakes, no two flowers, trees, or any other of the infinite number of life-forms'.

In considering systems of governance inspired by patterns of Nature, Cullinan proposes that the diversity of Earth's regulatory systems might be expressed through Earth Democracy. At present, constitutional democracies articulate the purpose of governance to be 'of the people, by the people, and for the people'. In our present circumstances, we may ask how well governance 'for the people' has worked. A short-term focus on human economic gains has placed Earth's biosphere, species, and ecosystems in jeopardy. Instead, as Cullinan understands it, diversified systems of Earth governance would be of the people and by the people, but *for* the whole Earth community.

Through an approach to governance called Earth Democracy, humanity's role is recontextualised within the Earth family and girded with a purpose that safeguards the wider Earth community. With roots in ancient societies, Earth Democracy is an emerging political movement that is gathering under banners of peace, justice, and sustainability. According to physicist and environmental activist Vandana Shiva, 'Earth Democracy connects the particular to the universal, the diverse to the common, and the local to the global'.[39]

To respect the particular, Earth Democracy emphasises local governance. Shiva describes it as a 'living democracy', a type of governance that 'grows like a tree, from the bottom up'. People who are grounded in a place, who know the plants and animals, seasons and signs, ecosystems and processes of that place on Earth are in the best position to speak and care for the lands, waters, and beings of that community. Shiva believes that localisation may pose 'an antidote to globalisa-tion', which has led to the loss of biological and cultural diversity through global economics, transnational corporations, and industrial agribusiness.

Earth Democracy proposes that decisions should be made at the most appro-priate level. Not every decision is made at the local level, according to Shiva. Instead, Earth Democracy is guided by the principle of subsidiarity, calling for decisions to be made at the lowest appropriate level of governance. Through subsid-iarity, local control would be denominated for urban air pollution, regional control would be appropriate for transboundary air pollution, and global control would be recognised for global atmospheric pollution.[40]

An example of Earth Democracy at the local level may be found in the Democracy Schools that have arisen from the efforts of Pennsylvania townships to keep corporate factory hog farms out of their communities. With assistance from Thomas Linzey and the Community Environmental Legal Defense Fund (CELDF), local groups drafted ordinances banning corporate actors from bringing such businesses into their communities, and then broadened home rule powers to grant constitutional rights to ecosystems while stripping corporations of constitutional rights.[41]

At the bioregional level, Earth Democracy supports efforts to institute forms of governance based on ecosystems. Bradley Karkkainen's proposal for 'collaborative ecosystem governance' asserts the need for decision-making at the ecosystem level and gives examples such as the 'watershed approach' to protection of aquatic ecosystems that is taking place in the Chesapeake Bay and Great Lakes Programs.[42] Because ecosystems such as the Chesapeake Bay do not fit within conventional governmental boundary lines, one important feature of this model is horizontal and vertical coordination across governments at the same level as well as across multiple tiers of government.

At the global and nation-state level, Earth Democracy can be expressed in ways that recognise our duty to future generations. The current evidence on global warming clearly demonstrates that actions taken now will have an impact on the systems and inhabitants of the world in the middle to latter half of the twenty-first century, as Nicholas Stern has pointed out. The severity and irreversibility of anticipated impacts of global warming mandate a response from the present generation. That response should match the scientific data that has been presented, requiring greenhouse gases to be reduced to the level that accords with the natural capacity of Earth to remove them from the atmosphere. Moreover, the looming extinction rates of other species put significant pressure on the existing moral and legal framework to expand consideration not only to future generations of human beings, but also to remote species and Earth systems.[43] Reconstructing law and governance along the lines of Earth Democracy has the potential to keep humanity from creating a 'garbage heap' for the diversity of life that will follow us.[44]

In these ways, Earth Democracy is not only an environmental philosophy; it is also a political philosophy. In assuming our duties to Earth, humanity also creates diverse democratic approaches to governance. Consequently, the preservation of the Earth community is linked with the reinvention of local, regional, and global governance.

Conclusion

Earth Jurisprudence seeks to shift the focus of jurisprudence from a narrow, anthropocentric perspective based solely on the welfare of humanity to an Earth-centred perspective that recognises the role of humankind within the Earth community. To make that shift, this essay has proposed that we need a depth of vision that appreciates the intrinsic value of Earth and all beings, systems, and entities in Nature, a clarity of vision to embrace our relationship with Earth as a trust, and a breadth of vision to support Earth Democracy in all forms of governance.

We have entered a pivotal time in the history of Earth, when the likelihood of global warming of at least two degrees Celsius will result in the compromise of all major ecosystems of Earth and the extinction of thousands of species. As the moral agents on this planet, humankind has the responsibility to recreate human institutions to meet this challenge. It is not too late for a renewal of systems of law and governance. The time is right for humanity to envision new systems of jurisprudence for the well-being of the entire Earth community. Earth Jurisprudence is in bud.

Acknowledgements

I am grateful for the philosophical groundwork of Thomas Berry and Cormac Cullinan, the spiritual leadership of Sr. Patricia Siemen, O.P., the insightful comments of CEJ members Mary Munson and Nicole Gerard, the painstaking editing of CEJ member Jane Goddard, and the helpful research assistance of Tim Martin, Erin Cox, Lisabeth Fryer, Erica Ashton, and Jimmy Davis.

I am also indebted to the many authors whose work informs this article, including S. de Beauvoir, *The Second Sex*, trans Howard M. Parshley, Vintage Books, London, first published 1949, 1989 ed; E. T. Freyfogle, *Bounded People, Boundless Lands: Envisioning a New Land Ethic*, Island Press, Washington, District of Columbia, 1998; Gaia Foundation and United Kingdom Environmental Law Association, *Wild Law: Ideas into Action—Where Can We Find Examples of Earth Jurisprudence in Practice?*, United Kingdom Environmental Law Association, 2008; H.F. Greene, 'Where is the Universe in the Universe Story?', *The Ecozoic: Reflections on Life in an Ecological Age*, no.1, 2008; Harvard Law Review Association, 'What We Talk About When We Talk About Persons: The Language of Legal Fiction', Note, *Harvard Law Review*, vol. 114, no. 6, 2001; P. Hawken, *Blessed Unrest: How the Largest Social Movement in History is Restoring Grace, Justice, and Beauty to the World*, Penguin Group (USA), New York, 2007; M. Hawkesworth, 'Confounding Gender', *Signs*, vol. 22, no. 3, 1997; O.A. Houck, 'Are Humans Part of Ecosystems?', *Environmental Law*, vol. 28, 1998; L.G. Kniaz, 'Animal Liberation and the Law: Animals Board the Underground Railroad', *Buffalo Law Review*, vol. 43, 1995; A.R. Light, 'The Waiter at the Party: A Parable of Ecosystem Management in the Everglades', *Environmental Law Reporter*, vol. 36, 2006; R. Luov, *Last Child in the Woods: Saving Our Children from Nature-Deficit Disorder*, Algonquin Books of Chapel Hill, Chapel Hill, North Carolina, 2008; A. Naess and G. Sessions, 'Platform Principles of the Deep Ecology Movement' in *The Deep Ecology Movement: An Introductory Anthology*, eds A. Drengson and Y. Inoue, North Atlantic Books, Berkeley, California, 1995; J. Naish, *Enough: Breaking Free from the World of More*, Hodder and Stoughton, London, 2008; J.R. Nash, 'Standing and the Precautionary Principle', *Columbia Law Review*, vol. 108, no. 2, 2008; B.G. Norton,'Future Generations, Obligations to' in *Encyclopedia of Bioethics*, vol. 2, ed W.T. Reich, Gale Cengage Learning, Farmington Hills, Michigan, 1995; J. Rawls, *Political Liberalism*, Columbia University Press, New York, expanded ed, 2005; K. Sale, *The Green Revolution: The American Environmental Movement, 1962-1992*, Farrar, Straus and Giroux, New York, 1993; E.F. Schumacher, *Small Is*

Beautiful: Economics as if People Mattered, Perennial Library, New York, 1989; H.D. Thoreau, *Walden*, ed B. Atkinson, Random House, New York, first published 1854, 1992 ed; V.I. Vernadsky, *The Biosphere*, ed P.N. Nevraumont, trans D.B. Langmuir, Copernicus, New York, first published 1926, 1998 ed; B.H. Weston and T. Bach, *Recalibrating the Law of Humans with the Laws of Nature: Climate Change, Human Rights, and Intergenerational Justice*, Vermont Law School, South Royalton, Vermont, and University of Iowa, Iowa City, 2009; C.F. Wilkinson, 'The Headwaters of the Public Trust: Some Thoughts on the Source and Scope of the Traditional Doctrine', *Environmental Law*, vol. 19, 1989; E.O. Wilson, *The Diversity of Life*, W.W. Norton and Co., Ltd, New York, 1992; M.C. Wood, 'Nature's Trust: Reclaiming an Environmental Discourse', *Virginia Environmental Law Journal*, vol. 25, 2007. All errors are my own.

Notes

1   D. Levertov, 'Beginners' in J. Chittister, *The Cry of the Prophet: A Call to Fullness of Life*, ed M.L. Kownacki, Benetvision, Erie, Pennsylvania, 2009, p. 55.

2   See C. Cullinan, *Wild Law: A Manifesto for Earth Justice*, Green Books, Devon, p. 18; R.F. Nash, *The Rights of Nature: A History of Environmental Ethics*, University of Wisconsin Press, Madison, Wisconsin, 1989.

3   T. Berry, *Evening Thoughts: Reflecting on Earth as Sacred Community*, Sierra Club Books, San Francisco, 2006, p. 17.

4   N. Stern, *Stern Review: The Economics of Climate Change, Executive Summary*, Cambridge University Press, Cambridge, UK, 2006; Intergovernmental Panel on Climate Change, 'Summary for Policymakers' in *Climate Change 2007: The Physical Science Basis: Contribution of Working Group I to the Fourth Assessment Report of the Intergovernmental Panel on Climate Change*, eds S. Solomon et al., Cambridge University Press, Cambridge, UK, 2007.

5   T. Berry, *The Dream of the Earth,* Sierra Club Books, San Francisco, 1988, pp. 44–45; Cullinan, *Wild Law*, pp. 85–86.

6   B. Swimme and T. Berry, *The Universe Story: From the Primordial Flaring Forth to the Ecozoic Era*, HarperCollins Publishers, New York, 1992, p. 74.

7   Berry, *Dream of the Earth*, p. 46.

8   Swimme and Berry, *Universe Story*, p. 76.

9   Ibid., p. 73.

10 Ibid., p. 74.

11 Swimme and Berry, *Universe Story*, p. 73.

12 Cullinan, *Wild Law*, p. 115 (quoting T. Berry, *The Origin, Differentiation and Role of Rights*, 2001).

13 I.M. Young, *Justice and the Politics of Difference*, Princeton University Press, Princeton, New Jersey, 1990, p. 99.

14 Fr J. Kavanaugh, 'Challenging a Commodity Culture' in *On Moral Business: Classical and Contemporary Resources for Ethics in Economic Life*, eds M.L. Stackhouse, D.P.

McCann and S.J. Roels, Wm B. Eerdmans Publishing, Grand Rapids, Michigan, 1995, p. 608.

15 K.E. Goodpaster, 'On Being Morally Considerable', *Journal of Philosophy*, vol. 75, no. 6, June 1978, p. 308; see also C. Palmer, *Environmental Ethics and Process Thinking*, Oxford University Press, New York, 1998, p. 63.

16 Palmer, *Environmental Ethics*, p. 63.

17 W.M. Hunt, 'Are *Mere Things* Morally Considerable?', *Environmental Ethics*, vol. 2, no. 2, 1980, p. 60.

18 C.D. Stone, 'Should Trees Have Standing? Toward Legal Rights for Natural Objects', *Southern California Law Review*, vol. 45, no. 2, 1972.

19 C.D. Stone, 'Should Trees Have Standing? Revisited: How Far Will Laws and Morals Reach? A Pluralist Perspective', *Southern California Law Review*, vol. 59, 1985, p. 1.

20 *Sierra Club v Morton*, 405 US 727, 742 (1972) (Douglas, J., dissenting); C. Stone, *Should Trees Have Standing? And Other Essays on Law, Morals, and the Environment*, Oxford University Press, New York, 25th anniversary ed, 1996, p. 13.

21 D.N. Cassuto, 'The Law of Words: Standing, Environment, and Other Contested Terms', *Harvard Environmental Law Review*, vol. 28, no. 1, 2004, pp. 84, 102.

22 504 US 555 (1992).

23 C.P. Gilkerson, 'Poverty Law Narratives: The Critical Practice and Theory of Receiving and Translating Client Stories', *Hastings Law Journal*, vol. 43, 1992, p. 920.

24 Earth Charter Commission, *The Earth Charter* (2000) <http://www.earthcharterinaction.org/ content/pages/Read-the-Charter.html> at 1 December 2009.

25 M.F. Belenky, L.A. Bond, and J.S. Weinstock, *A Tradition That Has No Name: Nurturing the Development of People, Families, and Communities*, Basic Books, New York, 1997, pp. 19–22.

26 J.E. Koons, 'Gunsmoke and Legal Mirrors: Women Surviving Intimate Battery and Deadly Legal Doctrines', *Journal of Law and Policy*, vol. 14, no. 2, 2006, pp. 683–85.

27 J.L. Griscom, 'On Healing the Nature/History Split in Feminist Thought' in *Feminist Ethics*, ed L.K. Daly, Westminster John Knox Press, Louisville, Kentucky, 1994, pp. 271–81; see also C. Merchant, *The Death of Nature: Women, Ecology, and the Scientific Revolution*, HarperCollins Publishers, New York, 1980.

28 E. Sahtouris, *Earth Dance: Living Systems in Evolution*, iUniverse, Lincoln, Nebraska, 2000, pp. 16–26.

29 T.N. Hanh, *Being Peace*, Parallax Press, Berkeley, California, 1996, pp. 45–46.

30 T. Berry, *Evening Thoughts: Reflecting on Earth as Sacred Community*, Sierra Club Books, San Francisco, 2006, p. 150.

31 A. Leopold, *A Sand County Almanac and Sketches Here and There*, Oxford University Press, New York, first published 1949, 1989 ed, p. 204.

32 Cullinan, *Wild Law*, p. 89 (quoting E. Goldsmith, *The Way: An Ecological View*, University of Georgia Press, Athens, Georgia, 1996).

33 T. Berry, *The Great Work: Our Way into the Future*, Bell Tower, New York, 1999, p. 173; see also P. Siemen, 'Weaving an Ethic of Right Relationships for the Earth

Community' in *Women Moving Forward Volume Three*, eds J.B. Bachay and R. Fernández-Calienes, BookSurge Publishing, Charleston, South Carolina, 2008, p. 78.

34   C.M. Rose, 'Joseph Sax and the Idea of the Public Trust', *Ecology Law Review Quarterly*, vol. 25, no. 3, 1998, p. 351.

35   146 U.S. 387 (1892).

36   Rose, 'Sax and the Idea of the Public Trust', p. 352.

37   P.E. Salkin, 'The Use of the Public Trust Doctrine as a Management Tool over Public and Private Lands', *Albany Law Journal of Science and Technology*, vol. 4, no. 1, 1994, p. 3; M.C. Jarman, 'The Use of the Public Trust Doctrine for Resource-Based Area-Wide Management: What Lessons Can We Learn from the Navigable Waters Trust?', *Albany Law Journal of Science and Technology*, vol. 4, no. 1, 1994, p. 8.

38   Berry, The Great Work, p. 149.

39   V. Shiva, *Earth Democracy: Justice, Sustainability, and Peace*, South End Press, Cambridge, Massachusetts, 2005, p. 1.

40   A. Rosencranz, 'The Origin and Emergence of International Norms', *Hastings International and Comparative Law Review*, vol. 26, no. 3, 2003, p. 310.

41   T. Linzey, 'Of Corporations, Law, and Democracy: Claiming the Rights of Communities and Nature', Twenty-Fifth Annual E.F. Schumacher Lectures, October 2005, <http://www.schumachersociety.org/publications/linzey_06.html> at 7 September 2009.

42   B.C. Karkkainen, 'Collaborative Ecosystem Governance: Scale, Complexity, and Dynamism', *Virginia Environmental Law Journal*, vol. 21, 2002, pp. 189–243.

43   Stone, 'Should Trees Have Standing? Revisited', p. 13.

44   J. Feinberg, 'The Rights of Animals and Unborn Generations' in *Philosophy and Environmental Crisis*, ed. W. Blackstone, University of Georgia Press, Athens, Georgia, 1974, pp. 64–65.

The Great Jurisprudence

Peter Burdon

> This abstraction called Law is a magic mirror, [wherein] we see reflected, not only our own lives, but the lives of all men that have been![1]

If the Western idea of law could be represented as a person, what would it look like? What physical and intellectual characteristics would it embody? To begin, the person would be old, to reflect the great lineage of our jurisprudence. Like the vast majority of its authors and interpreters it would be white, male and upper class. Further, our imaginary person would embody traditional Western values found in humanist philosophy, Christianity and modern liberal political philosophy. The intention of this thought experiment is to highlight that Western law is a reflection of our culture and the values and perceptions it embodies. Certainly, law is a significant description of the way in which a society analyses itself and projects its image to the world. It is a major articulation of a culture's self-concept, and represents the theory of society within that culture. Given the relationship between law and culture, it is not surprising that our jurisprudence reflects the characteristics cited above. Indeed, from the ancient Greeks to present scholarship, the canon of Western legal thought has been dominated by individuals who fit that description.

While these scholars have made great advances in legal thought and reasoning, in the last thirty years there has been a growing chorus of voices who contend that their unique perspective has been excluded or marginalised from orthodox legal theory. For ease of communication I will categorise these voices under the heading 'critical theory'. Representatives of 'critical theory' include critical legal studies, feminist jurisprudence, critical race theory, Marxist theory and queer theory. While critical theory is a house with many rooms, what unites theorists is a belief that society, and necessarily legal order, constructs and maintains a particular injustice. Further, critical theorists look at the ways in which the injustice can be undermined and ultimately eliminated. In some instances they also advocate an alternative direction for jurisprudence which responds to the critique presented.

The legal philosophy of Earth Jurisprudence represents an emerging branch of critical theory. Consistent with the widely accepted critique advanced in environmental philosophy, it contends that Western law and jurisprudence reflects an anthropocentric worldview. That is, a worldview which perceives human beings as the central and most important element of the Universe. In this paper, I will demonstrate this point through a critique of legal theory, property law and legal rights. Following this discussion, I will note that anthropocentrism no longer has any credibility in modern science and that human beings exist as one equal part of a broader Earth community. Given this shift in human understanding, Earth Jurisprudence seeks to articulate legal concepts that properly reflect our relationship and place in the Earth. This is a shift from 'anthropocentric' to an earth-centred, or 'ecocentric', theory of law. Further, Earth Jurisprudence recognises that both human beings and nature are relevant and necessary to the operation of law.

From these basic principles, Earth Jurisprudence advocates the existence of two types of law, which exist in hierarchical relationship. On top is the Great Law which represents the rules or principles of nature, which are discoverable by human beings and relevant to human-earth interaction. Underneath the Great Law is Human Law, which represents rules articulated by human authorities, which are consistent with the Great law and enacted for the common good of the comprehensive Earth Community. Regarding the interrelation between legal categories, two points are critical. First, Human Law derives its legal quality from the Great Law. Further, any law that ignores and transgresses the Great Law is considered a corruption of law and not morally binding on a population. This paper will extrapolate further these concepts to articulate theory of law that responds and accommodates the ecological imperative of the present age.

Anthropocentrism and the Law

Albert Einstein defined anthropocentrism as 'an optical delusion of human consciousness' where we come to regard 'humanity as the centre of existence'.[2] The *Macquarie Dictionary* provides more detail on the individual elements of anthropocentrism, defining it as an adjective that '[r]egards man as the central fact of the universe'; '[a]ssumes man to be the final aim and end of the universe'; and '[v]iews and interprets everything in terms of human experience and values'.[3]

While anthropocentrism has been thoroughly discredited by modern science, it is a perspective which has been promoted throughout the majority of Western history. For example, Aristotle noted: 'Plants exist for the sake of animals, the brute beasts for the sake of man – domestic animals for his use and food, wild ones (or at any rate most of them) for food and others accessories of life, such as clothing and various tools. Since nature makes nothing purposeless or in vain, it is undeniably true that she has made all animals for the sake of man'.[4] Similarly, the Hebrew Bible notes in Genesis 1:27–31: 'Be fruitful and increase in number; fill the earth and subdue it. Rule over the fish of the sea and the birds of the air and over every living creature that moves on the ground.'[5] Both classical philosophy and Christian theology are of central importance to the worldview of Western culture. Further, because law is a reflection of culture it also reflects these ideas. Phillip Allot notes:

> Society cannot be better than its idea of itself. Law cannot be better than society's idea of itself. Given the central role of law in the self-ordering of society, society cannot be better than its idea of law.[6]

For this reason it is no surprise that many of our law's most fundamental concepts and ideas imitate an anthropocentric worldview. This section will illustrate this point through a brief discussion of legal theory, property law and legal rights.

Legal Theory

Legal philosophy provides the foundation and intellectual base for positive law. As Karl Llewellyn notes, legal theory 'is as big as law – and bigger'.[7] Yet, despite their enormous variation and diversity, theories of law and justice in Western jurisprudence are predominantly anthropocentric. Mainstream theories such as natural law and legal positivism are concerned ultimately with human beings and human relationships. The concept of 'relationship' is extended from human beings, to communities, states, nations and elementary groupings operating within these categories.[8] However, only in rare circumstances is the concept extended to include nature or non-human animals. Further, when law is constructed as purposive and reference is made to the 'common good', as is common in natural law theory, the term is expressly limited to human good. Certainly, the hierarchical ordering of the human and non-human world is the unquestioned starting point for most theories of law.

Today, the dominant theory of law is legal positivism. Stated plainly; legal positivism asserts that it is both possible and valuable to produce a purely conceptual or purely descriptive theory of law, free from moral evaluation or external influence. Legal positivism claims that law is a science and like other scientific discourses, attempts to describe law from an objective perspective.[9] Positivism identifies and defines law through 'abstract' categories or doctrines, which it posits as authoritative rules applicable to the resolution of legal disputes. This legal philosophy expressly considers the influence of the non-human world as 'remote, inappropriate and unnecessary to the operation of law'.[10]

Further, legal positivism reflects the notion that human beings are self-validating and it is appropriate (and valid) for human beings to enact law in ignorance or even contrary to the needs of place and ecological principles. Cultural historian Thomas Berry comments on the self-validating nature of legal positivism, noting that law is 'framed for the advancement of the human with no significant referent to any other power in heaven or on Earth'.[11] As a result, Berry comments that '[h]umans have finally become self-validating, both as individuals and as a political community'.[12] Cormac Cullinan supports this point in his book *Wild Law*:

> Our secular legal philosophies almost universally deny that our jurisprudence needs to take account of any rules, norms or considerations that lie outside human society. Laws are generated entirely within our glass 'homosphere'. Our laws are understood literally, as laws unto themselves. All that matters are the legal convictions of the human community at the relevant time and the content of the written law.[13]

The conscious separation of human law from external factors has important consequences for the environment. For example, it enables sophisticated governance bodies such as the European Union to allocate greater fishing quotas than the fish stocks can sustain.[14] In Australia, despite recent regulatory measures[15], irrigators are still able to draw unsustainable amounts of water from the Murray Darling System to grow ecologically insensitive crops such as rice and cotton on arid land.[16] Perhaps most visibly, despite overwhelming evidence supporting anthropogenic climate change, it is perfectly legal for corporate bodies to pollute increasing amounts of carbon into the atmosphere.[17] Certainly, our legal philosophies provide no reason or mechanism for human laws to consider the role of place, space and nature in the creation of law. Law is considered a separate, higher authority and orthodox legal theorists 'do not see the need for any connection or continuity between our legal system and the Earth system'.[18]

Nature as Human Property

A second pertinent example of how our law reflects anthropocentric values is that our law defines nature as human property, which by definition can be used and exploited for human benefit. Eric T. Freyfogle notes:

> When lawyers refer to the physical world, to this field and that forest and the next-door city lot, they think and talk in terms of property and ownership. To the legal mind, the physical world is something that can be owned.[19]

While it is perhaps impossible for lawyers and legal scholars to regard nature as anything but property, the cultural nature of this view must be highlighted. In making this point, I am not suggesting that other cultures did not have concepts and rules regarding land use and access. The critical point is what ownership means and how property, as a legal concept and organising element, influences our relationship to the land. To begin this discussion it is instructive to note that the first sophisticated definition of property in western legal history was the Roman concept *dominium*.[20] In defining this concept, 11th century jurists noted that it was akin to 'lordship' and further noted that it was a sovereign, ultimate or an absolute right to claim title and thus possess and enjoy an item. While the institution of private property has never reflected such absolute language, the idea of dominion has been maintained in cultural narratives and 'lay' understandings of property.

The themes of control and possession are commonplace in our cultural narratives on property. Here the law's main message is that people are distinct from the land and its component parts. People are subjects and the land is merely an object, possessing little moral or legal worth. As Freyfogle notes, 'there is at work here a strict dichotomy of subject and object, legally worthy and legally worthless ... people are the ones who own and dominate and the land is the thing that is owned and dominated'.[21] In the last century, cultural narratives concerning property have been influenced heavily by the popularisation of liberal political theory.

The principles most commonly associated with liberal theory include freedom, toleration, autonomy, justice and individual rights. Of these ideals Jeremy Waldron notes, 'the deepest commitment of liberal political philosophy is to individualism' and providing freedom to fulfil individual potential.[22] Indeed, liberals hold that individual human persons are the most important factor in social and political matters. One may have an interest (and indeed many liberals do) in community, the environment and non-human animals but for a liberal, such interest is always secondary or derivative.

Under the influence of liberalism the 'idea' that nature was human property strengthened in pursuit of individual preferences and choices. Thus, from a property perspective, human beings are not only separate from nature, they are separate from each other.

Legal Rights

The implications of the preceding points are exacerbated further by the fact that our law places all value and rights in human beings. Berry notes:

> All rights have been bestowed on human beings. The other than human modes of being are seen as having no rights. They have reality and value only through their use by the human. In this context the other than human becomes totally vulnerable to exploitation by the human.[23]

Following from the first two points, nature and the non-human world are 'things' or 'objects' which exist for human benefit. While protective legislation does exist, this is the exception rather than the rule and only applies in circumstances of peril or danger. Further, protective legislation is often justified with reference to a human right to a clean environment or a non-right to be cruel or kill a particular species. One obvious consequence of ignoring the intrinsic value and legal rights of nature is that when environmental damage occurs on privately owned land, the law treats the problem as a property offence. Damage is measured and distributed for the benefit of the owner and not the land itself. In other circumstances where a property owner has not been directly harmed or perhaps does not wish to sue for damage to their property, it is very difficult to gain standing to sue for environmental protection.

In response to the anthropocentric focus of legal rights, some jurisdictions are beginning to legislate in favour of the rights of nature. However, the lack of uniform and consistent legislation on this issue leaves the natural world profoundly vulnerable to the needs of a growing industrial economy.

Earth Jurisprudence

In the previous section it was contended that Western law reflects an anthropocentric human–Earth relationship. The key question for this section is, how can law as an evolving cultural institution shift to reflect the relatively modern revelation that human beings are interconnected with and dependent upon a comprehensive Earth community?

Earth Jurisprudence is an emerging philosophy of law, first proposed by Thomas Berry in 2001. Berry was a persuasive critic of the anthropocentric paradigm and the myth of progress expounded by modern civilisation. In his important analysis 'Legal Conditions for Earth Survival' he argues that the present legal system 'is supporting exploitation rather than protecting the natural world from destruction by a relentless industrial economy'.[24] In 1987 Berry set about describing how human society could shift both its idea of law and its legal system toward an Earth-centred perspective. Most of his remarks are in outline to this shift, as witnessed in his early paper 'The Viable Human'. Berry notes:

> The basic orientation of the common law tradition is toward personal rights and toward the natural world as existing for human use. There is no provision for recognition of nonhuman beings as subjects having legal rights ... the naïve assumption that the natural world exists solely to be possessed and used by humans for their unlimited advantage cannot be accepted ... To achieve a viable human–earth community, a new legal system must take as its primary task to articulate the conditions for the integral functioning of the earth process, with special reference to a mutually enhancing human-earth relationship.[25]

The idea of 'mutual-enhancement' is fundamental to Earth Jurisprudence and is informed further by the concept of 'Earth community'. That is, human beings exist as one part of a community of life, which incorporates non-human animals and innate living systems. In regard to legal philosophy, the goal of mutual-enhancement necessitates that nature is not only relevant, but also appropriate and necessary to our idea of law.

To make this transition Berry and subsequent proponents of Earth Jurisprudence have posited the existence of two kinds of law, which exist in hierarchical relationship. On top is the Great Law, which represent the principles of nature which are discoverable by human beings through scientific method and relevant to human-earth relationships. Below the Great Law is Human Law, which represents binding prescriptions, articulated by human authorities, which are consistent with the Great law and enacted for the common good of the comprehensive Earth Community. Regarding the interrelation between these two types of law, two points are critical. First, Human Law derives its legal quality and power to bind in conscience from the Great Law. Further, because human beings exist as one part of an interconnected and mutually dependent community, a prescription that is directed to the comprehensive common good has the quality of law. In some instances, the content of law can be framed with reference to first principles of Great Law; for the rest, the legislator has the freedom of architect. Secondly, any law which ignores or transgresses the Great Law is considered a corruption of law and not morally binding on a population.

It will be clear to anyone familiar with legal philosophy that the basic structure and relationship between these different types of law share resemblance to
the Thomist and neo-Thomist natural law tradition. Theorists from within both
Earth Jurisprudence and natural law have noted the broad conceptual relationship
between the two theories. However, to date neither discipline has investigated this
relationship further. For Earth Jurisprudence to develop conceptually it is critical
that this analysis occurs. Further, the absence of such analysis has been lamented by
advocates of natural law theory. For example, Jane Holder writes that the 'absence
of matters of "physical nature" (as opposed to human nature) is striking' causing
her to reiterate (albeit in a different context) Lloyd Weinreb's denunciation of
'natural law without nature.'[26]

Legal Categories

It was noted above that the goal of legal positivism is the creation, implementation and study of law, free from external factors. According to this theory, 'there
is no law but positive law' and if one wants to discover what the law is, it can be
readily found in statutes, codes and cases. Further, statements of law are necessarily
kept separate from questions regarding what the law ought to be, its relationship
to morality or ecological principles. In contrast to this orthodox position, Earth
Jurisprudence advocates the existence of two types of law. This is similar to natural
law philosophy, which has advocated as many as four kinds of law existing in hierarchical relationship.[27] Prior to examining the two concepts promoted in Earth
Jurisprudence it should be noted that in this context the term *law* has analogous
meaning. From a linguistic and logical perspective, Human Law is law properly
speaking. Ontologically, however, the Great Law is the measure of human law and
by analogy is termed 'law'.

The Great Law

The first category of law advanced in Earth Jurisprudence is the Great Law. Berry
introduces this concept, noting that human society should recognise the 'supremacy
of the already existing Earth governance of the planet as a single, yet differentiated,
community'.[28] He notes further that an orientation toward the natural world 'should
be understood in relation to all human activities'[29] and that 'Earth is our primary
teacher as well as the primary lawgiver'.[30] Drawing on these comments, Cormac
Cullinan coined the term 'Great Jurisprudence' or 'Great Law' to help make sense of
this re-characterisation. Cullinan defines this term as 'laws or principles that govern
how the universe functions' and notes that they are 'timeless and unified in the
sense that they all have the same source'.[31] This law is manifest in the universe itself
and the examples provided by Cullinan include the 'phenomenon of gravity', 'the
alignment of the planets', the 'growth of planets' and the 'cycles of night and day'.[32]

It is important to pause and consider in more detail Cullinan's description
of the Great Law as representing the laws of nature. In particular, we need to

discern precisely what is a law of nature and in what sense they have meaning or relevance for human law. In introduction to the first question, it must first be noted that 'laws' play a central role in scientific thinking. Martin Curd notes that 'some philosophers of science think that using laws to explain things is an essential part of what it means to be genuinely scientific' and 'support for the view that scientific explanation must involve laws is widespread'.[33] Many also believe they are justified in trusting or relying on scientific inferences, because these predictions are based on established laws. In this view, our expectations regarding the behaviour of systems, materials and instruments is considered reasonable, to the extent that they are drawn from a correct understanding of the rules which govern them.

Yet, despite the critical importance laws have in science, there is little general agreement about the kind of things laws are that can do justice to all the attributes we commonly ascribe to them.[34] The existence of this dissonance presents a major challenge to Cullinan's description of the Great Law. Indeed, how can the law of nature influence human law if we are unable to provide an intellectually satisfying description of the former? Compounding this problem is the apparent lack of relevance many 'established' laws of nature have for human law. Indeed, it is difficult to imagine what specific application Newton's law of motion or Boyle's law of mass/pressure would have in human legal institutions. To be plain, I regard reference to the 'laws of nature' as too general and overwhelmingly broad to have relevance in Earth Jurisprudence. Indeed, consistent with other authors on Earth Jurisprudence, I contend that the Great Law should be defined with reference to 'first principles' uncovered in the scientific discipline of ecology.[35] This approach is also consistent with the current direction of environmental law[36] and helps distinguish between principles that are relevant to the regulation of human behaviour and ones that describe other phenomena in the universe.

The term ecology is derived from the Greek root *oikos* meaning 'house' or 'place to live'.[37] Thus, literally, ecology is the study of 'houses' or more broadly, 'environments'. It is the study of organisms 'at home'.[38] Eugene Odum notes further that because ecology is concerned especially with the make-up of groups of organisms, and with the functional processes on the lands, ocean and fresh waters, 'it is more in keeping with modern emphasis to define ecology as the study of the structure and function of nature, *it being understood that mankind is a part of nature*'.[39]

Focusing our discussion on ecology correlates in a substantial reduction of the range of potential disciplines and principles applicable to our theory. However, within ecology itself, there remain a considerable number of principles relating to ecosystems, energy transfer, biochemical cycles, species distribution, carrying capacity, community organisation, development and evolution. While I acknowledge the necessity of doing so, I will not attempt to provide an exhaustive list of the ecological principles that might be relevant to Earth Jurisprudence in this paper. However, one principle favoured by Berry and recognised in ecology as fundamental is interconnectedness. Odum describes interconnectedness as follows:

Living organisms and their nonliving (abiotic) environment are inseparably inter-related and interact upon each other. Any unit that includes all of the organisms (i.e. the community) in a given area interacting with the physical environment so that a flow of energy leads to clearly defined trophic structure, biotic diversity, and material cycles ... within the system is an ecological system or ecosystem.[40]

This principle is fundamental to human–Earth relationships and one that we transgress at our own peril. For these reasons it is regarded in Earth Jurisprudence as ontologically prior to human law and a principles which human beings can choose to conform too. This last point is critical – the hierarchy of legal categories described in Earth Jurisprudence does not represent a progression of logical neces-sities to which human beings must conform. Instead, the interaction of legal cate-gories depends on human beings consciously choosing to alter our law to confirm with the greater community that sustains and makes human life possible. Because of the role rational choice plays in Earth Jurisprudence, it is not subject to criticism under the banners 'noncognitivism' or what G.E. Moore termed the 'naturalistic fallacy'.[41]

Human Law

In Earth Jurisprudence, 'human law' is the essence of what is meant by the term law. It's meaning is largely consistent with that articulated in orthodox theory. That is, human law is a rule or prescription promulgated by the lawmaking authority of a community. To this basic definition, Earth Jurisprudence links human law to the 'common good' of the comprehensive Earth community, and holds that human law must be supported by the Great Law. Acknowledging these stand-ards in constructing human law is regarded as reasonable behaviour. Thus, an approximate definition of human law in Earth Jurisprudence could be expressed as follows: 'rules, supported by the Great Law, which are articulated by human authorities for the common good of the comprehensive whole.'

Importantly, this definition shares many similarities with legal positivism. Key areas of relationship include the recognition of human authority to make binding prescriptions for the community and the advancement of human law as a subject of separate consideration and a topic readily identifiable prior to any questions about its relation to external factors i.e. the Great Law. Further Earth Jurisprudence does not contest the known benefit of positive law in achieving social/common goods that require the deployment of state power or the co-ordination of public behaviour.

The dividing line between the two theories rests on several fine distinctions, which nonetheless carry theoretical significance. The most obvious difference is the appeal to 'higher law'. Further, Earth Jurisprudence views human law not as an object or entity to be studied dispassionately under a microscope. Instead, consistent with the writing of secular natural law theorist Lon Fuller, it views human law as a project, with a purpose.[42] Among other things, this includes to allow human beings to co-exist and flourish within society and the broader environment. This teleological understanding of law was first described by Aquinas in question 90, article two of his *Summa Theologica*. In answer to the question of whether law is always directed to the common good, Aquinas answers:

> Now the first principle in practical matters, which are the object of the practical reason, is the last end: and the last end of human life is bliss or happiness ... consequently the law must needs regard principally the relationship to happiness. Moreover, since every part is ordained to the whole ... and since one man is a part of the prefect community, the law must needs regard properly the relationship to universal happiness.[43]

On this account, law cannot truly be understood without understanding the ideal of 'common good' towards which it is striving. Further, as Fuller notes, to exclude the ideal from a theory of law on the basis of a 'separation of description and evaluation' is to miss the point entirely. The social practice and institution of law, 'is by its nature a striving towards' ideals such as common good. To support this argument, Fuller contrasted laws from other forms of governance such as managerial direction. Law is a particular 'means to an end' or a kind of tool. With this in mind, one can better understand the claim that rules must meet certain criteria connecting the means to the function, if they are to be accorded legal authority.[44]

Another way to understand this proposition is that legal authorities are not entirely free to create law. Instead, they must have knowledge and respond to factors which in orthodox legal philosophy are considered external to law, i.e. the Great Law. It should also be noted that in suggesting that laws are enacted for the common good of the comprehensive Earth Community, Earth Jurisprudence is introducing a moral component. Natural law writers such as Aquinas, Finnis and Fuller accept either implicitly or explicitly the relationship between purpose and morality in their respective theories. This recognition is maintained in Earth Jurisprudence. Importantly, this does not mean that human law is a conclusion from moral premises. Rather, following Berry, a lawmaker will be guided in promulgated law from the principles of ecology. In other words, the principles of nature and the moral idea of comprehensive common good are important for determining what the positive law should be. This is a fundamentally different claim from the contention that one ought to use naturalism to determine or describe what the positive law of a particular society currently is.

The Function of the Great Law
The Great Law has been described as ontologically prior and the measure of Human Law. Thus, for Human Law to attain legal quality and the power to bind a community in conscious, it must be consistent with the Great Law. Further, any purported law which is in conflict with the Great Law is regarded as a mere

corruption of law and not binding by virtue of its own legal quality. In this situation, Earth Jurisprudence upholds the moral right citizens have to disobedience and protest.

Legal Quality

In regard to the question of legal quality, Cullinan notes that ecological first principles such as interconnectedness should not be applied literally, as a rule or principle might. Instead, he notes that they can be understood as 'the design parameters within which those of us engaged in developing Earth Jurisprudence for the human species must operate'.[45] To understand whether a legislator has transgressed a principle such as interconnectedness, one must analyse the likely future consequence of a purported law. In some instances, such as the logging of old-growth forest, it will be relatively easy to establish that a 'first principle' such as interconnectedness has bee breached· In other cases, which involve some level of forecasting or where there is little information on the practice, process or perhaps ecosystem, the decision-making process will be more difficult. If it is determined that the purported law breaches this ecological first principle, it is a corruption of law and does not bind in conscience. First principles derived from ecology are also necessarily connected to the purpose of law, which is directed at the common good of the complete community.

While this reasoning may appear abstract, one can witness many similar statements in environmental philosophy and politics. For example, in 2007 former vice president of the United States Al Gore noted, 'I can't understand why there aren't rings of young people blocking bulldozers, and preventing them from constructing coal-fired power plants'.[46] These comments were followed in a 2008 address to the Clinton Global Initiative:

> If you're a young person looking at the future of this planet and looking at what is being done right now, and not done, I believe we have reached the state where it is time for civil disobedience to prevent the construction of new coal plants that do not have carbon capture and sequestration.[47]

In this example, we can presume that the proponent has applied for and received the relevant legal permits and licenses to carry out construction of a coal plant. Consistent with other large-scale projects there has likely been community consultation, opportunity for public comment and negotiation with stakeholders. However, in spite of these measures, because of the known ecological dangers caused by coal-fired power plants, and the risk it poses to the long-term common good, Gore questions its legal validity. More than this, he expresses his dismay that individuals are not positively breaking the law to stop the project. Certainly, we must we must question the value and legitimacy of any law that surpasses the ecological limits of the environment to satisfy the needs of one species. In this notion we witness the interrelation between the Great Law and Human Law.

To further make sense of this interrelation, it is useful to draw from the Thomist natural law tradition and this time, to the writing of John Finnis. Finnis distinguishes what he calls the 'focal' meaning of law from its secondary meaning.[48] Here the focal meaning of law refers to its ideal form and that actual law is a mere striving or approximation toward this form. Finnis argues that the central focus of law is the 'complete community', which he defines as 'an all-round association' that includes the 'initiatives and activities of individuals, of families and of the vast network of intermediate associations'.[49] Its purpose or point is to 'secure a whole ensemble of material and other conditions that tend to favour the realisation, by each individual in the community, of his or her personal development'.[50] Thus, the focal meaning of law is to secure the common good of human beings by co-ordinating the different goods of individuals within the complete community. For Finnis, this is the true purpose of law and it follows that any law that conflicts with this goal is not a law in the focal sense of the term. They are not true laws 'in the fullest sense of the term' and 'less legal than laws that are just'.[51]

The argument advocated by Finnis regarding the focal and secondary meaning of law is conceptually very useful for advancing a theory of Earth Jurisprudence. However, one should have misgivings about how 'complete' Finnis's notion of community is. His is a human community and centres on important social relations and the sharing of a common aim. In this construction, Finnis forgets the ground on which he stands, the air he breathes and the material his *opus* was printed on. More than this, by limiting his central case to the human community, he places in great jeopardy his notion of law's purpose. Indeed, individual and community goods are entirely dependent on a healthy and productive environment. If this comprehensive community flourishes we all flourish; if it falls we all fall. Importantly, Finnis does provide for the extension of his definition of complete community. Looking to the future he notes, '[i]f it appears that the good of individuals can only be fully secured and realised in the context of the international community, we must conclude that the claim of the national state to be a complete community is unwarranted'.[52] Following this thinking one step further, if the good of individuals and communities can only be secured by further extension to the comprehensive Earth Community, there is no reason why his reasoning could not similarly be extended.

The central case of law advanced in Earth Jurisprudence is truly a 'complete community'. Its point, or common good, is to secure the safety and future flourishing of this community. This is the true purpose of law, in the focal sense of the term and the only point of reference worth considering in pursuit of common good. It follows from this that laws that transgress the design parameters derived from ecological first principles and place in jeopardy the purpose of law, are not laws in the focal sense of the term. They lack legal quality. I will turn now to consider the consequence of this status.

Corruption & Civil Disobedience

Consistent with the interrelation of legal categories proposed in natural law, Earth Jurisprudence holds that a law that transgress first principles of nature and jeopardises the purpose of law, is not a law in the focal sense of the term. To be clear, Earth Jurisprudence does not invalidate human law in this manner. Instead, it provides a set of fundamental principles for a legal system that serves the true purpose of law and seeks to provide 'a rational basis for the activities of legislators, judges and citizens'.[53] Further to these functions, it also provides criteria for deciding whether citizens are morally bound to follow the law in so far as positive law may diverge from the ideal standards proposed in the Great Law. This function is connected with ideas on legal authority, obligation to obey the law and civil disobedience.

To begin, Earth Jurisprudence affirms the presumption of laws obligatory force. That is, individuals have a general obligation to obey the law as a result of the benefit they receive from it, including protection and material wealth. On this point, the modern social contract theory articulated by John Rawls remains influential. Briefly summarised, Rawls argued that a society is just if it is governed by principles which citizens would have agreed to in a state of ignorance of their individual position in society. Where a society is just or close to just there is, he says, a 'natural duty' for citizens to support its institutions.[54] Further, so long as the basic structure of society is reasonably just, Rawls argued that the duty to obey the law extends to obeying particular unjust laws – so long as they do not exceed certain thresholds of injustice. Rawls regarded the denial of basic liberties to cross this threshold and noted that if this occurred, one's *prima facie* duty to the law could be overridden and replaced with 'other more stringent obligations'.[55] Here, conscientious refusal to obey the particular laws is justified and, in the case of blatant injustice, civil disobedience may be justified.

Consistent with the basic framework of this theory, Earth Jurisprudence upholds the right to conscientious refusal and civil disobedience when a purported law contravenes first principles derived from ecology or places in jeopardy the comprehensive common good. While justified on a different basis, such a corruption of law sits alongside other violations such as the denial of civil liberties or discrimination on the basis of race or gender. Consistent with other investigations into legal authority, Earth Jurisprudence is a response to its times and the unique challenges it faces. Indeed, just as Rawls was informed by the growing civil rights movement in the United States, advocates of Earth Jurisprudence are informed by the increasing rate and intensity of environmental activism around the world. At present, our law and jurisprudence maintains an antagonistic stance toward these efforts. In response, Earth Jurisprudence seeks to establish an intellectual framework through which to understand this movement and advocates a legal philosophy that explicitly recognises the moral right we all have to protest laws that place the Earth community in peril.[56]

Conclusion: Re-thinking Law for the Ecological Age

In his fascinating study of the rise and fall of civilisations, biogeographer Jared Diamond suggests that one key reason past cultures have collapsed is 'a failure to respond adequately to a perceived problem, because of a reluctance to abandon deeply held values'.[57] Certainly, one central factor contributing to the present environmental crisis is our failure to understand and behave as members of the Earth community. Our law has been developed to facilitate a one-way exchange with the Earth and feed our ever-growing extractive industrial economy. Today, there is a great need to develop a jurisprudence that seeks to develop a mutually enhancing and beneficial human-earth relationship. As Berry notes, '[t]o be viable, the human community must move from its present anthropocentric norm to a geocentric norm of reality and value'.[58] Consistent with the philosophy expressed in Earth Jurisprudence, this is simply the recognition that the human community is a subsystem of a broader and primary earth system. As integral members of the Earth community, human beings need to act from within this comprehensive context, adapting Human Law to respect the Great law and learning once more to inhabit the Earth.

The classic German poet, Rainer Maria Rilke, expressed this simple message in his poem 'Wenn etwas mir vom Fenster fällt':

> How surely gravity's law,
> strong as an ocean current,
> takes hold of the smallest thing
> and pulls it toward the heart of the world.
> Each thing-
> each stone, blossom, child-
> is held in place.
> Only we, in our arrogance,
> push out beyond what we each belong to
> for some empty freedom.
> If we surrendered
> to earth's intelligence
> we could rise up rooted, like trees.
> Instead we entangle ourselves
> in knots of our own making
> and struggle, lonely and confused.
> So like children, we begin again
> to learn from the things,
> because they are in God's heart;
> they have never left him.
> This is what the things can teach us:
> to fall,

patiently to trust our heaviness.
Even a bird has to do that
before he can fly.[59]

Notes

1   O.W. Holmes, *The Speeches of Oliver Wendell Holmes,* Nabu Press, Charleston, 2010, p. 17.

2   A. Einstein quoted in Klaus Bosselmann, 'The Way Forward: Governance & Ecological Integrity', eds L. Westra, K. Bosselmann & R. Westra, *Reconciling Human Existence with Ecological Integrity*, Earthscan, New York, 2008, p. 319.

3   *The Macquarie Dictionary: Revised Edition*, Macquarie Library, London, 1985, p. 113.

4   Aristotle, *The Politics,* University Of Chicago Press, Chicago, 1985, p. 1256b.

5   For a classic critique of the relationship between Christianity and the environmental crisis see L. White Jnr, 'The Historical Roots of Our Ecologic Crisis', *Science*, vol. 155, 1967, p. 1203.

6   P. Allot, *Eunomia: New Order for a New World*, Oxford University Press, Oxford, 1990, p. 298.

7   K. Llewellyn, *Jurisprudence: Realism in Theory and Practice*, Transaction Publishers, New Jersey, 1962, p. 372.

8   N. Graham, *Lawscape: Property, Environment & Law*, Routledge, Oxon, p. 15.

9   J.W. Harris, *Legal Philosophies*, Oxford University Press, Oxford, 2002, p. 12.

10 Graham, *Lawscape*, p. 20.

11 Ibid., p. 13.

12 Ibid.

13 Ibid., p. 79.

14 C. Cullinan, *Wild Law: A Manifesto For Earth Justice*, Green Books, Devon, 2003, p. 74.

15 See the Murray Darling Basin Authority www.mdba.gov.au/.

16 Australian Conservation Foundation, Facts and Figures: Murray-Darling Basin, 2009, http://www.acfonline.org.au/articles/news.asp?news_id=122. Accessed 2 December 2009

17 See further www.un.org/climatechange/ & http://climate.nasa.gov/.

18 Cullinan, *Wild Law*, p. 79.

19 E.T. Freyfogle, *Justice & The Earth: Images for out Planetary Survival*, Free Press, New York, 1993, p. 49.

20 R. Pipes, *Property & Freedom,* Vintage, New York, 2000, p. xv. This early conception is important because as Joshua Getzler notes, 'Roman ideas about private and public property provide a kind of DNA of legal ownership, the intellectual structure within which most later legal thought has developed' J. Getzler, 'Roman Ideas of Land Ownership' in *Land Law: Themes and Perspectives*, eds, S. Bright & J. Dewer, Oxford University Press, Oxford, 1998, p. 81.

21 E.T. Freyfogle, 'Ownership & Ecology', *Case Western Law Review*, vol. 43, 1993, p. 1272.

22 Jeremy Waldron, 'Liberalism' in *The Shorter Routledge Encyclopaedia of Philosophy*, ed E. Craig, Routledge, New York, 2005, p. 570.

23 T. Berry, *The Great Work: Our Way Into the Future*, Bell Tower, New York, 1999, p. 72.

24 T. Berry, *Evening Thoughts: Reflections on Earth as Sacred Community*, Sierra Club Books, San Francisco, 2006, p. 107.

25 T. Berry, 'The Viable Human' in *Deep Ecology for the 21st Century*, ed G. Sessions, Shambhala, Boston, 1995, pp. 5–6.

26 J. Holder, 'New Age: Rediscovering Natural Law' in *Current Legal Problems*, ed M.D.A. Freeman, Oxford University Press, Oxford, p. 172.

27 Aquinas for example advocated the existence of eternal law, natural law, divine law and human law.

28 T. Berry, 'Forward' in C. Cullinan, *Wild Law*, p. 20.

29 T. Berry, *The Great Work*, p. 64.

30 Ibid.

31 C. Cullinan, *Wild Law*, p. 84.

32 Ibid.

33 M. Curd, 'The Laws of Nature' in *Philosophy of Science, The Central Issues*, ed M. Curd & J.A. Cover, W.W. Norton & Company, New York, 1998, 805.

34 Ibid.

35 A. Kimbrell, 'Recovery of Natural Law as a Paradigm for a New Jurisprudence' at Center for Earth Jurisprudence Symposium, *Framing an Earth Jurisprudence for a Planet in Peril*, Feb. 28–29, 2008 http://earthjuris.org/events_/02–08symposium/02–08symposium.html. Accessed 23 March 2009.

36 See R.O. Brooks, *Law and Ecology: The Rise of the Ecosystem Regime*, Ashgate, Aldershot, 2002.

37 C. Krebs, *The Ecological Worldview*, Univeristy of California Press, California, 2008, p. 2.

38 E. Odum, *Fundamentals of Ecology*, W.B. Saunders Company, Philadelphia, 1971, p. 3.

39 E. Odum, *Ecology*, Thomson Learning, London, 1966, p. 3.

40 Ibid.

41 G.E. Moore, *Principia Ethics*, Dover Publications, New York, 2004.

42 L. Fuller, *The Morality of Law*, Yale University Press, Yale, 1964, p. 53.

43 L. Fuller, 'Human Purpose and Natural Law' *Journal of Philosophy*, vol. 53, 1956, p. 697.

44 L. Fuller, *Morality*, p. 207.

45 C. Cullinan, *Wild Law*, pp. 84–85.

46 M. Leanard, 'Al Gore Calling for Direct Action Against Coal', *The Understory*, 2007, http://understory.ran.org/2007/08/16/al-gore-calling-for-direct-action-against-coal/. Accessed 1 February 2010.

47 (M. Nichols, 'Gore Urges Civil Disobedience to Stop Coal Plants', *Reuters*, 2010, http://
    uk.reuters.com/article/idUKTRE48N7AA20080924. Accessed 1 February 2010.

48 J. Finnis, *Natural Law & Natural Rights*, Oxford University Press, Oxford, 1980, p. 9.

49 Ibid., p. 147.

50 Ibid., p. 154.

51 Ibid., p. 279.

52 Ibid., p. 150.

53 Ibid., p. 290.

54 J. Rawls, *A Theory of Justice*, Harvard University Press, Cambridge, 1964, p. 3.

55 Ibid., p. 350.

56 Natural Resource Defense Council, 'Hostile Environment: How Activist Judges
    Threaten Our Air, Water, and Land', 2001, http://www.nrdc.org/legislation/hostile/
    hostinx.asp. Accessed 1 February 2010.

57 The collapse of Easter Island is an obvious example, where the community deforested
    the land so that the timber could be used for transporting stones. The stones were
    being built to construct a giant statute for worship. See further J. Diamond, *Collapse:
    How Societies Choose to Fall or Succeed*, Penguin Books, New York, 2006, pp. 79–119.

58 T. Berry, *Viable Human*, p. 8.

59 Rainer Maria Rilke, 'Wenn Etwas mir vom Fenster Fällt' in Anita Barrows & Joanna
    Macy, *Rilke's Book of Hours,* Riverhead Books, New York, 1996, p. 116.

Part Two

Inspiration for Earth Jurisprudence

Section 1: Science and Nature

The greatest inspiration for Earth Jurisprudence is the natural world. In order to align our law and governance structures to promote a mutually beneficial relationship with nature, we must develop a deep understanding of our intimate connection and dependence on the natural world. This section will introduce the reader to key principles in holistic science, such as Gaia theory, phenomenology and the principles of interconnectedness, differentiation, and autopoiesis. Following this, it will outline activities to promote a personal connection to place. Implicit in this study is that nature must be understood in the head and felt in the heart.

Gaia and Earth Jurisprudence

Stephan Harding

The idea that nature has rights is an anathema to mainstream Western culture, which for the last 400 years or so has seen the cosmos as nothing more than a vast dead machine which can be exploited without let or hindrance for the exclusive benefit of the human species. As a result, there is now virtually no doubt that our culture has unleashed a massive crisis upon the world. We are changing the very climate of the Earth, we are driving millions of species into extinction and we are eroding the social ties that bind us into healthy families and communities. There are many metrics that can help focus our minds on the immensity of what is happening – here are just two. Firstly, because of our burning of fossil fuels, the Earth has not experienced atmospheric carbon dioxide levels as high as today's for about 740,000 years, and secondly, we are wiping out our fellow species with a ruthlessness that beggars the imagination – every day we exterminate some 100 species around the world at a rate about 1000 times faster than the natural background rate of extinction. The crisis is so colossal that some eminent scientists, such as James Lovelock and James Hansen, warn that we could face the collapse of civilisation within a matter of decades.

It is now widely accepted that we have brought this dangerous predicament upon ourselves through our pursuit of unlimited economic growth powered by the burning of fossil fuels and the ruthless exploitation of wild ecosystems. But there is a deeper cause that lies not in these outer actions, but in the very way that we have been taught to see the world ever since we were children. This is a worldview so dangerous that it has led us to wage an unwitting war on nature that we cannot possibly win – we are under the spell of a perspective that we must quickly modify if we are to have any viable future on the Earth.

Our death-dealing worldview is simply this: that, for us, our great turning world is no more than a vast dead machine full of 'resources' that have value only when they are converted into money. We think that mountains, forests, and the great wild oceans are all dead things that we are free to exploit as we wish without let or hindrance. Our culture values only quantities such as weight, height, and money in the bank. We have been taught to disregard qualities – to believe that the sense of elation we feel in the mountains or the calm we experience by a sunlit lake are merely our own idiosyncratic subjective impressions that tell us nothing real about the world. Furthermore, we have been persuaded to think that good citizenship involves buying more and more material goods so that the global economy can grow.

What if our relationship to nature is dysfunctional because of an unbalanced psychological development both individually and within the culture as a whole? And if this is indeed the case, is there anything we can do about it? It was C.G. Jung, the great Swiss psychologist, who observed that we all have four psychological functions, or 'ways of knowing', which operate as pairs of opposites: Intuition and sensing, and thinking and feeling. Intuition gives insight into the nature and deeper meaning of things, whilst sensing yields a direct apprehension of the world around us through the substrate of our physical bodies. Thinking interprets what is there through the exercise of logic and reasoning, whilst feeling helps us to ascribe positive or negative value to phenomena and situations – ultimately this is the sphere of ethics. Thinking and feeling are evaluative, whilst sensation and intuition are perceptive. Jung discovered that each of us has a dominant function, whilst the opposite function remains largely unconscious and undeveloped. The other two functions are only partially conscious, generally serving the dominant function as auxiliaries. Of course, this typology suffers from the limitations of all models, but Jung found it useful enough to say of it that it 'produces compass points in the wilderness of human personality'. Mental and physical health in Jung's therapeutic approach requires the conscious development of the neglected function together with an awareness of the four functions in oneself so as to achieve a well-rounded personality.

By applying Jung's insight to our culture, it is easy to spot the fact that we are suffering from an over-development of a particularly dangerous style of thinking that became immensely persuasive during the scientific revolution in the sixteenth and seventeenth centuries. It was at this time that philosopher/mathematician René Descartes and others made a convincing case that there is a fundamental ontological divide between the rational human soul and a soulless material universe, which was seen as nothing more than a vast mechanical contrivance there for us to dominate and control with impunity through the exercise of pure analytical reasoning.

The mechanistic worldview, useful as it is in certain limited ways, is at odds with a more ancient sensibility that saw the Earth and the entire cosmos as sentient beings worthy of reverence and respect. Our ancestors, and indeed many indigenous people to this day, sensed that they lived within a great psyche, the psyche of the cosmos itself – the *psyche kosmou*, as Plato called it, or, in Latin, the *anima mundi* – the 'soul of the world'. In this view matter is sentient to its deepest roots – everything is capable of experience, including subatomic particles, mountains, entire ecosystems, and indeed the Earth itself. Father Thomas Berry brilliantly expresses this insight when he says that 'the universe is not a collection of objects but a communion of subjects'. According to this 'panpsychist' or 'animistic' perspective, our own consciousness has its roots within the atomic mode of sentience, for it must somehow be entangled in (or perhaps emerge out of) the complex and intricate communion that takes place amongst the atoms that make up our physical bodies. We can only become truly sensitive to the living qualities of the world when we combine these rational arguments with our intuition, sensing and feeling. These tell us that every speck of matter has intrinsic value irrespective of any use that we might put it to, and that nature's subtle qualities – her colours, sounds and textures – are a kind of language or 'text' that constantly speaks to us of our animate surroundings. Thus, in this view, everything in nature has rights simply because it exists.

Strangely enough, the idea that the Earth is alive has come back to the modern world in an unlikely arena – within science itself. In 1972 James Lovelock, the great British scientist, proposed that our planet consists of a tightly coupled set of complex feedbacks between life, rocks, air and water that gives rise to the emergent ability of the planet as a whole to regulate its own surface conditions within the narrow limits suitable for life. Inspired by William Golding, Lovelock chose to his name his theory of a self-regulating Earth after Gaia, the ancient Greeks' animistic divinity of the Earth. A key insight from Gaia theory for us to ponder is that we humans are not in charge of the planet – that we are, after all, not the most important of her species. Instead, in the words of Aldo Leopold, the great American environmentalist of the last century, we are 'just plain members of the biotic community'– special in our own unique ways, but no more special in principle than the trees, the great whales, the teeming denizens of the microbial realm or the millions of other species that populate our newly stricken world.

Perhaps it is time to counter our dangerously outdated 'mechanistic' worldview with a more fruitful, more soulful science-based idea in tune with the wisdom of our indigenous ancestors that inspires us to uncover our deep indigenous connection with the earthly community of rocks, atmosphere, water and living beings – with the animate Earth that enfolds us. We can do this by learning how to relate holistically to the animate Earth through our four ways of knowing. We use our reason to study the Earth as an emergent self-regulating system that has kept our world habitable since the appearance of life some three and a half thousand million years ago thanks to the tumultuous and multitudinous interactions between living beings and the atmosphere, rocks and water that surround them. This is the essence of Gaia theory, which teaches us that we live symbiotically within a vast, evolving, sentient creature that has been charting her yearly path around the sun for thousands of millions of years, evolving her capacity for keeping her crumpled surface suitable for life as her biodiversity has increased over geological time. We can ponder the consistencies or otherwise between Gaia theory and the theory of natural selection, we can build mathematical models of the carbon cycle coupled to an active biota, and we can look at how the Earth could respond to climate change as a fully integrated complex system consisting of life coupled to its abiotic 'environment'. But we speak of the players in this 'system' not as dead cogs in a static machine, but as animate beings with particular and oftentimes peculiar personalities. Carbon atoms are the placid Swedes of the chemical world; oxygen atoms are its passionate Italians – for if the world is truly ensouled then even atoms are 'persons' in the most rudimentary sense of the word.

Then we go further. We can use this rational knowledge to fuel our intuitive sense of connection to the whole community of nature by recreating Gaia's long and complex evolutionary trajectory in our imaginations and by engaging in rigorous meditative explorations of her tightly coupled feedbacks. We deliberately connect with the qualities of rocks, atmosphere, oceans, clouds, individual organisms and entire ecosystems by spending quiet time savouring their essences much as we would that of a poem or a piece of music. As we deepen our perceptual abilities, we find a remarkable degree of agreement with each other in what we discover by means of this more phenomenological approach to nature. In addition, we work with exercises that help to shift our everyday perceptual frameworks.

I offer you one such 'meditation'. Just try it. I guarantee that it will give you an unexpected depth of belonging to the Earth. From the point of view of our mainstream culture you will be 'doing nothing', but in fact you will be engaging in highly subversive act – the demolition of our suicidal and vastly destructive mechanistic worldview.

> Lie on your back on the ground outside in as peaceful a place as you can find, in the forest perhaps, or by the roaring sea. Relax and take a few deep breaths. Now feel the weight of your body on the Earth as the force of gravity holds you down.
>
> Experience gravity as the love that the Earth feels for the very matter that makes up your body, a love that holds you safe and prevents you from floating off into outer space.
>
> Open your eyes and look out into the vast depths of the universe whilst you sense the great bulk of our mother planet at your back. Feel her clasping you to her huge body as she dangles you upside down over the vast cosmos that stretches out below you.
>
> What does it feel like to be held upside down in this way – to feel the depths of space beyond you and the firm grip, almost glue-like grip of the Earth behind you?
>
> Now sense how the Earth curves away beneath your back in all directions. Feel her great continents, her mountain ranges, her oceans her domains of ice and snow at the poles and her great cloaks of vegetation stretching out from where you are in the great round immensity of her unbelievably diverse body.
>
> Sense her whirling air and her tumbling clouds spinning around her dappled surface.
>
> Breathe in the living immensity of our animate Earth.

Do this again and again, at every available opportunity. Let yourself be 'Gaia'ed' by the great round sentience of our living world. Deeply experience what it feels like to meld with the great wild body of our animate Earth in this way. See how this simple act brings you into a deeply felt relationship with the whole of life.

Experiences such as these lead us into the realm of ethics as we deeply question our own lifestyles and those of our society in the light of our deepening connection with the personhood of the Earth – this is the 'deep ecology' approach of the great Norwegian philosopher Arne Naess. When we do this, we sense that is wrong to seriously harm the great turning world within which we live. This realisation gives us the energy and insight to change our lifestyles in beneficial ways. Research by people such as David Reay at the University of Edinburgh has shown that we can make a massive difference in our personal lives thanks to simple acts such as: turning down our heating in winter by just one degree Celsius; using our cars less; composting organic waste; avoiding flying; driving at or below speed limits; eating locally produced food; reducing, reusing and recycling; and by turning off all standbys and transformers. Also, we can think carefully before we buy anything new. Could we buy it second-hand, or even do without? We could become involved in strengthening our local communities, and find satisfaction in talking, telling stories and making music together rather than in working so mindlessly hard to buy the mostly useless consumer products promoted by the mass media for filling the gaps in our lonely lives. All of this doesn't seem like much, but if enough of us consume less in these ways we will make a huge difference, thereby removing the need for several new power stations in the UK.

Connecting with the Earth, consuming less and developing a sense of community can also give us the will and energy to work for change at the societal level. In this domain perhaps the most important thing to do is to agitate for an economy that is in a steady state, rather than working blindly for one which seeks to grow by extracting more and more of the Earth's finite resources from her ancient crumpled surface. Those of us touched by the animate Earth feel the urge to work towards creating an economy in which the things that grow are the development and deployment of renewable technologies, the restoration of degraded ecosystems, the recreation of vibrant local communities and economies, and the adoption of ecologically diversified farming practices. Policies inspired by this kind of 'intelligent growth' would also stimulate those non-material things that can grow without limit – spirituality, creativity, depth of community and simple living. These are, after all, the sources of our deepest satisfactions and our sense of well-being.

So how can we promote intelligent growth? We can begin by consuming less in our personal lives, as outlined above. But we can also lobby government (via our MPs) to take climate change seriously by setting a ceiling on greenhouse gas emissions through the implementation of a rigorous carbon rationing system, by giving significant tax breaks and other incentives for implementing energy saving measures, by funding massive research efforts into renewable energy and by developing ecologically sound ways of food production, building and transportation.

We can also agitate for the acceptance of the Earth Jurisprudence approach within mainstream culture, in which we recognise that each entity has rights given to it by existence, not by the arbitrary whims of human lawmakers. Thus, since 'personhood' is everywhere, atoms and molecules have rights too. We need to realise that the atoms in coal, oil, gas and other fossil fuels have rights to stay in the ground, as do mineral deposits and water in aquifers. If we humans want to extract some of these atoms and molecules for our own purposes, then we will need to use our best science to work out how many of them we can extract without destabilising the integrity of the Earth, but we will also need to ask our best shamans to ask the atoms and molecules for their permission to be exhumed for our benefit. And once we have used them to make our products, we must make sure that they are recycled, and that they eventually return to the Earth's crust when their work on our behalf has been done.

If we don't immediately make these radical changes we will almost certainly invoke the fearsome wrath of Gaia. Her law is that any being that destabilises her climate will experience feedbacks from the whole 'system' that will curtail the activities of that being. So we have a choice. We can carry on with business as usual and live in rightful fear of Gaia. Or we can learn to love her hills, her wild forests and her oceans as we love a cherished grandmother. Perhaps only then, motivated by this love of all earthly things, will we find the inspiration for mending our ways and for massively reducing our impact on the great animate being that gave us birth.

Bibliography

Abram, D., *The Spell of the Sensuous*, Vintage, New York, 1997.

Berry, T., *The Great Work*, Bell Tower, New York, 2000.

Harding, S.P., *Animate Earth: Science, Intuition and Gaia*, Green Books, Devon, 2009.

Jung, C.G., *On the Nature of the Psyche,* Routledge, New York, 2001.

Lovelock, J.E., *The Revenge of Gaia*, Penguin Books, New York, 2006.

Lovelock, J.E., *Gaia: Medicine for an Ailing Planet*, Gaia Books, London, 2005.

Sessions, G. (ed). *Deep Ecology for the 21st Century,* Shambala, Boston, 1995.

Wilson, E.O., *The Future of Life,* Alfred A Knopf, New York, 2002.

Eco-Centric Paradigm

Peter Burdon

In *The Great Work* Thomas Berry writes that the deepest cause of the environmental crisis 'is found in a mode of consciousness that has established a radical discontinuity between the human and other modes of being'.[1] This mode of consciousness is termed anthropocentrism, defined by Albert Einstein as an 'optical delusion of human consciousness' where we come to regard 'humanity as the centre of existence'.[2] Anthropocentrism is the view that human beings are the final aim and end of the universe and that the universe exists to satisfy the needs and desires of human beings.[3]

Anthropocentrism has its origins in Stoic philosophy and Christian theology.[4] Throughout Western history, it has enjoyed widespread acceptance amongst the intellectual and ruling classes of society. However, alongside this worldview a different perspective has continually surfaced and challenged the intellectual foundation of anthropocentrism. Perhaps the first significant challenge occurred in 1543 when Polish astronomer Nicolaus Copernicus published *On the Revolution of Celestial Orbs*. In this book, Copernicus presented a new view of the cosmos and humanity's place in it. Contrary to the dominant view of his age, Copernicus posited the sun at the centre of the universe and held that the Earth rotated on its own axis.[5] In 1610 Galileo Galilei substantiated this theory in *Sidereus Nuncius* where he presented qualitative evidence on the circulation of planets within Earth's galaxy. This combination of ideas fuelled a revolutionary change in conventional cosmology and 'demolished the intellectual and moral order of the western world'.[6] Indeed, writing at this time, British poet John Donne lamented,

> The new Philosophy calls into doubt,
> The Element of fire is quite put out;
> The Sun is lost, and th' Earth, and no mans wit
> Can well direct him where to look for it ...
> 'Tis all in peeces, all coherence gone;
> All just supply, and all Relation.[7]

The Copernican revolution marked a significant shift in the worldview of Western culture. However, it did not overthrow the existing anthropocentric paradigm. Indeed, this perspective was strengthened during the scientific revolution, when human beings took up a new mantle, 'claiming power not as part of the web of creation, not even as caretakers, but as masters of a cosmic machine'.[8] In the centuries that followed, a host of scientific advancements challenged this perception once again. Charles Darwin's *The Origin of Species* was the biological equivalent of the Copernican revolution. Significantly, Darwin posited human beings into a 'long-running family saga that includes apes and chimpanzees'.[9] While Darwin 'shoved the human species off its pedestal'[10] succeeding generations of evolutionary biologists such as Stephen Jay Gould showed that natural selection does not necessarily lead to increasing levels of complexity or greater intelligence.[11] Indeed, in opposition to Aristotle's 'Great Chain of Being'[12], Gould illustrated there is 'no splendid evolutionary ladder leading steadily onwards and upwards towards Homo Sapiens'.[13] As David Suzuki notes:

> Darwinian evolution has cast us as the children of chance, creatures with enough self-awareness and wit to recognise ourselves as a kind of cosmic joke. From Copernicus to Darwin to the reflections of modern eminent scientists, in the Western world Homo Sapiens has undergone a relentless diminuendo, ending up as just another species that happened to evolve way out in the heavenly boonies.[14]

In the modern era, the environmental crisis is another important event, which forces us to confront anthropocentrism and our continued exploitation of the Earth. This crisis has already begun to put human survival at risk, as well as many other living systems.[15] In *The Universe Story*[16] Berry and Brian Swimme contend that a third revolution is upon us and involves a shift from the terminal Cenozoic to the Ecozoic era.[17] Berry notes:

> The Cenozoic period is being terminated by a massive extinction of living forms that is taking place on a scale equalled only by the extinctions that took place at the end of Palaeozoic around 220 million years ago and at the end of the Mesozoic some 65 million years ago. The only viable choice before us is to enter into the Ecozoic period, the period of an integral community that will include all the human and non-human components that constitute planet Earth.[18]

Berry and Swimme prioritise three interrelated ideas as being integral to the Ecozoic era. The first is communion (or interconnectedness) which contends that the universe[19] is an integrated whole, rather than a hierarchical collection of dissociated parts. The second idea is differentiation, which recognises that the universe is ordered by diversity, complexity, variation and multiform nature. Finally, the ecological paradigm recognises that the universe is characterised by autopoiesis, subjectivity, self-manifestation and interiority.

My intention in this paper is to expand and analyse the framework established by Berry & Swimme and provide a launch pad for further study. Indeed, while many authors comment freely on the ecozoic paradigm, few have expanded its internal ideas and noted the rich literature that supports the concept. To fulfil this objective, I will focus primarily on scientific disciplines. In doing so, I do not seek to advance science over other disciplines or ways of understanding the universe.

Indeed, as Jerome Revetz has noted, science is profoundly shaped by value commitments and biases.[20] Instead, I choose this mode of expression because it remains a powerful and persuasive means of communication in Western culture. Indeed, detached from mechanistic assumptions, science is a superlative tool for investigating the nature of the physical world – it is also the distinctive discovery of our Western culture. Speaking to this point, Freya Matthews comments, '[w]e may be discontented or disappointed with its findings, and we may wish to supplement scientific method with other investigative techniques, but if a new worldview is to attain legitimacy and take root in this culture, it must ultimately have the sanction of science'.[21] I will begin this discussion with an analysis of the interrelatedness of all life.

Communion

> That the Universe is a communion of subjects rather than a collection of objects is the central commitment of the Ecozoic.[22]

Relationship is the essence of existence.[23] This point was noted by thirteenth century mystic, Hildegard of Bingen, who wrote, 'everything that is in the heavens, on the earth and under the earth is penetrated with connectedness, penetrated with relatedness'.[24] Western science has been playing catch-up and attempting to provide empirical proof of mystical insights like this for centuries. Berry notes that the scientific advancements that have occurred over the past three centuries 'might be considered among the most sustained meditations on the universe carried out by any cultural tradition ... science has given us a new revelatory experience'.[25] This revelation can be witnessed in the modern understanding of the origin of the Universe. Indeed, we now have empirical evidence that the Universe was brought into being through a single energetic event or primordial fire. When this fire sparked, each element in the universe was connected to everything else. At no point in the future was this connection broken for, as Swimme notes, 'alienation of a particle is a theoretical impossibility'.[26] Swimme notes further:

> That which blossomed forth as cosmic egg fifteen billion years ago now blossoms forth as oneself, as one's family, as one's community of living beings, as our blue planet, as our ocean of galaxy clusters ... To enter the omnicentric unfolding universe is to taste the joy of radical relational mutuality. For we know this body of ours could have been a giant sequoia. We know in a simple and direct way that we share the essence of and so easily could have been a migrating pelican ... we reach the conviction that we could have been an asteroid, or molten lava, or a man, or a woman, or taller or shorter, or angrier, or calmer, or more certain, or more hesitant, or more right or wrong.[27]

Communion is profoundly revealed in many other scientific disciplines. Many writers focus on theoretical physics to make this point.[28] In addition, there is no greater or more convincing source than the science of ecology, which is the study of relationships, energy transfers, mutualities, connections and cause/effect networks within natural systems.[29] Ecology provides solid and observable facts to the merely theoretical findings in physics and has enriched the way we think about the natural world by two new concepts, community and network. These concepts shift our perception of an ecological community from hierarchy and toward 'an assemblage of organisms, bound into a functional whole by their mutual relationships'.[30] A further achievement of ecology is the understanding that most organisms are not only members of ecological communities, but are also 'complex ecosystems themselves, containing a host of smaller organisms that have considerable autonomy and yet are integrated harmoniously into the functioning of the whole'.[31]

This point can be illustrated in every level of nature.[32] Consider for a moment a forest that you are familiar with. In your mind, enter the forest and feel the cool air provided by the canopy above. Walk over and sit at the base of one of the large trees that sit before you. Your nose is alerted to a strong smell at the base of the tree. You reach down and pick up a truffle that is growing freely amongst some of the other trees of the forest. You observe that trees with truffles at their base are larger and greener than those without. This is because truffles extract water and minerals from the soil and dispense them over the roots of their host. While you reflect on this reciprocal relationship, a Longfooted Potoroo[33] (or another medium size mammal of your choice) hops into view and stops to rest beside you. It bends down to eat the truffles and before leaving excretes on the base of another nearby tree. The spores of the truffle, now coated in rich organic matter, can begin to regenerate and thereby enhance the overall health of the forest.

In this example, forest, truffle and Potoroo – three very different life forms – are all bound together in a remarkable web of relationships. By entering into the ecosystem, we as human beings represent a further layer of interdependence. While portrayed as a passive observer, at the most basic level we are breathing air and returning carbon dioxide to the trees. We may pick the truffles for cooking and return our compost to further enrich soil quality. It is easy to see our interconnectedness to nature in unspoiled nature. It is much harder to recognise that even a city is nature. Indeed, every single component of a city comes from nature and will be re-absorbed back into the Earth in time.

The concept of network has become increasingly prominent in ecology. This has provided a new perspective on the 'so-called hierarchies of nature'.[34] Since living systems (at all levels) are networks, we can visualise them as webs of relationships interacting in a network fashion with other systems. One key consequence from this evolution in knowledge is that it reverses the Cartesian mechanistic scientific method and holds that natural systems cannot be properly understood by isolated analysis. Indeed, the properties of the parts 'can only be understood within the context of the larger whole'.[35] Further, as quantum physics has illustrated, there are no parts at all.[36] What we refer to, as a 'part' is actually an inseparable web of relationships. Consequently, the shift from parts to a whole 'can also be seen as a shift from objects to relationships'.[37] Indeed, as Saint-Exupery noted centuries earlier, 'man is but a network of relationships, and these alone matter to him'.[38]

Differentiation

Nature abhors uniformity.[39]

Nature is ordered by differentiation.[40] This quality can be witnessed in elementary particles, atomic beings, in the radiant structures of the physical world and the complexities of the galaxy and planetary systems. The biosphere is an intricate tapestry of interwoven life forms and even the seemingly desolate arctic tundra is sustained by a complex interaction of diverse flora and fauna. Certainly, biodiversity is the greatest wonder and mystery on Earth.[41]

Each element of the universe is new and distinct from all other structures. Indeed, the interactions that control elementary particles are distinct from those involving atoms, which are again different from the dynamics involved with stars and galaxies.[42] While one can find similarities within levels, the generic form of the equations on one level is particular to that level and 'each level of the universe is a distinct world'.[43] When combined with the concepts, network and community, the notion of differentiation further challenges the anthropocentric idea that the Earth exists for human beings. Indeed, it illustrates that life is not singular and focused on one species. Life is characterised and receives strength and resilience from the diverse interactions of the Earth community.

Perhaps the best description of Earth's diversity comes from the eminent Harvard biologist Edward O. Wilson. In his classic text *The Diversity of Life* Wilson notes:

> How many species of organisms are there on Earth? We don't know, not even to the nearest order of magnitude. The number could be close to 10 million or as high as 100 million ... with the help of other systematists, I recently estimated that the number of known species of organisms, including all plants, animals and microorganisms, to be 1.4 million. This figure could easily be off by a hundred thousand ... evolutionary biologists are generally agreed that this estimate is less than a tenth of the number that actually live on Earth.[44]

Over the last forty years, biologists have begun exploring the deep sea, a vast domain that stretches 300 million square kilometres. This environment is amongst the most inhospitable places on earth and yet it is home to hundreds of thousands of undocumented species.[45] On land, the great naturalist-explorer William Beebe noted while sitting in a rainforest canopy, 'yet another continent of life remains to be discovered, not upon earth, but one to two hundred feet above it'.[46] In Columbia's Choco region, half the species remain undocumented and of those a large portion lack a scientific name. On average, two new bird species are discovered each year[47] (sometimes after they are extinct). Even new kinds of mammals are discovered on occasion. For example, in 1988 a new lemur was found in Madagascar, a monkey from Central Africa and a new Muntajak deer from western China. In 1990, a primate named the black-faced lion tamarin was located on a small coastal island, just 65 kilometres from the city of Sao Paulo. Even the order of Cetacea, which consists of the largest animals on Earth, is not completely known.[48]

Among the least known groups are fungi and bacteria. Speaking first to fungi, there are '69,000 known species but 1.6 million thought to exist'.[49] The range is even greater with bacteria, of which a mere 4000 species are recognised. Wilson notes:

> the vast majority of bacterial types remain completely unknown, with no name and no hint of the means needed to detect them. Take a gram of ordinary soil, a pinch held between two fingers, and place it in the palm of your hand. You are holding a clump of quartz grains laced with decaying organic matter and free nutrients, and about 10 billion bacteria. How many bacterial species are present? Take one-millionth of the pinch of soil and spread it evenly over nutrients poured into standard culture dishes. If each and everyone of the bacteria in this near-invisible solid sample could multiply, we would expect to see over 10,000 little colonies growing on the nutrient surface, one from each bacterium. But they cannot and we do not. We get only between 10 and 100 colonies.[50]

Studies such as this illustrate the vast diversity in the natural world. We are a long way from even making an educated guess on the diversity of natural systems, especially complex ecosystems such as rainforests, wetlands, and oceans. Yet, 'life' is greater than human beings and is made possible by millions of organisms, most of which we have never seen, never mind studied in a meaningful way.

Autopoiesis

The biosphere is structured by autopoiesis.[51] The term autopoiesis stems from the Greek root *auto* (self) and *poiein* (making) and refers to life's continuous production of itself. Our biosphere is autopoietic in the sense that it maintains itself.[52] Its vital 'organ' is the atmosphere. Importantly, the term 'autopoiesis', as first described by Chilean neurobiologist Humberto Maturana, applied strictly to *living* entities on Earth.[53] Berry and Swimme use poetic license and analogy to extend this concept further to include galaxies and the universe.[54] Commenting on their use of the term, the authors note,

Autopoiesis refers to the power each thing has to participate directly in the cosmos-creating endeavour. For instance, we have spoken of the autopoiesis of a star. The star organises hydrogen and helium and produces elements of light. This ordering is the central activity of the star itself. That is, the star has a functioning self, a dynamic or organisation centred within itself. That which organises this vast entity of elements and action is precisely what we mean by the star's power of self-articulation.[55]

This is clearly a broader use of the term than conventionally employed in science. It is not necessary to go to this extent to show the relevance and importance of autopoiesis on Earth. For example, oceanographic studies illustrate that salts such as sodium chloride and magnesium sulphate are continuously eroded from continents and carried into the ocean through river systems. Chemical studies suggest that salt should have accumulated to levels that make life impossible for non-bacterial organisms.[56] However, this has not occurred and salt-sensitive organisms continue to thrive in ocean ecosystems.[57] To account for this occurrence, oceanographers have argued that seafaring microorganisms have been sensing and stabilising ocean salinity.[58] In this example, life has responded to prolong its own survival.

This connects to another hypothesis, which contends that Earth has a homeostatic feedback mechanism, which maintains optimum conditions for life. The first scientist to study how this mechanism operates was Vladimir Ivanovich Vernadsky (1863–1945). Vernadsky inspired the development of the Gaia theory by James Lovelock and Lynn Margulis in the 1970s.[59] Importantly, while Vernadsky described both organisms and minerals as 'living matter' the Gaia hypothesis describes Earth's surface, including the ocean and crustal rocks as 'alive'.[60]

In *Gaia: A New Look at Life on Earth,* Lovelock explained the Gaia hypothesis, noting that the 'entire range of living matter on Earth from whales to viruses and from oaks to algae could be regarded as constituting a single living entity capable of maintaining the Earth's atmosphere to suit its overall needs and endowed with faculties and powers far beyond those of its constituent parts'.[61] Lovelock went on to define Gaia 'as a complex entity involving the Earth's biosphere, atmosphere, oceans, and soil; the totality constituting a feedback of cybernetic systems which seeks an optimal physical and chemical environment for life on this planet'.[62] From this initial hypothesis, Lovelock claimed the existence of a 'global control system' of ocean salinity, atmosphere composition and surface temperature. In explaining this point, he notes:

> Seen in all its shining beauty against the deep darkness of space, the Earth looks very much alive. This impression of life is real. Only a planet with abundant life and able to retain its water and regulate its unique atmosphere and climate could appear so different from its sister planets, Mars and Venus, both of which are dead. Of course the Earth is not alive like an animal, able to reproduce itself and have its progeny evolve in competition with other animals. It is a superorganism, alive like the great ecosystems or some giant tree, the largest life form we yet know. I think it wrong of science to deny the status of life intermediate between inanimate matter and a sentient organism, yet greater and longer lived than most organisms.[63]

As explained by Margulis, the Gaia hypothesis takes a slightly different approach. To explain the Gaia theory, she notes that 'the temperature of the planet, the oxidation state and other chemistry of all of the gases of the lower atmosphere (except helium, argon, and other nonreactive ones) are produced and maintained by the sum of life'.[64] Importantly, Margulis emphasises that the Gaia hypothesis is a biological idea and rejects the term 'organism' in her interpretation. She notes

> Those who want Gaia to be an Earth goddess for a cuddly, furry human environ-
> ment find no solace in it. They tend to be critical or to misunderstand. They can buy
> into the theory only by misinterpreting it …Earth is an ecosystem, one continuous
> enormous ecosystem composed of many component ecosystems. Lovelock's posi-
> tion is to let the people believe that Earth is an organism, because if they think it is
> just a pile of rocks they kick it, ignore it, and mistreat it. If they think Earth is an
> organism, they'll tend to treat it with respect. To me, this is a helpful cop-out, not
> science.[65]

Despite these differences, Lovelock & Margulis agree on the fundamental tenants of Gaia science and in particular the notion that the Earth's surface is alive. In this context, autopoiesis refers to the aggregate, emergent properties of the gas-swapping, gene-trading, growing and evolving organisms on Earth. Indeed, just as relations among the body's cells regulate temperature and blood chemistry, so planetary regulation occurs in response to interaction among the Earth's living inhabitants.[66] Life extends across and enlivens the planet – Earth in a very real sense is alive. This is not a metaphysical or spiritual claim, but instead a physiological truth. Each breath connects living organisms with the biosphere. The biosphere also breathes and this is marked by the decrease in carbon dioxide from the light side of the globe, the changing of the seasons and photosynthetic activity.[67] Considered to its greatest physiological extent, life is the planetary surface. Indeed, as Margulis and Sagan note, 'earth is no more a planet-sized chunk of rock inhabited with life than your body is a skeleton infested with cells'.[68]

Conclusion

This brief description of modern science presents a description of the Earth that is characterised by communion, differentiation and autopoiesis. Certainly, these (and many other) revelations have the potential to move society beyond the outdated and harmful anthropocentric paradigm. The assimilation and application of these ideas is the *great work* before the present generation. As individuals, we must learn to assimilate this revelation and look at the universe with fresh eyes. As a culture, we must evolve our social institutions toward a vision of Ecozoic era. Above all, we must live with a deep respect and humility for our place in the cosmos and celebrate our existence in the great matrix of life. This idea of 'celebration' might appear out of place, but for Berry and Swimme, it sits as the single expression for the entire universe. They note:

> The awesome aspect of the universe is found in qualitatively different modes of
> expression throughout the entire cosmic order but especially on the planet Earth.
> There is no being that does not participate in this experience and mirror it forth in

some way unique to itself and yet in a bonded relationship with the more comprehensive unity of the universe itself. Within this context of celebration we find ourselves, the human component of this celebratory community. Our own special role is to enable this entire community to reflect on and to celebrate itself and its deepest mystery in a special mode of conscious self-awareness.[69]

Notes

1   T. Berry, *The Great Work: Our Way into the Future*, Bell Tower, New York, 1999, p. 4.

2   Albert Einstein, quoted in K. Bosselmann, 'The Way Forward: Governance and Ecological Integrity', *Reconciling Human Existence with Ecological Integrity,* eds L. Westra, K. Bosselmann & R. Westra (2008) p. 319.

3   *The Macquarie Dictionary: Revised Edition*, Macquarie Library Pty Ltd, Dee Why, New South Wales, 1985, p. 113

4   See Berry, *Great Work*, p. 136.

5   E. McMullin, *The Shorter Routledge Encyclopaedia of Philosophy*, ed. Edward Craig, Routledge, New York, 2005, p. 150.

6   D. Suzuki, *The Sacred Balance: Rediscovering Our Place in Nature*, Allen & Unwin, Crows Nest, New South Wales, 1997, p. 13.

7   John Donne, quoted in Suzuki, *Sacred Balance,* 1997, p. 14.

8   Suzuki, *Sacred Balance,* 1997, p. 13.

9   Suzuki, *Sacred Balance,* 1997, p. 13.

10   Suzuki, *Sacred Balance,* 1997, p. 13.

11   S.J. Gould, *The Richness of Life: The Essential Stephen Jay Gould*, W.W. Norton and Company, New York, 2006.

12   See T. Burns et. al., 'Hierarchical Evolution in Ecological Networks', in *Theoretical Studies of Ecosystems: The Network Perspective*, eds M. Higashi & T. Burns, Cambridge University Press, Cambridge, 1991.

13   Gould, *Richness of Life*, 2006.

14   Suzuki, *Sacred Balance,* 1997, p. 15.

15   Note that in 2009, over 20 million people were officially classified as environmental refugees. For a robust account see L.R. Brown, *Plan 4.0: Mobilizing to Save Civilization*, World Watch Institute, New York, 2009.

16   T. Berry & B. Swimme, *The Universe Story: From the Primordial Flaring Forth to the Ecozoic Era*, Harper Collins Publisher, San Francisco, 1992, pp. 71–81.

17   Science divides the Phanerozoic eon into the Palaeozoic, Mesozoic and Cenozoic eras. Other books which describe paradigm shift are T. Kuhn, *The Structure of Scientific Revolutions*, University of Chicago Press, Chicago, 1996; F. Capra, *The Turning Point: Science and the Rising Culture*, Bantam Books, Toronto, 1983; P. Sorokin, *Social and Cultural Dynamics*, Porter Sargent Publishers, Boston, 1970; L. Mumford, *The Transformations of Man*, Harper Torchbooks, New York, 1956; D. Korten, *The Great Turning*, Berrett-Koehler Publishers, San Francisco, 2006; E. Laszlo, Macroshift: Navigating the Transformation to a Sustainable World, Berrett, San Francisco, 2001.

18 T. Berry, 'The Ecozoic Era', 1999 <http://www.earth-community.org/images/The%20 Ecozoic%20Era.pdf> accessed 1 March 2009.

19 I refer to the universe as the most comprehensive and inclusive point in the discussion. My analysis applies equally to the Earth, bioregions and microanalysis.

20 See J. Revetz, *Scientific Knowledge and its Social Problems*, Transaction Publishers, New Jersey, 1971. Revetz noted that science has long been a totem of European secular culture. In the century-long struggle between the ideologies of science and those of religion, the central claim was that science uniquely had truth, as opposed to nonsense studies like theology and metaphysics. However, values enter science through the selection of the problem to be investigated – the choice of the problem, who makes the choice and on what grounds. Society, the political realities of power, prejudice and value systems will influence even the 'purest' science. Values also play an important role in determining what is actually seen as a problem, what questions are asked and how they are asked. Other critics such as Hilary and Steven Rose have reinforced these arguments.

21 F. Mathews, *The Ecological Self*, Barnes and Nobles Books, Maryland, 1991 p. 49. Indeed, while many writers draw from the vast reservoir of Eastern wisdom to articulate scientific principles, ultimately salvation must come from within our own culture and its traditions. Speaking to this point the Carl G. Jung notes, 'Of what use to us is the wisdom of the Upanishads or the insight of Chinese Yoga, if we desert the foundations of our own culture as though they were errors outlived and, like homeless pirates, settle with thievish intent on foreign shores', C. Jung, *The Secret of the Golden Flower*, Harcourt Brace & Company, San Diego, 1962, p. 144.

22 Berry & Swimme, *Universe Story*, p. 243.

23 For an excellent introduction to this idea see Stuart Kauffman, *The Origins of Order*, Oxford University Press, New York, 1993.

24 Hildegard of Bingen quoted in M. Fox, *Meditations with Hildegard of Bingen*, Bear & Company, Vermont, 1982, p. 41.

25 T. Berry, *The Dream of the Earth*, Sierra Club Books, San Francisco, 1988, p. 18.

26 Berry & Swimme, *Universe Story*, p. 77.

27 B. Swimme, *The Hidden Heart of the Cosmos*, Orbis Books, New York, 1999, p. 111.

28 See for example F. Capra, *The Web of Life: A New Scientific Understanding of Living Systems*, Anchor Books, New York, 1996; F. Capra, *The Tao of Physics*, Flamingo, London, 1982; Mathews, *Ecological Self*.

29 G. Synder, *A Place in Space: Ethic, Aesthetics and Watersheds*, Counterpoint, Washington, 1995, p. 75.

30 Capra, *Web of Life*, pp. 33–34

31 Ibid., p. 34.

32 See R. Hopkins, *The Transition Handbook*, Green Books, Devon, 2008, pp. 60–61. Hopkins describes an excellent game 'The Web of Resilience' which can be used in small groups to demonstrate this point.

33 An Australian marsupial now classified as rare.

34 Capra, *Web of Life*, p. 35.

35 Ibid., p. 37.

36 Capra, *Tao of Physics,* p. 142.

37 Ibid., p. 37.

38 Saint-Exupery quoted in N. Evernden, *The Natural Alien: Humankind and the Environment*, University of Toronto Press, Toronto, 1993, p. 43.

39 Berry, *Great Work,* p. 149.

40 Berry & Swimme, *Universe Story,* p. 73. The classic text on biodiversity is Edward O. Wilson, *The Diversity of Life*, W.W. Norton and Company, New York, 1992.

41 E.O. Wilson, 'Editors Forward' in E.O. Wilson & F. Peter, *Biodiversity*, National Academy Press, Washington, 1988, p. v.

42 Berry & Swimme, *Universe Story,* p. 73.

43 Ibid.

44 Wilson, *Diversity*, pp. 132–133.

45 Ibid., p. 141.

46 William Beebe, quoted in Wilson, *Diversity*, p. 140.

47 Wilson, *Diversity*, p. 147.

48 Ibid., 148.

49 Ibid., 142.

50 Ibid.

51 Berry & Swimme, *Universe Story,* p. 75.

52 For an excellent paper on this topic see P. Lyon, 'Autopoiesis and Knowing: Reflections on Maturana's Biogenic Explanation of Cognition' in *Cybernetics & Human Knowing*, vol. 11, no. 4, 2004.

53 See for example H.R. Maturana, 'Autopoiesis, Structural Coupling and Cognition: A History of These and Other Notions in the Biology of Cognition' vol. 9, no. 3–4, 2003.

54 Berry & Swimme, *Universe Story*, p. 75.

55 Ibid.

56 L. Margulis & D. Sagan, *What is Life?*, Simon & Schuster, New York, 1995, p. 27.

57 Ibid.

58 Ibid.

59 J. Lovelock & L. Margulis, 'Biological Modulation of the Earth's Atmosphere', in *Icarus*, vol. 21, 1974.

60 J. Lovelock, *Gaia: The Practical Science of Planetary Medicine*, Gaia Books, London, 1991, p. 38. Note that autopoiesis only applies to living matter. However, Lovelock's inclusion of rocks, ocean etc into this larger category is explained as follows: 'With loving organisms the intensity of life varies from part to part. Your hair, your nails, and the outer layers of your teeth contain no living cells, yet they are undeniably a part of you. So it is with the atmosphere, the oceans and the crustal rocks of the Earth: They are parts o the organism in which life is thinly dispersed ... the seas and the rocks are usually seen as lifeless in themselves. I however, see them as essential parts of a larger organism'.

61   J. Lovelock, *Gaia: A New Look at Life on Earth*, Oxford University Press, Oxford, 1979, p. 18.

Suzuki, *Sacred Balance,* 1997, p. 145.

62   Lovelock, *Gaia,* p. 18.

Suzuki, *Sacred Balance,* 1997, p. 145.

63   J. Lovelock, *Gaia: The Practical Science of Planetary Medicine*, Gaia Books, London, 1991, p. 36.

64   L. Margulis, 'Gaia is a Tough Bitch', 2007, http://www.edge.org/documents/ThirdCulture/n-Ch.7.html accessed 18 November 2008.

65   Ibid.

66   Margulis & Sagan, *What is Life?,* p. 28.

67   Ibid.

68   Ibid.

69   Berry & Swimme, *Universe Story,* p. 264.

Place as Inspiration

Joel Catchlove

The following essay emerged from a workshop developed by Joel Catchlove and Richard Smith, at Wild Law, Australia's first conference on Earth Jurisprudence, held in October 2009 at Woodhouse, in the Adelaide Hills, South Australia. Woven throughout the essay are activities adapted for the workshop that are intended to cultivate awareness of place.

Activity – Mapping Home

To introduce the workshop's focus on place, participants are given a sheet of paper and markers and invited to draw a map or representation of their home. In doing so, they are invited to consider where they are from, the unique characteristics of their home, their town, their country of origin or even their lives so far. They may also like to consider mapping their journey to the workshop venue, and their experience of the venue so far. Participants then introduce themselves to other participants using their maps. Maps can be collected for display.

Earth Jurisprudence is nested within a broader shift towards place-based ecological understanding, characterised by interconnected, global movements like bioregionalism, permaculture, and place-based education. Purely technical amendment of law is not sufficient unless it is coupled with a deep cultural and personal shift that understands our connections and interdependence with all life. Underlying this shift is the need to re-establish a deep relationship with the world and the specific places we inhabit. Freya Mathews urges a re-investing of matter and place with an 'animating principle', 'so that to live in a particular place will be to enter into relationship with it, a relationship that can come to claim us so powerfully that the place in question becomes "home", our anchor to the world'.[1] For writers like Robert Thayer, we have become 'in certain fundamental ways, homeless', and the rediscovery of our home bioregion, or 'life-place', is central to a 'regenerative future'.[2] As Christine King notes, 'the problems we face today, as well as solutions, are no longer essentially technical, but have to do with the way we relate to each other and the world around us'.[3]

For just over 170 years, industrial civilisation has been settled on my home bioregion, the Adelaide Plains. Yet already the Plains express a microcosm of the ecological crises experienced across the continent. Rising temperatures, degradation of land and water, and biodiversity loss, together with declining water supplies, exists in sharp contrast to the history of Kaurna occupation of the Plains. For over 40,000 years, the Kaurna Aboriginal nation, one of over 120 Aboriginal nations across Australia, has lived on the Adelaide Plains in a 'semi-sedentary lifestyle nurturing the places and passing on environmental knowledge'[4] through a unique and enduring connection to place. As Markus highlights, the contrast between the lifestyle of Australian Indigenous cultures and the lifestyle of our own settled, industrial culture represents a 'fundamental shift in attitudes towards nature'.[5] Where the former recognises and is responsive to the unique limits of place, the second 'challenges nature' and 'manipulate[s] the environment ... often far beyond its capacity'.[6] These are sentiments shared by Mathews, who asserts that the environmental crisis has emerged because industrialised cultures view the world as 'lifeless clay, for us to convert into things for our own use, convenience and diversion'. Within such a view of the world, it becomes impossible for us to truly dwell in the world, 'all places are interchangeable' and 'all places, like all things, can be converted into "property" '.[7]

While our European-based calendar carves the seasons of the Plains into neatly mathematical three-month chunks, the Kaurna calendar offers one profound illustration of the depth of place-based understanding. In contrast to the abstracted rigidities of the European calendar with its four seasons, the four to six seasons of the Kaurna calendar expresses a sense of time as 'subjective, multi-dimensional and determined by the nuances of nature'.[8] Each of the seasons emerge from specific seasonal, ecological or cosmological events, depicting a deep awareness of environmental change and triggering seasonal activities and movements. The understanding of time and place demonstrated by the Kaurna calendar, and the extreme longevity of Aboriginal cultures highlights that these place-based cultures are perhaps the only model of ecological sustainability we have.

Activity – Touch Hunt

To introduce the idea of building awareness and encouraging tactile experiences with nature, pairs of participants are given a list of 'touches' to find in the environment surrounding them (they might include scratchy, rippled, smooth, rough, crunchy, tickly, cold, bendy or others). The group then reconvenes and pairs are invited to share their discoveries.

Activity – Blindfold Walk

The Blindfold Walk further extends the tactile experiences of the Touch Hunt, and engages other senses in building awareness of place. In pairs, one participant is blindfolded and led on a short walk, with the guide exposing their blindfolded companion to a range of sensory experiences with the surrounding environment. The guide should consider experiences that engage touching, smelling, listening, and even tasting, as appropriate! After a few minutes, the blindfold is removed, and the formerly blindfolded participant tries to retrace their journey. Roles are then swapped. The group reconvenes and participants are invited to share interesting or surprising experiences and discoveries.

Without a relationship with place, a culture cannot appreciate and respond to its sustainable ecological limits. For Kirkpatrick Sale, to become 'dwellers in the land', 'the crucial and perhaps only and all-encompassing task is to understand place, the immediate specific place where we live'.

> The kinds of rocks and soils under our feet; the source of the waters we drink; the meaning of the different kinds of winds; the common insects, birds, mammals, plants, and trees; the particular cycles of the seasons; the times to plant and harvest and forage – these are the things that are necessary to know. The limits of its resources; the carrying capacities of its lands and waters; the place where it must not be stressed; the places where its bounties can best be developed; the treasures it holds and the treasures it withholds – these are the things that must be understood. And the cultures of the people, of the populations native to the land and of those who have grown up with it, the human social and economic arrangements shaped by and adapted to the geomorphic ones, in both urban and rural settings – these are things that must be appreciated.[9]

This, says Kirkpatrick, is the essence of bioregionalism.

In asserting that connection to place is essential for environmental and social sustainability, bioregionalism offers a powerful model for the development of an ecological culture. Bioregionalism accepts that radical and fundamental change is necessary to cultivate a sustainable society, rather than merely 'fine-tuning and reforming industrial civilisation's present practices'.[10] In its vision that 'free people, freely associating, informed by the biological and geological truths of their home-places ... bring the best within themselves, along with the health of their lands, to full flowering',[11] bioregionalism reaches beyond connection with the landscape to reimagine social relationships.

In his synthesis of bioregional approaches, Doug Aberley articulates shared philosophies that express an urge towards the decentralisation of power on a bioregional level (territories defined by distinct biological, geographic and cultural characteristics) governed through autonomous, democratic, and participatory processes.[12] Political and cultural processes are measured by their ability to achieve social and ecological justice, within which we can imagine a diversity of expressions of Earth Jurisprudence, each shaped by the uniqueness of its home place. Bioregional economies are shaped towards local self-reliance, and the preservation and restoration of diversity and ecological health, and are supported by 'intricate networks of federation', 'woven on continental, hemispheric and global bases to ensure close association with governments, economic interests and cultural institutions in other bioregions'.[13]

The relocalisation of economies is a characteristic particularly evident in a spectrum of grassroots movements from the climate change and peak oil-based Transition Towns and Post-Carbon Network, to the broader movement for local food. Place-based movements like bioregionalism hold their roots in the practices and traditions of our ancestors and traditional cultures across the world.[14] However, they do not seek to displace the unique and enduring relationship of indigenous peoples with the landscape, but rather to move towards a mutually supportive, parallel culture of sustainability.

Expressions of bioregionalism are already emerging. These expressions are as diverse as the bioregions that sustain them, as collections like *Home! A Bioregional Reader* so powerfully demonstrate. In translating the bioregional vision into action, McGinnis suggests a 'place-based initiation process' that opens 'human senses and sensibilities to the surrounding landscape' and articulates 'one's connections with others' in the human and non-human community through local political processes, developing ecological literacy and self-education.[15] The process of learning place is supported by a wealth of resources. Nature journaling is a powerful way to open one's senses and learn place through the observation and documentation of natural patterns, forms and cycles.[16] Self-directed courses like the Wilderness Awareness School's *Kamana,*[17] emphasise the development of observation and awareness, and can be adapted to contexts outside of North America. Likewise, Oregon's Northwest Earth Institute offers discussion courses for small groups to explore literary perspectives on topics such as 'Reconnecting with Earth' and 'Discovering a Sense of Place'.

In its commitment to exploring local ecology and cultural histories, democratic decision-making and induction into community processes, place-based education strongly echoes the central principles of bioregionalism.[18] Now spreading across schools in the United States and the inspiration for powerful educational projects in Adelaide, South Australia,[19] place-based education grounds 'learning in local phenomena and students' lived experience'. It recognises that 'human learning once occurred within the context of specific locales', but has now shifted so that 'teachers direct children's attention away from their own circumstances and ways of knowing and toward knowledge from other places that has been developed by strangers they most likely will never meet'.[20] In an environmental context, 'knowledge from other places' may look like young children learning about global crises such as climate change or deforestation, which, rather than creating 'empowered global citizens', can instead cultivate hopelessness and disempowerment.[21] Instead, argues David Sobel, 'what is important is that children have an opportunity to bond with the natural world, to learn to love it and feel comfortable in it, before being asked to heal its wounds'.[22] Such an approach to learning to love the natural world is echoed by Richard Louv, who details how the most significant childhood factors in developing ecological consciousness for naturalists and environmental activists was wild, exploratory, independent play in natural environments, rather than mandatory, adult-regulated activities.[23]

Such immersive education provides opportunities for children to 'give back to others in ways that validate their own existence', while enabling them to become 'skilful and confident about their capacity to shape their own lives in ways that will benefit themselves and their children and grandchildren'.[24] By adopting students' neighbourhoods as a textbook, place-based education emphasises bioregional knowledge before expanding towards the continental and global. As McGinnis notes, 'bioregionalism is the ground you walk'.[25] Nested within culture, community and ecology, North American studies have also demonstrated that place-based education increases student enthusiasm, engagement and academic performance across disciplines.[26]

Activity – Seeing Ecology

Ecology is the study of natural systems and the relationships that sustain and connect all life. In the 1970s, ecologist Barry Commoner developed as series of 'laws of ecology' which can help us explore fundamental ecological realities and to develop a place-based understanding of sustainability. Commoner's five principles are:

1. Everything is connected to everything else
2. Everything has to go somewhere
3. Everything is always changing
4. There is no such thing as a free lunch
5. Everything has limits

In five small groups, participants receive a principle, with a short explanation (look online for this). Groups then explore the surrounding area for an example of that ecological concept. Using their example, they then draw, write about or develop a short performance of their ecological concept and present it back to the group.

Permaculture offers one of the most profound and practical engagements with developing a relationship with place. Developed in the 1970s by Australians David Holmgren and Bill Mollison, permaculture is a philosophy for the design of sustainable human habitats that mimics ecological principles. In its recognition of small-scale, localised economies and governance structures that reflect natural systems, permaculture is closely connected to bioregional and indigenous cultural resurgence movements.[27] The first of the permaculture principles articulated by Holmgren emphasises is to 'observe and interact', with an emphasis on protracted observation before initiating the smallest possible interventions. This foundational principle asserts that all the 'knowledge that we need to create and manage low energy human support systems can come from working with nature',[28] echoing place-based education's employment of the schoolyard, local community and ecosystems as the 'textbook'.[29]

Through developing a deep understanding of place to cultivate 'an abundance of food, energy and fibre for the provision of local needs',[30] permaculture returns us to the most fundamental of our connections to the earth. Food is one of the most intimate and fundamental ways we express our relationship with the world. Exploring the health of our food system illuminates our crisis of disconnection and highlights the need to root ourselves in place. As chef and author Jessica Prentice describes, food is a powerful medium for expressing webs of connections with 'place and season, history, and people'.[31] Agrarian philosopher Wendell Berry asserts that eating is an 'agricultural act' that 'ends the annual drama of the food economy that begins with planting and birth'.[32] This understanding of the relationship between eating and the land has largely been lost, and we must rediscover the knowledge that 'eating takes place inescapably in the world,' and that 'how we eat determines ... how the world is used'.[33] Through the medium of growing food, permaculture offers a framework for developing an understanding of soils, seasons, rainfall and climate that offers a yield and encourages constant adaptation towards a sustainable, place-based culture. If we are seeking a relationship with place, then there is nothing more intimate than cultivating food and taking the harvest of a particular place into our bodies.

From permaculture to place-based education and beyond, expressions of bioregional thinking and action are as diverse as the bioregions that sustain them. In the same way that we exist within a complex web of relationships that nourish and sustain our existence, Earth Jurisprudence is nested within a network of interrelated movements that emphasise relocalisation and connection to place as a response to an array of environmental and societal crises. At the heart of these movements is a commitment to our home landscapes. Such a relationship with place is fundamental not only in the building of an Earth Jurisprudence but also in the cultivation of a truly sustainable culture.

Notes

1   F. Mathews, 'Becoming native to the city', in *Changing Places: Re-imagining Australia*, ed J. Cameron, Longueville Books, Double Bay, p. 197.

2   R.L. Thayer, *LifePlace: Bioregional Thought and Practice*, University of California Press, Berkeley, 2003, p. 1.

3   C. King, 'Nature as a Guide for Localisation: Applying Ecological Systems Principles to Development Practice', *New Community Quarterly*, vol. 7, no. 3, pp. 12–21.

4   D.S. Jones, 'The Ecological History of Adelaide 2: The Adelaide Plains Environment and its People Before 1836', in *Adelaide, Nature of a City: The Ecology of a Dynamic City From 1836 to 2036*, eds C.B. Daniels & C.J. Tait, BioCity: Centre for Urban Habitats, Adelaide, 2005, p. 54.

5   N. Markus, *On Our Watch: The Race to Save Australia's Environment*, Melbourne University Press, Carlton, 2009, p. 29.

6   Ibid., p. 29.

7   Mathews, *Native to the City*, p. 197.

8   Ibid., p. 59.

9   K. Sale, *Dwellers in the Land: The Bioregional Vision*, New Society Publishers, Gabriola Island, 1991, p. 42.

10 S. Mills, 'Foreword', in *Home! A Bioregional Reader*, eds V. Andruss, C. Plant, J. Plant & E. Wright, New Society Publishers, Gabriola Island, 1990, p. vii.

11 Ibid., p. vii.

12 D. Aberley, 'Interpreting Bioregionalism: A Story From Many voices', in *Bioregionalism,* ed. M.V. McGinnis, Routledge, Oxon, 1999, pp. 36–37.

13 Ibid., pp. 36–37.

14 M.V. McGinnis, *Bioregionalism*, Routledge, Oxon, 1999, p. 2.

15 Ibid., p. 8.

16 C.W. Leslie & C.E. Roth, *Keeping a Nature Journal: Observing, Recording, Drawing the World Around You*, Storey Publishing, North Adams, 2003.

17 J. Young, *Kamana One: Exploring Natural Mystery*, Owlink Media, 2001.

18 G.A. Smith, 'Place-based Education: Learning To Be Where We Are', *Phi Delta Kappan*, vol. 83, no. 8, pp. 584–594.

19 B. Comber, & H. Nixon, 'Children Reread and Rewrite Their Local Neighbourhoods: Critical Literacies and Identity Work', in *Literacy Moves On: Using Popular Culture, New Technologies and Critical Literacies in the classroom,* ed. J. Evans, David Fulton Publishers, London, 2004, pp. 127–148.

20 G.A. Smith, 'Place-based Education: Learning To Be Where We Are', *Phi Delta Kappan*, Vol. 83, No. 8, p. 586.

21 D. Sobel, *Beyond Ecophobia: Reclaiming the Heart in Nature Education*, The Orion Society, Great Barrington, 1996, p. 9.

22 Ibid., p. 10.

23 R. Louv, *Last Child in the Woods: Saving Our Children From Nature-Deficit Disorder*, Algonquin Books of Chapel Hill, Chapel Hill, 2008, pp. 150–151.

24 G.A. Smith, 'Place-based Education: Learning to be Where We Are', *Phi Delta Kappan*, Vol. 83, No. 8, pp. 593–594

25 McGinnis, *Bioregionalism*, p. 9.

26 D. Sobel, *Place-based Education: Connecting Classrooms and Communities*, The Orion Society, Great Barrington, 2005, p. 25.

27 D. Holmgren, *Permaculture: Principles and Pathways Beyond Sustainability*, Holmgren Design Services, Hepburn, 2002, p. 173.

28 Holmgren, *Permaculture*, p. 16.

29 Sobel, *Place-based Education,* p. 25.

30 Holmgren, *Permaculture*, p. xix.

31 J. Prentice, *Full Moon Feast: Food and the Hunger for Connection*, Chelsea Green Publishing Company, White River Junction, 2006.

32 W. Berry, 'The Pleasures of Eating' in *The Art of the Commonplace: Agrarian Essays of Wendell Berry,* ed. N. Wirzba, Counterpoint, Berkeley, 2002, p. 321

33 Berry, *Pleasure of Eating*, p. 324

Section 2: Theology and Philosophy

Law is a product of culture and an important articulation of a society's self-concept. Judeo-Christian theology and philosophy have played a central role in the development of Western law for over two thousand years. Moreover, theology and philosophy represent an important source of values and ideas for individuals and communities. This section will consider leading ideas from both sources that seek to move beyond anthropocentrism and promote a genuine relationship with nature. Further, it will consider a fundamental obstacle to Earth Jurisprudence, which is that values cannot be derived from facts, and an important shift in knowledge from truth to approximate description.

Beyond Dominion and Stewardship

Gloria L. Schaab

In his well-known and frequently cited 1967 essay 'The Historical Roots of Our Ecologic Crisis', historian Lynn White Jr. levels a pointed critique at the Christian tradition with regard to its negative impact on the environment and on ecology. Focusing on Christianity as 'the most anthropocentric religion the world has seen', White insists that Christianity established a dualism between humanity and the natural world that gave humanity, by divine will, the right and duty to exploit nature for human ends.[1] In response to this critique, I would suggest that there are multiple interpretations of the environment often overlooked in environmental discourse through which the Christian tradition may be focused. In this essay, I survey six sets of Christian interpretive lenses on the environment. Three of these lenses do produce the distorted Christian visions of the environment which have been blamed in large measure for the environmental crises we experience in our world today. The other three lenses, however, produce what I deem an authentic Christian vision of the environment that has the potential to promote the intrinsic value and full flourishing of the environment. After peering through these six lenses, I appropriate the work of theologian Elizabeth Johnson to propose three approaches to the environment consistent with the Christian models I set forth.

The Critique

Criticism of the Christian approach to the environment stems from visions focused through three predominant lenses: (1) divinely commanded dominion over creation based on the creation stories of Genesis 1 and 2, (2) anthropocentrism rooted in the doctrine of the human being as *imago dei*, and (3) emphasis on other-worldliness promoted by both mysticism and eschatology.

Dominion

The critique begins with the epic poem that constitutes the priestly version of the biblical story of creation in Genesis 1. After creating male and female, 'God blessed them, saying: 'Be fertile and multiply; fill the earth and subdue it. Have dominion over ... all the living things that move on the earth".[2] It is a particular interpretation of this command to subdue and to have dominion that seems to have given license to the human community to ravage and despoil the natural environment. It enables human beings to look upon the environment as having only instrumental value – that is, as valuable solely in terms of what it supplies the human being.

This stands in opposition to the environment as invested with intrinsic value – that is, as valuable in and of itself as indispensable, purposeful, and irreplaceable life forms. A dominion perspective has led to overconsumption and to overpopulation; to pollution and despoliation; to ozone depletion and to global warming; to species extinction and to human starvation.

Imago Dei

In the same passage from the book of Genesis, after the orderly unfolding of the cosmos from primordial chaos, 'God said: 'Let us make [humans] in our image, after our likeness. Let them have dominion over ... all the creatures ... every seed-bearing plant ... and every tree".[3] In the second Genesis account of creation, the older Yahwist version, it says, 'God formed the man out of the clay of the ground and blew into his nostrils the breath of life, and so man became a living being ... Then, God said: 'It is not good for the man to be alone. I will make a suitable partner for him'. So God formed out of the ground various wild animals and various birds of the air, and ... whatever the man called each of them would be its name'.[4]

In both scenarios, the human sits atop the hierarchy of created being. In the first account, human priority arises from being created in the image and likeness of the Creator and thus transcending the rest of creation. In the second account, the human attains priority by being created first in the order of creation, with all other creatures being made for the sake of the human. Moreover, in the Jewish tradition, to name is to have power over; hence, naming each creature in the second account is tantamount to the having the dominion granted by the first. The effect of human priority in the story of creation is an overarching anthropocentrism in Christian theology that makes the needs and the good of the human being the central principle upon which ethical decisions are based. Some Christian theologians have attempted to use the notion of *imago dei* to redeem the Christian tradition with regard to the environment by proposing that the human being in the image of God shares God's creative love and responsibility for the sustenance of the planet. This love and responsibility translates to the notion of human *stewardship* for creation. Nevertheless, even the understanding of the steward in the household of God implies a hierarchical and utilitarian relationship between humanity and the rest of creation, a relationship which, in view of human sinfulness, can easily slip into dominion in the face of competing claims.

This World or the Next?

The dualism between the rest of the natural world and human beings in biblical creation stories was mirrored in a dualism between matter and spirit in Greek philosophy. Grounded in Platonic thought, dualism proposes that two distinct principles of being exist in the cosmos: the principle of matter (body) and the principle of spirit (soul). In this dualism, the realm of the spirit-soul was considered superior to the realm of matter-body. As a result, such Christian Platonism does not promote the vitality of the natural world, but the glory of the world to come. This emphasis on the *new* creation instead of on this *present* creation inevitably has consequences for the environment, since the denial of the goodness and value of the natural world predictably results in its misuse. Moreover, when fuelled by Christian apocalypticism, this lens looks toward the destruction of the present world in order that God may bring forth a new heaven and a new earth for those who are faithful. Clearly, neither consequence bodes well for the preservation and full flourishing of the environment.

Redeeming the Tradition

In view of these distorted perspectives on the environment, one is tempted to ask, 'Is the Christian tradition redeemable? Does it have any good news for the environment?' I would strongly submit that it is redeemable and it does have good news! To support this position, I propose three alternative views of the environment from the Christian tradition. I submit that these lenses – the relational, the sacramental, and the incarnational – lead to an ecological vision of 'a flourishing humanity on a thriving earth in an evolving universe, all together filled with the glory of God'.[5]

Relational Lens

Grounded in the doctrine of creation, the relational approach invites human beings to view the environment as an integral part of the relationship that exists between God as Creator and the cosmos as creation. Rather than viewing the concept of creation solely in terms of the original *creatio ex nihilo* of the cosmos, the understanding of God as Creator must augmented by notions of a dynamic and ongoing creative relation between God and world, consistent with the insights of evolutionary science concerning ongoing creativity in and through the very stuff of the material world. Theologically, this ongoing creative relation between God and the cosmos has been termed *creatio continua*. If God is understood as Creator and the process of creation has been ongoing throughout cosmic history and into the cosmic future, then this reveals a perpetual creative relation of God to the cosmos. God is thus understood as *semper Creator* who is directly involved in the continuing processes of the cosmos in and through the inbuilt creativity of the universe.

This perennial relationship of God to creation is effectively symbolised through the panentheistic model of God–world relation. In this model, God is the circumambience in which the environment lives, moves, and has its being and the world-as-a-whole is conceived within God, as a child is conceived in its mother's womb. Like the female procreativity that best reflects it, in the panentheistic model of God–world relation, God as mother transcends and is distinct from the creation dwelling within. Nevertheless, the life and energies of God as mother are immanent within creation, promoting its life, sustenance, and growth. This ongoing relationship plainly demonstrates that the cosmos shares in God's very life and thus has enduring importance to God. Thomas Aquinas calls this relationship between God and creation 'participation'. Aquinas states, 'Because the divine goodness could not be adequately represented by one creature alone, God produced many and diverse creatures ... Thus the whole universe together participates in divine goodness more perfectly, and represents it better, than any single creature whatsoever'.[6]

Because of this common source and participation in God, a profound kinship exists among all members of creation. Such kinship has the capacity to engender in humanity a spirit of humility and interdependence toward the environment. Realisation of God's ongoing and integral relationality to the environment accentuates the intrinsic rather than instrumental value of the cosmos, denounces the despoliation of the environment through human choices and behaviour, and makes plain that God's creative relation and intention extends not simply to the full flourishing of humanity, but also to the natural environment itself. Imaging the cosmos within God and God within and around the cosmos, moreover, graphically demonstrates that what is experienced by creation is acutely experienced by God. This fosters ethical responses which promote values and actions that protect, sustain, and enhance the environment. Clearly, these are values and actions in keeping with the love of a Divine Mother for the developing life within her, demanding abundance of life for the child of her womb.

Sacramental Lens

Theology – quite literally, the study of God – is essentially an exploration into the mystery of God and into the mystery of those things which God created, namely the natural world. Nevertheless, theology's exploration into the mystery of God does not follow the same course as one might use to enter into the mysteries of the natural world. Unlike the natural world, God is Spirit[7] and as Spirit cannot be seen or touched or heard in the same ways in which the physical elements of the natural world can.

If this is the case, how can persons come to truly *know* God? The Christian tradition teaches that persons of faith can come to know God because God has chosen to reveal Godself to and through what God has given life. Based in his belief in Jesus Christ and in positive revelation, the Jesuit theologian Karl Rahner taught that God's very nature is that of free and self-communicating love and that the very existence of the cosmos and its creatures is a result of this divine freedom and love. Moreover, if God is self-communicating love, then that presupposes an 'other' *to* whom and *through* whom God communicates in love. According to Rahner, God communicates *to* human beings *through* all that God has created, both human and non-human. Therefore, those seeking to explore the mystery of the God must begin in those places through which God reveals Godself. Through the sacramental lens on the environment, this revelation is discovered in cosmic creation.

'In the beginning when God created the heavens and the earth, the earth was a formless wasteland, and darkness covered the abyss, while a mighty wind swept over the waters. Then God said, "Let there be light" and there was light. God saw how good the light was'.[8] God spoke – God communicated Godself – and creation came into being. While most theologians regard this creation story in Genesis as a form of sacred allegory or myth, it nonetheless expresses an important step into the mystery of God. It reveals that God is the source of all creation and that all of creation is the self-expression of God. Therefore, one can come to know God by attending carefully to God's self-communication through the natural world. Furthermore, because the natural world continues to change and develop in ever more surprising ways, God's creative self-communication is unceasing through the continuous creativity of our evolving cosmos.

In *Summa Theologica* Aquinas formalised this understanding of how God as Creator can be known through what God creates. Aquinas presented this argu-ment: 'When an effect is better known to us than its cause, from the effect we proceed to the knowledge of the cause. And from every effect the existence of its proper cause can be demonstrated, so long as its effects are better known to us ... Hence the existence of God, in so far as it is not self-evident to us, can be demonstrated from those of His effects which are known to us'.[9] In his *Summa Contra Gentiles,* Aquinas summarised his ideas in a simpler fashion: 'There is some manner of likeness of creatures to God ... [Thus] from the attributes found in creatures we are led to a knowledge of the attributes of God'.[10]

Because of this likeness or 'analogy of being', inferences can be drawn about the attributes and purposes of God based on objects and relationships in the natural order. This is because, in creating the world, God who is Being itself, shares 'being', i.e. 'existence', with creation. Because the natural world shares in the being of God, creation can mediate God to us. Consequently, we can express our understanding of God in words and images drawn from the natural world despite the fact that God exceeds anything that can be said or imagined. This capacity, moreover, has received a new dynamism in the dialogue between Christian faith and evolutionary science.

In his book *Theology for a Scientific Age,* scientist-theologian Arthur R. Peacocke explores the nature and characteristics of the evolving cosmos to see what it can reveal about the nature, characteristics, and purposes of God.[11] Since the cosmos displays unity and diversity in its many structures and life forms, Peacocke infers that its Source must be both essentially one and unfathomably rich. Given that the cosmos has order and regularity, this suggests a rationality underlying creation. The universe, however, also demonstrates constant change. Therefore, God must act not only to sustain the cosmos, but also to continuously create new forms throughout the passage of time. Ultimately, from this continuous creativity and from the very stuff of creation, the human person emerges. In view of the emergence of human beings, Peacocke inferred that God who is the source of such personal beings must be a personal Being in Godself. And since human persons are by nature relational and purposeful in their actions, then God in whose image persons are made must be relational and purposeful in Godself.

This revelatory nature of the universe described by Peacocke is expressed through the concept of *sacramentality*. In the Christian tradition, 'The essence of a sacrament is the capacity to reveal ... God, by being what it is ... [by] being thoroughly itself'.[12] Hence, 'By its nature, a sacrament requires that it be appreciated for what it is and not as a tool to an end'.[13] Consistent with this, the sacramental lens on the environment sharpens the vision of the intrinsic value of the environment sighted through the relational lens. Moreover, the understanding of the cosmos as sacrament is consistent with the panentheistic model of God–world relation described above. In his writings, Peacocke used the triad 'in, with, and under' to refer to the action of God in panentheistic relation to the cosmos, the same triad used by Martin Luther to describe Christ's sacramental presence in the Eucharist. As sacrament, the cosmos is that through, in, with, and under which God comes and discloses Godself.

Undoubtedly, human exploitation and abuse of the natural world often serves to obscure and obliterate the revelatory power of the environment. Eco-theologian Thomas Berry bemoans the fact that 'losing the richness of life around us will impoverish our sense of the God whose being is symbolically revealed to us through the extravagant diversity and beauty of nature'.[14] However, a profound awareness of the sacramental character of nature, symbolised in the Christian tradition not only in Eucharist, but in the waters and fire of baptism and the oils of anointing, has the capacity revive this sensibility and to inspire contemplation, preservation, and revitalisation of the environment.

Incarnational Lens

Informed by scripture and the mystery of Christ, the incarnational lens on the environment invites humanity to imagine '"the Word made flesh" as not limited to Jesus of Nazareth but as the body of the universe'.[15] Proposed by theologians like Sallie McFague, this metaphor of the universe as God's word-made-flesh results in a radical reinterpretation of the God–world relationship. Moreover, when seen against the backdrop of the Genesis account which uses the image of God's word calling all creation into being, this view sees the word of God made flesh in all the matter that makes up the universe.

In her book *The Body of God*, McFague proposes a radically incarnational vision of the God–world relationship, which, like the sacramental perspective above, contends that all of creation mediates divine reality and communicates the way in which God and the world relate. Integrating biblical, Christological, and scientific understanding of evolutionary processes, McFague reflects on the mystery of the incarnation. 'Were we to imagine', McFague suggests, '"the Word made flesh" as not limited to Jesus of Nazareth but as the body of the universe ... might we not have an ... awesome metaphor for both divine nearness *and* divine glory'.[16] Within the pantheistic paradigm that McFague also appropriates, her incarnational metaphor results in a radicalisation of divine immanence and transcendence. In the model of the universe as the body of God, divine transcendence is 'radically and concretely embodied ... *in* the differences, in the concrete embodiments, that constitute the universe'.[17] As a result of God's embodied transcendence, moreover, God is immanently present in and through all bodies, an image prevalent in both the Hebrew and Christian traditions that take seriously the mediation of God by the world. From this incarnational viewpoint, 'At one level our model – the universe as God's body – moves us in the direction of contemplating the glory and grandeur of divine creation ... while at another level it moves us in the direction of compassionate identification with and service to the fragile, suffering, oppressed bodies that surround us'.[18]

Because of its core doctrine of incarnation, McFague considers Christianity uniquely suited to embrace the model of the world as God's body, understood in shape and scope through what McFague terms 'the Christic paradigm'. Based upon the story of Jesus of Nazareth, this paradigm suggests that the direction of creation is toward fullness of life and inclusive love for all, especially those needy and vulnerable. Moreover, according to McFague, in an ecological age, the oppressed, the needy, and the outcast must include the 'new poor', that is, non-human beings and the environment itself.

In the context of the incarnational model, humanity's bond with all the bodies of the natural world is rooted not only in our common creation story and in the common stuff of our material existence, but also in the immanence of God's very self within the cosmos God has created. This understanding of incarnation refutes a view of the universe centred on humanity; it counters the hierarchy of the spirit over the body and rejects human dominion over creation. The incarnational model motivates Christians to live in right relationship with the larger whole of which we are a part; to care for the community of life with understanding, compassion and love; and protect and restore the integrity of Earth's ecological systems. Moreover, they do so with the understanding of the teaching of the Matthean Jesus expanded by the incarnational paradigm of the natural environment, 'whatever you do to the least of mine you do to me'.[19]

Practical Ramifications
Having re-envisioned the Christian interpretation of the environment through the lenses of relationality, sacramentality, and incarnation, I would like to conclude this paper by summarising three forms of Christian praxis toward the environment consistent with the three visions I described above. Proposed by Elizabeth Johnson in her essay 'God's Beloved Creation', these three praxes are the contemplative, the ascetic, and the prophetic. Contemplative praxis toward the environment invites persons to view the world lovingly, rather than with utilitarian arrogance. It calls

us to appreciate the beauty of creation and to be awed by its mystery. This praxis is particularly suited to those who view the environment through *sacramental* lenses and see God's presence in simple, earthy things. As Johnson notes, 'The earth, with all its creatures, is the primordial sacrament, the medium of God's gracious presence and blessing'.[20] Ascetic praxis toward the environment from what I considered a *relational* viewpoint encourages human beings to practice discipline in using Earth's resources and to make ecologically and environmentally responsible choices. 'The morality of our actions', Johnson claims, must be measured by 'one stringent criterion' – 'whether or not these [actions] contribute to a sustainable earth community'.[21] Finally, from an *incarnational* approach, Johnson describes a prophetic praxis toward the environment. This praxis recognises the ongoing destruction of the environment as a sign of sinfulness and challenges Christians to action for justice on behalf of Earth. This is a form of justice that entails a conversion from anthropocentrism to 'ensure vibrant life in community for all' and that calls for political action and structural transformation to insure the vitality of the environment. As Johnson powerfully insists, 'A flourishing humanity on a thriving earth, in an evolving universe, all together filled with the glory of God: such is the theological vision needed in this critical age of Earth's distress'.[22]

Notes

1   Lynn White, 'The Historical Roots of Our Ecologic Crisis', *Science,* vol. 155, 1967, pp. 1203–1207.

2   Gen. 1:28.

3   Gen. 1:26–31.

4   Gen. 2: 19–25.

5   Elizabeth A. Johnson, 'God's Beloved Creation', *America,* vol. 184, no. 13, 2001, p. 12.

6   Thomas Aquinas, *The Summa Theologica of St. Thomas Aquinas* (henceforth *ST*), Second and Revised Edition, trans. Fathers of the English Dominican Province; available from http://www.newadvent.org/summa/1047.htm (Accessed 14 February 2007), I.47.1.

7   Cf. Jn. 4: 24.

8   Gen. 1: 1–4a.

9   Aquinas, *Summa Theologica,* I.2.1.

10 Thomas Aquinas, *Summa Contra Gentiles,* from *The Jacques Maritain Center,* University of Notre Dame; available from http://www2.nd.edu/Departments/Maritain/etext/gc1_33.htm (Accessed 26 August 2009), 1.33.

11 Arthur R. Peacocke, *Theology for a Scientific Age: Being and Becoming: Natural, Divine and Human,* Augsburg Fortress, Minneapolis, 1993, pp. 99–134.

12 Michael J. Himes and Kenneth R. Himes, 'The Sacrament of Creation: Toward an Environmental Theology', *Commonweal,* vol. 117, 1990, p. 45.

13 ibid., p. 46.

14 Thomas Berry, in John Haught, 'Ecology and Eschatology', in *And God Saw That It Was Good,* ed. Drew Christiansen and Walter Grazer, U.S. Catholic Conference, Washington, DC, 1996, p. 55.

15 Sallie McFague, *The Body of God: An Ecological Theology,* Fortress, Minneapolis, 1993, p. 131.

16 ibid.

17 ibid., p. 155.

18 ibid., p. 135.

19 Mt. 25:40.

20 Johnson, 'God's Beloved Creation', p. 10.

21 ibid.

22 ibid., p. 12.

'And finally ... humans!'

Jason John

So ends the story of life on Earth, to all intents and purposes, for those who begin it with, 'In the beginning God'. A good many Christians, Jews, and Muslims (and many humanists besides) live as if humans are the pinnacle, the climax, the whole point of life on Earth.

I write as a 'professional Christian' who tries not to live that way. I am an ordained minister in the Uniting Church in Australia, working as an ecominister, based in Bellingen, New South Wales. I am also an honours graduate with a double major in zoology, and a postgraduate qualification in environmental studies. I accept the scientific rewriting of the Earth story in which humans are absent from its final chapters. I also accept the scientific rewriting of the *beginning* of the story, in which humans are absent.

Humans are a very brief chapter, or perhaps paragraph, in the middle of the story of life on Earth.[1] Yet until recently the Jewish, Christian and Islamic[2] stories understood humanity and our place on Earth through biblical stories which give the pivotal place to humanity, created 'a little lower than the angels,' of all species on Earth the only ones created in God's image.

Until recently? Admittedly, even now many or most Christians believe that the story of humanity literally begins in Genesis 1–3.

So despite being a science-loving minister, whose PhD argued that Christians should celebrate the fact that humans are only a short-lived part of the story of life[3], I want to start by exploring the creation stories in Genesis, and explore what it means to be in relationship with the rest of life on Earth. This question is central to the Great Work of Thomas Berry and underlies the shift required in Earth Jurisprudence.

You may notice that there are *two* creation stories in Genesis, though this will surprise even many non-literalist Christians, who are used to harmonising them into one. Genesis 1:1–2:3 describes a seven day creation. Its context, as a relatively late story written during exile in Babylon, is fascinating but beyond our scope. Genesis 2:4–2:25 is the much older, earthier, story of Adam and Eve. As Norman Habel[4] emphasised in a series of public talks in 2005, the two stories present *very different* ways of understanding our place on Earth, and have very different implications for something like Earth Jurisprudence.

The first story: Genesis 1 (dominion, fear and dread)

This is the passage classically blamed for the destruction of the environment by Judeo-Christian societies. I am persuaded by arguments that it is, rather, the passage most co-opted by people in Judeo-Christian societies who were going to plunder the environment anyway: 'God said to them, "Be fruitful and multiply, and fill the earth and subdue it; and have dominion over the fish of the sea and over the birds of the air and over every living thing that moves upon the earth."' (Genesis 1:28)

The story culminates in Genesis 9 with the tale of Noah and the ark, where God blesses Noah and his sons:

> God blessed Noah and his sons, and said to them, 'Be fruitful and multiply, and fill the earth. *The fear and dread of you shall rest on every animal of the earth,* and on every bird of the air, on everything that creeps on the ground, and on all the fish of the sea; into your hand they are delivered. Every moving thing that lives shall be food for you; and just as I gave you the green plants, *I give you everything.*'

The shudder that many hopefully experience when reading this story recedes a little when we remember that this is a story, written in Babylon during exile, which is primarily aimed at asserting an egalitarian vision in which *all humans*, not just the king of Babylon, are created in God's image.[5] *All people* have been given the Earth and its creatures for food, not just the king of Babylon. *All humans*, not just the foreign king, have the divine right to exercise dominion.

Also, in earliest Christian times, the statement that humans are created in the image of God was emphasised in a Roman society where many humans were considered to be of no value whatsoever.[6] In both its original setting and subsequent use, the motivation was to widen the egalitarian circle. There is a trajectory being established that we will come back to later.

Nonetheless, in our day it has been cited as a divine sanction, even command, to exercise dominion or control of the planet, as the only ones of any great worth, since we alone bear God's image. In 1992, for example, Hugh Morgan, an Australian businessman and head of a mining company, had this to say:

> bridges, railway stations, even large dump trucks and front end loaders. These things are the results of carrying out the injunction given in the first chapter of Genesis 1, verse 28.[7]

As Ann Wansbrough points out, however, it appears that Morgan, the managing director of Western Mining Corporation, 'reflects his own approach to life and to nature, rather than a serious study of the scripture'.

So if you encounter someone who wants to argue that Genesis delivers all things to humans as property, and gives us dominion (and therefore the notion of Earth Jurisprudence is unthinkable), the best strategy may be to simply agree (perhaps mentioning its original egalitarian purpose if you get the chance). But don't stop there! Quickly move them on to the second story. If you let them cite the Bible as some kind of authority, now they will have to let you.

The second story: Genesis 2 (to serve and protect)

Genesis 2:4–2:25 is *a completely different story!* Committed literalists will have a hard time accepting this one, but ask them to bear with you. This is a much more ancient story, and reads more like many of the Aboriginal Dreaming stories which we have in Australia. God wanders around in the garden in the cool of the evening, and wonders where Adam and Eve are after the talking snake leads them astray.

If people are committed to understanding the world through the Bible, your task is to get them to adopt *this* story, rather than Genesis 1, as their story. Or at least to get them to listen to it seriously.

Note first that Adam is a play on the Hebrew for earth (*adamah*). So 'The Adam' is an earth creature. From the earth God creates an earthling. From the humus God draws forth a human. The word play emphasises our connection to the earth, which is promising from an Earth Jurisprudence perspective.

This androgynous earthling is then given the task of tilling and keeping the land. At least that's the way us agricultural, property obsessed types like to translate it. God gives us the garden to farm it and keep it. I'm willing to bet that every garden variety Western Christian who reads about keeping the land reads it in terms of private property, at least subconsciously. We certainly act as if we do. We can imagine that this was a very useful text for those wanting to dispossess indigenous peoples in newly invaded countries, who weren't 'doing anything' with their land.

In the rest of the scriptures, however, 'till' is overwhelmingly translated as 'serve'. And 'keep' has the meaning of 'keep safe,' or protect.[8] As written in Genesis 2:15, '[t]hen God put the earth creature in the garden of Eden to serve and protect it'.

God then sets limits on human aspirations: we can eat anything we like, but not the tree of the knowledge of good and evil. Imagine if that was our starting point for Earth Jurisprudence. We are here to serve and protect life on earth, accepting our limitations. Not a bad story. Except, of course, it's wrong.

Wrong, if we consider it to actually describe the role and place of humans on Earth. Neither of these 'small stories,' which assume that humans have been present for the whole history of God and life, are literally our creation stories.

We now have a very big story, of which we are a tiny part, which does that much better. Thank God! We are saved from taking Genesis 3:16–19 about painful childbirth, patriarchy and weeds literally.

But what does a lawyer care about the big story? Or a judge? Or a politician? Or Hugh Morgan? What do any of us care? Most of us act as if life is a small story. Humans have always been around, we're the *pièce de résistance* of evolution.

Anything which predates writing is old, and agriculture began practically at the dawn of time itself.

The two small stories in Genesis speak to us because they cover the part of history where we assume everything happened (from the start of human agriculture onwards). Of the many creation stories to choose from, they hold sway because they were privileged when the Roman Empire subsumed Christianity, and because they stretch 'all the way back' to the dawn of agriculture.

Despite a revival of interest in Aboriginal Creation stories around the world, none of them will ever be adopted in the West, deliberately or even subconsciously, to the extent that Genesis 1 and 2 have been.

So we have two to pick from. Dominion, fear and dread, or serving and protecting. If you are reading a book about Earth Jurisprudence, or wild law, I'm betting you'd prefer to see people embracing the latter. How do we encourage them to do that? I can only speak within the Christian tradition, where Jesus challenges his companions to be servants, not dominators. Jesus said,

> You know that the rulers of the Gentiles exercise dominion over them, and their great ones wield power over them. It will not be so among you; but whoever wishes to be great among you must be your servant, and whoever wishes to be first among you must be your slave; just as the Son of Man came not to be served but to serve, and to give his life as a ransom for many.[9]

So which of the two competing Creation stories are more Christian? Is it more Christian to seek dominion and control? Or to seek to serve and protect, especially if we think of ourselves as the greatest species?

Jesus continues, '[t]he greatest among you will be your servant. *All who exalt themselves will be humbled*, and all who humble themselves will be exalted'[10].

Which seems to parallel – unintentionally – the warning in Genesis 2 of not accepting our limits by eating the fruit of knowledge.

Can the billions of people who claim that Genesis is part of a book which has authority in their lives, and guides their understanding of human nature, and the relationship between God, humans and other animals, be persuaded that *Jesus's* teaching about servanthood and dominion are relevant to the *Genesis* stories about servanthood and dominion?

I honestly don't know. But you could try!

Because, at the small story scale, we see Genesis 1 and 2 being played out around the world, not as God ordained lifestyles, but as ones which humans have chosen, in making themselves gods. These two stories are at war whenever some humans want to treat an intact ecosystem, or even a farm, as a commodity, and others see it as something which must be protected, which may have needs which are more important than human greed.

In the small scale: millennia, century or lifetime, our ability to devastate ecosystems and also our ability to pull back have been repeatedly documented. Pull back?

Those who have travelled according to the myth of dominion actually seem to need to be pulled back (by laws) or pushed back (by servants and protectors).

Again, those of you reading this know at least as much about that as I do.

So here it ends. With a choice between two paths, two mythological stories. If I have done nothing more than show you that both paths exist in the scriptures, and if you can do nothing more than persuade some of the tens of millions of powerful, rich Christians in the world that *both* paths are biblical, and that Jesus's followers ought to lean to the latter one, then perhaps something has been achieved. If those many people out there whose love for other creatures has cut them adrift from their dominion exercising, human-focussed Christian communities, perhaps they will draw some strength from knowing that the second path exists. That to walk the path of serving and protecting life on Earth is a fully biblical, fully Christian path. That wild law and Earth Jurisprudence are legitimate Christian pursuits, even a Christian responsibility.

A world in which law was based on Genesis 2 (and a recovery of the Sabbath laws, rewritten for our context) would be a wild law kind of place. A place where earthlings asked first how they could best serve and protect the garden in which they lived, and from which they had been drawn forth. Where they created laws to help them do that, given that it was too late to turn back time and spit out the fruit of the knowledge of good and evil.

Feel really free to leave it there. Most Christians do.

The third story: life-centred Christianity

Increasingly, however, Christians are exploring a third story. One which weaves in the insights of cosmologists, evolutionists, geneticists, psychologists and the like.

I'd like to show you some of this story. I should warn you that I accept a lot more of the scientific versions of the story than many other evolution-accepting Christians are inclined to. For example, the famous Catholic evolutionary theologian Denis Edwards[11], and the previous Pope[12], place significant restrictions on what evolution is allowed to tell us about ourselves. Whilst they readily abandon the historicity of Genesis, they doggedly preserve its theology, especially the notion that humans alone are created in the divine image, with an unbridgeable spiritual gulf between us and other animals.

Their story still ends, in other words, with 'and finally, humans!'

If there is an unbridgeable gap, or as Pope John Paul put it, an, 'ontological leap'[13], then there is plenty of scope to undermine Earth Jurisprudence, whatever else may be said about the value of creation. Since humans are a separate category, it would always make sense that we have separate laws. And since 'man [sic] is the only creature on earth that God wanted for its own sake,'[14] those laws will doubtless always privilege us.

Let me instead take you for a wander down a fully evolution-embracing path. One which, in the face of evidence about our true place in the story, is thoroughly biocentric, or life-centred.

A path on which we understand that the central story of the universe is between God and life, not God and *Homo sapiens*. Where we come to see our place in the universe as being much smaller than we think it to be. Where we do not so much think like a mountain, as think like the God who was here before the mountains were even formed. Who for billions of years had a relationship with life before we ever arrived.

Where we discard the theology that creates an intrinsic distinction between us and the other animals. And where, with this more humble view of our place in this universe and in God's heart, we make room for the other creatures which we share this planet with, so that we can all have the relationship with God which God desires.[15]

The big story affirms one part of the Genesis 2 story, and upends almost everything else.

We are earthlings. Humans from humus. Like all other species, we evolved from the elements, we are nourished by the earth, via plants, via the placenta until our birth. We return to dust when we die.

But we are not specially created in the image of God. There is no biological discontinuity between us and other creatures. There is no basis, apart from papal decree, to argue that we are spiritually or psychologically discontinuous from other animals, especially when we remember all the recently extinct *Homo* species. Even just amongst our peer species we find degrees of intelligence, love, compassion, tool use, culture and morality. Everything which has traditionally been used as evidence that humans are distinct, and have a distinct soul, have been shown instead to exist on a continuum.

But we cannot attribute pieces of the image to each species, or degrees of the image. Instead, if there is an image of God, then it is seen in life in all its fullness. All its evolutionary past, ecological present, and unknown future. We are a temporary part of the image of God.

If Life is the image of God, then, according to Christian tradition, it should follow that Life has dominion on Earth. And indeed this is true. Long before humans evolved, living creatures shaped their environments, consciously and unconsciously. Without the 'dominion' of microbes, oxygen consumers like us would never have evolved. No animal has total dominion on Earth, but it belongs much more with the bacteria and worms than it does with us. Humans are definitely an optional accessory.

This applies as much to the more benign 'serve and protect' story. Life got on fine without us for three billion years. It will continue long after we are gone.

We are not dominators, or servers, protectors, stewards, or co-creators, any more than millions of others species have been.

So what are we?

We are a tiny tributary of the stream of life on Earth, and perhaps the only one, that is able not only to be loved by God, but to love God back. To sit back and observe the product of three billions of years of evolution, and wonder at it. Revere it. Love it.

Loving life is very Christian, but sounds too much like a soft-drink advertisement, so I prefer to use Schweitzer's term, reverence. Reverence for life (including our own). That's our job. Reverence for the God of life. Love God. Love yourself. Love your neighbour as yourself. In this world in which humans are literally the neighbours of all other species on the planet, this summary of Jesus's message now arguably applies to *all* our neighbours.

Heresy?

Jesus was labelled a heretic when he suggested that neighbour meant not the pious Jew next door, but the worthless, despised Samaritan who wasn't worth the time of day (Luke 10:25ff). Noting in passing that it was a lawyer who triggered Jesus's parable in the first place, we see here a radical extension of Foreigner Jurisprudence, which I think begins a trajectory which comfortably continues to Earth Jurisprudence.

All creatures are our neighbours. Neighbours have rights not to be unduly impinged upon by our activities. Good fences (and laws) make good neighbours.

Geographically and ecologically we are called to broaden our understanding of neighbour as we revisit Jesus' teachings. Genetically we are challenged to rethink 'family'. Sociologically we are warned against continually talking about all humans as 'us' and 'we' when we discuss ecology. Not all humans are equal. That sounds ridiculously obvious, but note how often articles about the environment talk about what 'we' are doing to the planet. Instead, there is a continuum of wealth, or power, measured as the ability to seize dominion here in the short term. When we revisit Jesus's teachings on the rich and the poor, we see this as not being solely about human interests, but rich humans and their impact on, and attitude towards, the sea of poor humans and other animals.

Since animals are part of the image of God, our neighbours, then what Jesus said about release for the captives, is relevant to the millions of captives in factory farms and chemical labs. Do to others what you would want them to do to you[16].

But we don't, do we? 'Do to others …'. Not many of us anyway, and not all of us always.

Writing this on my 39th New Years Day, the proposition that Earth Jurisprudence might become an actual *reality*, rather than a nice idea, seems like so much 'blah'. And I even went to bed early, and sober! Copenhagen seems to have failed, and it was only concerned with the consequences of climate change for humans. Is the paper which this article, and this book, is printed on worth cutting down a tree for? Won't it just end up as bookshelf filler or toilet paper?

I've hit Matthew Fox's *via negative (negative path)*, the way of despair and grief, again.[17] None of you will be a stranger to it. If we remember it, we may be able to empathise with those whom we ask to lay aside their precious dominion narrative. To die to a part of themselves. To renounce a story which justifies their sense of privilege and entitlement. Hopefully Fox's *positive* way – the Genesis 2 alternative – will give them the strength to pass through the negative way. Maybe they will get as far as his *creative* way. The creation of a new story out of the evolutionary and ecological insights we now have.

Exploring life-centred Christianity has been for me, life-giving. With varying degrees of success I have travelled along the fourth of Fox's ways: the way of transformation. Any of us connected to this book – as readers or contributors – bear a similar hope for humanity. Whether our hope is a flickering candle or a blinding beacon, we hope for laws to help humans transform our relationship with our neighbours, which bring freedom to the captives and good news to the poor. We hope for sight for those blinded to the fact that they are so possessed by technology and the fear of death that they have cut themselves off from Life, and damaged everyone else in the process.

We hope, I think, that Western culture and all it has influenced is not evil, but immature. A teenager. Discovering its power, but ignorant of its responsibilities. Self-obsessed. Full of dynamic energy and innovation and passion, whilst blind to its impact on others around it. Not just refusing to accept limits, but failing to really grasp that they even exist.

Jesus always steered away from labelling the characters in his parables, to make people think. If you get to the end of the story wondering why on earth it is there, you can check this footnote[18].

Here is my final story of hope, from Luke 15:11ff,

> Then Jesus said, 'There was a man who had two sons. The younger of them said to his father, "Father, give me the share of the property that will belong to me". So he divided his property between them. A few days later the younger son gathered all he had and travelled to a distant country, and there he squandered his property in self-indulgent living. When he had spent everything, a severe famine took place throughout that country, and he began to be in need. So he went and hired himself out to one of the citizens of that country, who sent him to his fields to feed the pigs. He would gladly have filled himself with the slop that the pigs were eating; and no one gave him anything. But when he came to his senses he said, "How many of my father's hired hands have bread enough and to spare, but here I am dying of hunger! I will get up and go to my father, and I will say to him, 'Father, I have sinned against heaven and before you; I am no longer worthy to be called your son; treat me like one of your hired hands'." So he set off and went to his father. But while he was still far off, his father saw him and was filled with compassion; he ran and put his arms around him and kissed him. Then the son said to him, "Father, I have sinned against heaven and before you; I am no longer worthy to be called your son". But the father said to his slaves, "Quickly, bring out a robe – the best one – and put it on him; put a ring on his finger and sandals on his feet. And get the fatted calf and kill it, and let us eat and celebrate; for this son of mine was dead and is alive again; he was lost and is found!" And they began to celebrate.

'Now his elder son was in the field; and when he came and approached the house, he heard music and dancing. He called one of the slaves and asked what was going on. He replied, "Your brother has come, and your father has killed the fatted calf, because he has got him back safe and sound". Then he became angry and refused to go in. His father came out and began to plead with him. But he answered his father, "Listen! For all these years I have been working like a slave for you, and I have never disobeyed your command; yet you have never given me even a young goat so that I might celebrate with my friends. But when this son of yours came back, who has devoured your property with prostitutes, you killed the fatted calf for him!" Then the father said to him, "Son, you are always with me, and all that is mine is yours. But we had to celebrate and rejoice, because this brother of yours was dead and has come to life; he was lost and has been found." '

May the younger brothers amongst us look around at the shit we're in, and return humbly to reality. May we seek a way back to the family we thought we could transcend and ignore. May the Earth welcome us back graciously, and may any older brothers amongst us find it in their hearts to celebrate rather than condemn.

May those of you committed to pursuing laws which help the younger brother find his way back, go in peace: with the serenity to accept what cannot yet be changed, the courage to change what you can, the wisdom to know the difference, and the humility to ask for help on the journey.

Notes

1   P.D. Ward and D. Brownlee, *The Life and Death of Planet Earth: How the New Science of Astrobiology Charts the Ultimate Fate of Our World,* Times Books, New York, 2003, pp. 106, 142.

2   Because the Jewish, Islamic and Christian faiths share the Genesis 1–3 creation stories, and because all three faiths have members who approach the scriptures literally and mythically, much of what I say in this article will be applicable to Judaism and Islam.

3   The thesis is at http://catalogue.flinders.edu.au/local/adt/public/adt-SFU20051212.182616/. A less paper consuming versions at http://ecofaith.org/articles/.

4   Norman Habel is a lecturer at Flinders University, South Australia, and coordinator of the Season of Creation.

5   M. Brett, 'Earthing the Human in Genesis 1–3' in *The Earth Story in Genesis* eds N.C. Habel and S. Wurst, Sheffield Adademic Press, Sheffield, 2000, pp. 73–83.

6   P. Singer, *Animal Liberation: A New Etics for Our Treatment of Animals,* Thorsons, London, 1991, p. 191.

7   A. Wansbrough, *Environment and Compassion: Caring for our Earth,* ELM, New South Wales, 1996, p. 5.

8   Rainbow Spirit Elders, *Rainbow Spirit Theology: Towards An Australian Aboriginal Theology*, HarperCollins, Melbourne, 1997, p. 79; N. Habel, *Bible Studies on the Readings for a Season of Creation* from http://www.seasonofcreation.com/studies/forest/Bible%20Studies%20-%20Forest%20Sunday.pdf, 2004, pp. 2–4.

9   See Matthew 20:25, parallel in Mark 10:43.

10 Matthew 23:11

11 D. Edwards, *The God of Evolution: A Trinitarian Theology*, Paulist Press, New Jersey, 1999, p. 11.

12 Pope John Paul II, 'Message to the Pontifical Academy of Sciences: On Evolution' from http://www.ewtn.com/library/PAPALDOC/JP961022.HTM, 1996.

13 Pope John Paul II, *God of Evolution*.

14 Pope John Paul II, *God of Evolution*.

15 Adapted from a short video at http://ecofaith.org/jason/ Find the link at the bottom of the web page. See also www.thankgodforevolution.com. My thesis contains a fairly comprehensive and critical review of others' journeys (http://ecofaith.org/articles). Some of my own thoughts, filtered through several years of ecoministry, are available at http://ecofaith.org/books.

16 For one example of trying to apply this in worship, see under multimedia at http://www.ecofaith.org/worship/.

17 M. Fox, *Original Blessing: A Primer in Creation Spirituality*, Bear & Company, New Mexico, 1983.

18 I see rich Westerners as the teenage younger son, and the Earth community as the father. Today for the first time I saw the older brother as those other cultures who have remained much more connected to the family, and perhaps those of you us who escaped the younger brother's destiny. More connected cultures and environmentalists face the temptation of seeing the 'younger brother' as irredeemable: awaiting the moment when he gets what's coming to him.

Cosmology and Earth Jurisprudence

Herman F. Greene

The reason for Earth Jurisprudence is to provide a legal response to the planetary ecological crisis. If there were no such crisis, there would be no need for Earth Jurisprudence. Thus, Earth Jurisprudence must be considered as the study of laws affecting human-Earth relations with intent to respond to this crisis and provide a viable future for humans and other life.

Earth Jurisprudence: The Great Law, Rights of Nature, and Natural Law

A distinguishing feature of Earth Jurisprudence is that it begins with the 'Great Law' of how the planet works, and seeks to conform human laws to this Great Law so that human activity will be coherent with it and life will flourish.[1] Another distinguishing feature of Earth Jurisprudence as developed by Thomas Berry is that nature has rights. For Berry those rights are derived from 'existence itself'[2], and are determined by existence[3].

By claiming that rights originate in existence and the universe, Berry appeals to natural law, a theory that the there is a 'law or body of laws that derives from nature ... binding upon human actions apart from or in conjunction with laws established by human authority'.[4] Some treat natural law as being concerned with the source of law's authority. John Finnis states:

> The fulcrum and central question of natural law theories of law is: How and why can law, and its positing in legislation, judicial decisions, and customs, give its subjects sound reason for acting in accordance with it? How can a rule's, a judgment's, or an institution's legal ('formal,' 'systemic') validity, or its facticity or efficacy as a social phenomenon (*e.g.,* of official practice), make it authoritative in its subject's deliberations?[5]

Berry wanted more. He wanted natural law to be the guide to the rights of nature. This paper concerns problems related to the arguments that the rights of nature are (i) derived from existence, and (ii) determined by existence.

Cosmology as the Authoritative Ground of Earth Jurisprudence

Berry recognised the need for an authoritative ground for the rights of nature. For him, this meant cosmology. He wanted to present a functional cosmology, one that would ground and guide human affairs. He felt this was what was missing from the work of leading social thinkers of his day:

All of these writers fail ultimately in judging the present and in outlining a program for the future because none is able to present data consistently within a functional cosmology. Neither humans as a species nor any of our activities can be understood in any significant manner except in our role in the functioning of the earth and of the universe itself. We come into existence, have our present meaning, and attain our destiny within this numinous context, for the universe in its every phase is numinous in its depths, is revelatory in its functioning, and in its human expression finds its fulfillment in celebratory self-awareness. Neither the psychological, sociological, nor theological approaches is adequate. The controlling context must be a functional cosmology.[6]

His cosmology is behind the legal principles he articulated for Earth Jurisprudence. And therein lies the problem because modern thought is on the whole acosmic.

Modernity and the End of Cosmology

In *The Wisdom of the World: The Human Experience of the Universe in Western Thought,*[7] Rémi Brague gives an account of the state of cosmology in the West. He distinguishes cosmology from cosmography, our physical map of the universe, and cosmogony, the story of how things came to be. Cosmology to Brague, 'as is implied by the word *logos*, is not that of a simple discourse, but an account of the world in which a reflection on the nature of the world as a world [as some kind of existing reality with common features throughout] must be expressed'.[8] Cosmology is reflexive, which is not always the case with cosmography or cosmogony. Because it is reflexive, cosmology requires an experiencing subject and 'must therefore necessarily imply something like an anthropology. [It] encompasses a reflection on the way in which man can fully realise what he is – an ethics'.[9]

He begins his account with the 'experienced cosmology' of humans before the 'Axial Age' of 600–200 BCE. Then there was no concept of 'world', no word designating all of reality in a unified way. It was not until the Greeks that a sense of 'cosmos' arose, one that encompassed humans and the universe, one where humans would grapple with who they are and what they should be from the nature of the 'world'.

The Greek word for world was '*kosmos*'. 'Pythagoras was the first to call "*kosmos*" the encompassing of all things … because of the order *(taxis)* that reigns in it.'[10] The world had a moral order that governed both nature and humans. In Plato:

> Good is the supreme principle. Good exercises its sovereignty over physical reality, but it equally rules the conduct through which the human individual turns his soul into a coherent whole (ethics) and gives the polis where his humanity must come to its fulfilment the unity without which the polis must fall (politics).[11]

The other great model of the cosmos in antiquity was the Abrahamic model carried forward in the sacred texts of the Hebrews, Christians and Muslims. Brague summarises this model as follows:

> The world is created by a good God, who affirms at every stage of creation that which he has just freely brought into being is 'good,' indeed in his ordered edifice 'very good' (Genesis 1). But the phenomena that seem most sublime within the physical world are not those of the highest level. They are in fact of lesser value compared with man, whom they serve. Man, therefore, is not meant to govern himself according to the phenomena of the world but must seek elsewhere for a model of behaviour. In the final analysis, that model is God himself and God manifests himself less through his creation than through a more direct intervention. He can either give the world his law, as in Judaism and Islam, or he can indeed enter into that world through incarnation, as in Christianity.[12]

These two models, one seeing the cosmos as ordered goodness from which humans are derivative, the other seeing nature and humans as independently created with nature being subservient to humans and all of creation being of a lower order than the world of the divine, have intertwined with each other in western thought.

A different model entered Western thought in the wake of the scientific breakthroughs in the sixteenth century and afterwards. Brague calls this 'the end of the world', a return to the pre-Axial Age 'absence of world' but in a different sense. 'The image of the world that emerged from physics after Copernicus, Galileo, and Newton is of a confluence of blind forces, where there is no place for consideration of the Good'.[13] The world was no longer a whole, but a result of disparate forces. Cosmology gave way to cosmography – the stars, for example, no longer reflected the order of heaven, an ethical model to which one was to adapt oneself, but lacked any significance until some new theory might account for the facticity of their existence[14].

Cosmology also gave way to cosmogony, as a focus on theories to account for the origins of nature became more important than the truth expressed in it. To the extent that post-Copernican science revealed a truth about nature, it was of its moral indifference. '[Consequently,] cosmology lost its relevance in two ways ... on the one hand, its ethical value was simply neutralised as the cosmology was considered amoral; and on the other hand it was more seriously discredited as being immoral'.[15] From this acosmic vantage point, good was no longer understood to be in nature, it had to be introduced by humans

> by force, by taking nature against the grain ... inside the only realm that [was] within the scope of human action ... the earth. Modern technology defines itself through the undertaking of domination, through a plan to become, according to the famous epigram of Descartes, the 'master and possessor of nature'.[16]

The philosopher E. Maynard Adams gives further insight into the 'immorality' of the universe in modern thought. In Adams's view, the modern scientific account of the world eliminated value and meaning concepts as categories that described or explained reality. He observed science became such a powerful tool for manipulating and controlling nature and yielded such benefits that only those aspects of knowledge that promoted these capacities were regarded as veridical. Chief among these aspects was the idea that an event 'is caused by the environmental, elemental, or antecedent factual conditions that necessitated it'.[17] This is the 'naturalistic' concept of causation that undergirds modern science. It stands in opposition to the teleological concept of causation, the idea that something happens for the realisation of an end (for what ought to be).

For the universe or existence to provide guidance as to what ought to be and ultimately what laws ought to be with respect to human–Earth relations and the authority for those laws as Berry advocates, there must be a commonly held sense of a meaningful, purposeful universe – a new cosmological awareness.

This leads to the question how could such a cosmology come into being or what language would best allow such a cosmology to be recognised, such that it might be the authoritative basis of, and a functional guide to, Earth Jurisprudence as it is in Berry's formulation?

Following Berry, I will pursue this question with reference to the universe (accounts of the nature of the universe) and ecology (accounts of the processes of life on Earth).

Cosmology and the Universe
With reference to the universe as a whole, here are six potential sources of such a cosmology: phenomenology, ontology, post-modern science, process thought, indigenous traditions and the 'universe story'.

Phenomenology
In a conversation I had with Thomas Berry in the winter of 2008, he explained cosmology by reciting his poem 'It Takes a Universe'. The poem begins this way:

> The child awakens to a universe.
> The mind of the child to a world of wonder.
> Imagination to a world of beauty.
> Emotions to a world of intimacy.
> It takes a universe to make a child.[18]

These are the primordial experiences that make us human – without 'wonder', we have no mind; without 'beauty', we lack imagination; without 'intimacy', we have no emotional bonding. These are given to us by virtue of being Earthlings. Berry says that if we grew up on the moon, our minds, imaginations and emotions would be as barren as the moon.

Phenomenology concerns the examination of our experience without consideration of what is objective reality or purely subjective response. So to say one becomes phenomenologically aware of a meaningful universe, means that one finds this awareness to be present in one's experience. Who, as a child or as an adult, has not had moments of awakening to a world of wonder, beauty and intimacy? Intuitively we know that such a universe has meaning.

Post-Modern Science

It is possible, however, for one to become disconnected from this awareness or not to trust it. We are our minds as well as our awarenesses. We see things in the world that don't correspond to wonder, beauty and intimacy. We are required to interact with our world to sustain ourselves and even to overcome threats to our existence. We may come to explain the world in a way that lacks meaning. Such an explanation was given by modern science beginning in the fifteenth century CE – the world was understood as objects in motion that were subject to mechanistic laws of causation. In the biological world, Darwin introduced ideas that the evolution of life was determined by blind forces of random mutation and natural selection.

The science of the twentieth and twenty-first centuries, however, offer the possibility of understanding the universe that is consistent with categories of meaning. Instead of a clockwork universe, relativity theory and quantum mechanics describe a dynamic universe with an element of uncertainty. Post-modern physical cosmology gives an account of a universe that has a story, a universe that has evolved through time with surprising developments. Instead of reductionistic biology with bottom-up understandings, biology now includes ecology and emergence theories that have top-down understandings as well. Under the Gaia theory, Earth as a whole is a kind of self-regulating organism. Evidence of this is that despite dramatic changes over millions of years in the climactic conditions on Earth and kinds and dispersion of species, the concentration oxygen in Earth's atmosphere has been 20.7 per cent.[19]

Some refer to this twentieth century science as post-modern because it differs from the mechanistic science of modernity. This science still seeks to discover patterns or laws of nature, but does so in a more holistic, interdependent, dynamic way.

Ontology

For the most part, Berry was not interested in a philosophical understanding of the universe, he was interested in an *experienced* cosmology. Yet beneath his evocative poetry of the universe, he held to the understanding of ontology or being, of that which gives the universe its existence.

In conversations I had with Thomas Berry, he would make the following statements:

- In the phenomenal world, the universe is the text without context.
- Everything that exists within the universe can only be understood with reference to the universe.
- The universe is self-referent in its being, but not self-explanatory.[20]

What Berry means by these statements is that we have no direct knowledge or experience of anything other than this universe. Thus, every being, other than the universe, is explained by its origin, role and function in the universe and its relation to other beings in the universe. The givenness of the universe in our experience, however, does not explain why it is the way it is or why there is anything at all.

For Berry, why the universe is the way that it is and why there is something rather than nothing must be explained by something beyond the phenomenal world (the world of our experience). Our knowledge of the nature of this originating source, the divine, is derived, however, from our knowledge and experience of the universe in the phenomenal order. Thus, he explained in conversation, while in the order of existence the divine precedes the universe, in the order of our knowledge and experience as human beings the universe precedes the divine.[21]

Further in conversation Berry explained the reason there is a universe is that the nature of the divine is love, and being love means giving of oneself and sharing with others. In the beginning, the divine gave of itself the universe.[22] The nature of the divine is reflected, not in any one part of the universe, but in the universe as a whole and necessarily therefore in every diverse part and dynamic that constitutes the universe. One of Berry's favourite quotations was the following from Thomas Aquinas:

> For He brought things into being in order that His goodness might be communicated to creatures, and be represented by them; and because His goodness could not be adequately represented by one creature alone, He produced many and diverse creatures that what was wanting to one in the representation of the divine goodness might be supplied by another. For goodness, which in God is simple and uniform, in creatures is manifold and divided; and hence the whole universe together participates in the divine goodness more perfectly, and represents it better than any single creature whatever.[23]

While some would argue knowledge that the universe originated in divine goodness is theological, Berry insisted that it was cosmological. By this he meant that it could be known by natural knowledge, that is through reason and experience. The experience would be of the types discussed in the preceding sections on phenomenology and post-modern science. And from this experience one would be led by reason to knowledge that the universe originated in goodness.

Process Metaphysics

Much of modern philosophy in the West has been an attempt to reconcile philosophy with modern science. For example, logical positivism in the early twentieth century held that nothing was true unless it was empirically verifiable or could be arrived at through logic, and analytic philosophy has considered what can meaningfully said in humanistic fields that is not mere confusion of language. This subjugation of humanistic categories of understanding to those of naturalistic science led E. Maynard Adams to say, as discussed above, that value and meaning categories were eliminated from the descriptive/explanatory framework for understanding the universe.

For there to be an intellectual defence for the reinstatement of these categories, there must be a philosophy that encompasses our humanistic experience and the understandings of contemporary science in a coherent, logical system. Such a system is referred to in philosophy as 'metaphysics'. In philosophy, cosmology is branch of metaphysics. Therefore, in a philosophical sense, as opposed to a physical science sense, cosmology must be sustained by metaphysics.

One notable effort to establish such a metaphysics is process metaphysics. The fundamental aspects of process metaphysics are creativity, organic change over time, interiority (pan-experientialism) and interdependence. As so understood, process thought is not new. There have been many philosophies that are based on process, including Heraclites of Ancient Greece and Lao Tzu of China and in the twentieth century Sri Aurobindo in India, and Henri Bergson and Teilhard de Chardin of France.

Among the process philosophers, Alfred North Whitehead stands out. He described his magnum opus, *Process and Reality: An Essay in Cosmology*[24], as 'specu-lative philosophy' and defined this as 'the endeavour to frame a coherent, logical necessary system of general ideas in terms of which every element of our experience can be interpreted'.[25] Whitehead felt that no philosophy, including his own, would ever complete this endeavour. Yet, he contributed much on which to build, and may prove to be a seminal influence globally on post-modern philosophy.

Whitehead's metaphysics is a 'philosophy of organism'. A brochure of the International Process Network explains, in brief, his philosophy this way:

> The central claim of Whitehead's 'process-relational philosophy of organism' is that the world is not made up of independent material objects. In contrast to Cartesian philosophy, in which the world is thought to be made up of 'static substances,' dependent only on themselves for their existence, Whitehead's philosophy depicts the dynamic interrelatedness of the multitudes of entities which compose the world. In this way of thinking, each entity requires others in order to exist, and each is thoroughly engaged in creative life-processes of becoming.[26]

Not mentioned in this summary, but also important are Whitehead's ideas that the true things (*res vera*) in the world are events and these events are internally related in a manner that is analogous to feeling or experience, as well as externally related in a manner that is analogous to efficient causation. Further, Whitehead understood a general tendency in the universe toward beauty by which he meant the complex harmonisation of feeling.[27]

Indigenous Traditions

Indigenous traditions are characteristically cosmologically grounded. In these tradi-tions nature has a language, and humans communicate with nature. The tree speaks, the bird speaks, the wind speaks, the river speaks, the cloud, the sky and the sun speak. They are brother, sister, mother, grandmother, father and grandfa-ther. There is one spirit common to all. Humans are part of nature. Humans can no more go out into nature than they can go out into their bodies. Humans do not 'make' laws, rather they discover them. Human well-being depends on main-taining harmonious order in obedience to the laws that govern all of nature. The role of law, as expressed by the shaman, is to mediate between the order and spirit of the universe and the daily life of humans.

> While our scientific understanding of the world is far different from the understand-ings that informed indigenous traditions, the traditions bring us to awarenesses of the universe and nature that are obscured by science. It is a nature that is alive, active, intimate, immediate, engaged, fully relational, moral in purpose (in fairness, also capricious and terrifying) and sustaining.[28]

The Universe Story

A sixth potential source of cosmology of a meaningful and purposeful universe is referred to in Berry's work as the 'universe story'. In *The Universe Story*, Berry and co-author Brian Swimme[29] presented the scientific account of the evolutionary development of the universe based on Big Bang physical cosmology and evolu-tionary biology, but they also presented an interpretation of this scientific account in humanistic terms that drew from phenomenology, process metaphysics, indig-enous traditions and classical religious and humanistic traditions.[30] Some find adequate meaning in the scientific account itself (without the humanistic interpre-tation) with its magnificent complexity, improbable events, transitions, interrelated-ness and actualities and its concepts of evolutionary development and emergence.[31]

Ecology as a Functional Cosmology

With reference to Earth only, ecology was for Berry a functional cosmology.[32]

Ecology is a nonreductionistic approach to understanding life on Earth and how it is sustained by dynamic interrelationships. According to Edward Goldsmith, ecology understands '[l]iving things [as] differentiated parts of the hierarchy of natural systems that make up the ecosphere, and the ecosphere has a critical struc-ture which enables it to maintain its homeostasis in the face of environmental challenges and to provide each of its subsystems with an optimum environment'[33].

He also states 'Ecology Seeks to Establish the Laws of Nature'[34], by which he means:

> To study the structure and function of the ecosphere and its constituent natural systems is to seek out their pattern. The general features of this pattern are relatively non-plastic, which is another way of saying that they are subject to constraints – in this case, that particular set of constraints required to ensure that their behaviour will serve to maintain the stability of the ecosphere. It is these constraints that we must refer to as the laws of nature or Gaian laws.[35]

These Gaian laws are not fixed, rather they are dynamic and are hierarchically emergent. Violating the laws destabilises the system.

Ecology is not, like the potential sources of cosmology in the preceding section on 'Cosmology and the Universe', a reflexive understanding of the characteristics of the cosmos. Rather it is a practical analysis of how life systems in the world maintain their viability and vitality. Both, however, restore the sense that we are part of a meaningful, purposeful world. They are both necessary for a truly functional cosmology.

Does Existence Determine the Rights of Nature?

The final question covered in this paper is, if we accept such a cosmology of meaning and purpose, can cosmology determine the rights of nature in a legal sense?

Laws have to be able to apply to broad groups of people of diverse opinions and orientations and must for the most part be self-enforcing. For example, a tax system cannot operate without public cooperation – there is no public audit or enforcement mechanism that can compel compliance in general. And for there to be such compliance, the public has to accept the authority of the regulatory agency and that the laws are just and serve a public purpose. In an acosmic world, it is difficult to establish the rights of nature – nature tends to be viewed from the perspective of what is good for humans. In a cosmology of meaning and purpose, however, every being has meaning and purpose. Such a cosmology does determine the rights of nature at least from the standpoint of the authority of laws relating to the rights of nature.

This leaves open the question of whether cosmology prescriptively determines the rights of nature? This is question will only be answered in part here. To say that every being has rights, does not address the question of the relative rights of beings. One can say that the right of each particular human to exist is protected by law, but one cannot say the right of each particular bacteria or worm to exist is protected by the law.

A cosmology of meaning and purpose leads to a biocentric, geocentric or universe-centric orientation, but it does not change the fact that humans make the judgments concerning the rights of nature. In a legal sense, neither the universe, nor Earth, nor any other-than-human creature can represent itself. In this representative function humans will weigh the rights of nature and of humans. Nevertheless, to operate from a biocentric, geocentric or universe-centric orientation expands the context in which these judgments are made. Further, to the extent humans learn ecology, 'listen' to animals and plants and Earth, develop empathy and compassion for animals and plants and Earth, and come to understand and sense Earth from a Gaian perspective as having organic and living qualities, humans are more able to represent the Earth community in law and accept the authority and purpose of laws that concern the rights of nature.

Such an orientation doesn't determine laws any more than all the diverse and conflicting principles of acosmic modern understandings determine laws. As always the development of laws is a deliberative and collaborative balancing of interests. From such an orientation, however, the interests include the entire community of life and the well-being of the human community is understood within the well-being of the Earth community as a whole.

Notes

1 C. Cullinan, *Wild Law: A Manifesto for Earth Justice*, Green Books, Devon, UK, 2002, pp. 75–78.

2 'The primary supposition here is that the interdependence of every mode of being on every other mode of being requires humans to recognise that every being has rights derived from existence itself'. T. Berry, 'Legal Conditions for Earth Survival', in *Evening Thoughts*, ed Mary-Evelyn Tucker, Sierra Club Books, San Francisco, 2006, p. 109.

3 'Rights originate where existence originates. That which determines existence determines rights'. ibid., p. 149.

4 *American Heritage Dictionary of the English Language*, 3rd edn., Houghton Mifflin Company, Boston, MA, 2002.

5 J. Finnis, 'Natural Law Theories', in *The Stanford Encyclopedia of Philosophy*, ed Edward N. Zalta, 2008, viewed 5 January 2010, <http://plato.stanford.edu/archives/fall2008/entries/natural-law-theories/>.

6 ibid., p. 87.

7 R. Brague, *The Wisdom of the World: The Human Experience of the Universe in Western Thought*, trans. Theresa Lavender Fagan, University of Chicago Press, Chicago, 2004.

8 ibid., p. 4.

9 ibid., p. 5.

10 ibid·, p. 19, quoting Aëtius, *Placita*, 2.1.1., P. 327, DK 14 A 21.

11 ibid., p. 32.

12 ibid., p. 60.

13 ibid., p. 185.

14 ibid., p. 186.

15 ibid., p. 194.

16 ibid., p. 209.

17 E.M. Adams, 'Is Science Really Compatible with Religion?' *The Ecozoic Reader*, vol. 2, o. 3, 2002, p. 28.

18 T. Berry, 'It Takes a Universe', quoted in interview by Caroline Webb, 'The Mystique of the Earth', *Caduceus*, Spring, 2003, p. 7, viewed 21 June 2007 <http://www.caduceus.info/archive/59/59_archive.htm>.

19 This is the 'Gaia hypothesis' first developed by Dr. James Lovelock. See J. Lovelock, *Gaia: A New Look at Life on Earth*, 3rd ed., Oxford University Press, New York, 2000 (first published in 1979).

20 Principle 2 of Berry's Principles of Earth Jurisprudence: 'Since it has no further context of existence in the phenomenal order, the universe is self-referent in its being and self-

normative in its activities. It is also the primary referent in the being and the activities of all derivative modes of being'. Berry, *Evening Thoughts*, p. 149

21 This is what Berry means when he says 'the universe is the primary revelation of the divine'. T. Berry, *The Great Work*, Bell Tower Books, San Francisco, 1999, p. 91.

22 The concept is that of divine 'plenitude' and of 'being itself' as the source of all things in existence without which they would not have come into existence and would not be maintained in existence. For a discussion of the place of divine plenitude in Berry's thought, see A.M. Dalton, *A Theology for the Earth: The Contributions of Thomas Berry and Bernard Lonergan*, University of Ottawa Press, Ottawa, Canada, 1999, pp. 40–45.

23 T. Aquinas, *Summa Theologica* I, 47, 1, quoted in part, for example, by Berry in *The Great Work*, p. 71.

24 A.N. Whitehead, *Process and Reality* (Corrected Edition), eds David Ray Griffin and Donald W. Sherburne, The Free Press, New York, 1978.

25 ibid., p. 3.

26 International Process Network brochure 2006, available from Herman Greene, 2516 Winningham Road, Chapel Hill, NC 27516, USA.

27 'The teleology of the Universe is directed to the production of Beauty'. A.N. Whitehead, *Adventures of Ideas*, The Free Press, New York, 1967, p. 12. For Whitehead, each event, or actual occasion, in the universe comes to be in ferment seeking a final resolution or satisfaction of its 'prehensions'. The occasion begins with a subjective aim, some unrealised potential relevant to the situation in which it arises. There is an element of self-creation or freedom in the occasion, one not to be explained but only to be observed. The occasion, while much is given to it, is its own creation. This reaching to achieve some new realisation of form, to achieve the unrealised potential, is its teleology.

28 For a contemporary description of such a universe, which draws in part on indigenous traditions, see S. Harding, *Animate Earth: Science, Intuition, and Gaia*, Chelsea Green, White River Junction, VT, 2006.

29 B. Swimme and T. Berry, *The Universe Story: From the Primordial Flaring Forth to the Ecozoic Era*, HarperSanFrancisco, San Francisco, 1992.

30 H. Greene, 'Where is the Universe in the Universe Story?', *The Ecozoic*, No. 1, 2008.

31 *See, e.g.*, U. Goodenough, *The Sacred Depths of Nature,* Oxford University Press, New York, 1998; E.O. Wilson, *Concilience: The Unity of Knowledge*, Alfred A. Knopf, New York, 1998.

32 Berry, *The Great Work*, p. 84.

33 E. Goldsmith, *The Way: An Ecological World-View*, University of Georgia Press, Athens, GA, 1998.

34 Title of Chapter Two, ibid.

35 ibid., p. 12.

Extracting Norms From Nature:
A Biogenic Approach to Ethics

Pamela Lyon

We do not find our ethical premises in our biological nature, or under cabbages.

– Peter Singer (1981) *The Expanding Circle*

The challenges of refashioning our systems of governance and law to reflect the interests of the biosphere are many and large. Not least is how to persuade people to expand their sphere of concern beyond not only all other human beings and animals recognisably like ourselves – such as chimpanzees, other great apes and mammals, whose behaviour can be more or less comfortably benchmarked against our own – but further still to embrace all living things. This is a big ask. A growing number of philosophers are concerned with just this problem.[1] Until now, I have not been one of them. I am not a professional ethicist, bio- or otherwise. I am a philosopher concerned with biological cognition – understanding what the mind is, what it does, how it works and how it might have evolved. Anthropocentric bias is common in my line of work. My intention here is to see whether the alternative perspective I devised for reframing questions and issues in cognitive science – the biogenic approach – might work for reframing bioethical issues in the service of Earth Jurisprudence.

One of the greatest philosophical obstacles to continuing Thomas Berry's 'Great Work' is the truism that values cannot be derived from facts. Paraphrasing Hume, one cannot extract *ought* from *is*. No matter how many empirical facts you may have to hand, they do not amount, in and of themselves, to a prescription for ethically correct action in a given situation. Whether in the observance or the breach, it is an individual's ethical framework, not simply the facts of the matter, which guide that individual's action. An ethical framework is a set of assumptions, ideally a reasonably rational and coherent set of assumptions, which may draw on empirical facts but are, at bottom, human interpretations of whatever facts are invoked. These interpretations are grounded in predominantly human concerns. Needless to say, this presents a serious problem for those seeking a principled basis for Earth Jurisprudence.

In *Sociobiology: The New Synthesis*, a manifesto equally influential and incendiary, Harvard zoologist Edward O. Wilson suggested, as others had done before

him, that we *can* derive rules for human behaviour from empirical facts, specifically the facts of biology. More precisely, Wilson claimed that the more we come
to know about biology, and particularly the cooperative and altruistic behaviour of
social animals including (but not limited to) ourselves, the clearer it will become
that human ethics derive from our biological nature, which – like the nature of
every living thing – is a product of natural selection and other evolutionary forces.[2]
At first glance, sociobiology might seem an alternative to an anthropocentric ethics,
but it is not. While the focus of his lifelong study encompasses the most routinely
self-sacrificing organisms on the planet – ants – Wilson sees the origins of human
beneficence not in deep evolutionary history but in the requirements of a particular
kind of group living. The 'biological nature' through which the sociobiologist will
discover the origins of ethics is not that nature humankind shares with all living
things but, rather, the special nature that arises from clan and tribal life, whose
closest manifestation elsewhere in the animal world is found among our nearest
primate relatives, chimpanzees.[3]

Peter Singer responded to Wilson's strong claims for sociobiology in the ethical
domain in *The Explanding Circle: Ethics and Sociobiology*, which argues for the
enlargement of the sphere of ethical concern. One would expect no less of the
author of *Animal Liberation*. Of the several ways in which Wilson believes biology
can be relevant to ethics, Singer takes issue at greatest length with the strongest
and 'potentially most significant' claim: that biologists might one day 'discover
ethical premises inherent in man's biological nature'.[4] Singer's counter-argument is
principally Hume's, known in modern moral philosophy (following G.E. Moore) as
the 'naturalistic fallacy'. Arguments from facts to values are fundamentally invalid,
Singer writes, because, 'Facts, by themselves, do not provide us with reasons for
action. I need facts to make a sensible decision, but no amount of facts can make up
my mind for me.'[5] Singer enlists two main examples to make his point. First, given
a $500 windfall, no amount of facts will help me choose between the following
two options: give the money to a best-practice charity to alleviate suffering among
the impoverished Valod tribespeople of India, or use it to make my family happy.
Second, Singer cites the well-exercised philosophical example of the charging bull.

> [T]he fact that the bull is charging does not, by itself entail the recommendation:
> 'Run!' It is only against the background of my presumed desire to live that the
> recommendation follows. If I intend to commit suicide in a manner that my insur-
> ance company will think is an accident, no such recommendation applies.[6]

In this chapter I will argue that the ethicist is mistaken in an important sense,
but the biologist, too, misses a crucial point. While Singer is entirely correct that no
amount of facts will tell me what to do with my $500, he – and most other moral
philosophers – are wrong in presuming that the naturalistic fallacy is exceptionless,
and that the gulf between fact and value is everywhere and always unbridgeable.
Wilson's error, I will argue, is that he neglects the most obvious ways in which

many (perhaps most) biological facts have an intrinsically normative dimension, what the moral philosopher Philippa Foot calls 'natural normativity'.

I will argue that most biological facts imply values in important ways – they are inescapably norm-laden – precisely because of the nature of living organisation. I will argue that the intrinsic normative dimension of biological life provides one avenue for advancing the notion that the interests of the biosphere should be represented in systems of governance and law.

The Biogenic Approach

When I began studying in 1995, I was deeply puzzled by the absence of a truly evolutionary approach to the problem of *what cognition is* in Anglophone philosophy of mind. Cognition was then generally, and confidently, regarded as computation over representations in the brain, although it was difficult to say (and still is) what exactly 'computation' means, much less what a 'representation' is. More remarkably, there was nothing close to consensus about what sorts of organisms are cognitive and what sorts aren't. Opinions spanned the spectrum from the most complex organisms to the simplest. Beyond the species boundary of *Homo sapiens*, viewpoints mainly were proffered on the basis of 'intuition' benchmarked against what we know, or surmise, about human cognition. This high degree of uncertainty about how the central object of inquiry is even to be *identified*, much less explained – after hundreds of years of investigation – struck me as very possibly unique in the history of science but seemed an especially bad look in a modern life science.[7]

Scientific knowledge of life processes advanced astonishingly in the twentieth century, which began in ignorance of the fundamental units of inheritance and ended with the sequencing of the human genome. Virtually all of this progress was made possible by understanding how biological functions and processes work first in very simple organisms (bacteria, yeast, worms, fruit flies), then in more complex organisms (fish, frogs, mice) and ultimately in primates, including humans. Almost everything we know about the basic mechanisms of genetics, development, how the constituents of the body are made, sensory signalling and information processing, circadian timing, memory and innumerable other biological processes was – and is – acquired this way. Yet when I mentioned this to colleagues I was told that simple organisms were 'reflex machines' and had nothing interesting to reveal about cognition.[8]

It struck me that what separates most of cognitive science from most of biology – and might explain the former's failure to make much theoretical sense of a mountain of data now so vast 'it almost inhibits meaning'[9] is a methodological assumption. Cognitive science proceeds largely on the basis that the features of human cognition can tell us most about the biological function of cognition (whatever it might be) and thus should be the starting point both for enquiry and for generalising to other species, where possible. I call this the *anthropogenic* approach to cognition, because investigation originates or begins (*genesis*) with humankind (*anthropos*). That this assumption dominates the sciences focused on comprehending the most cherished facet of human experience is entirely understandable, even if it may also constitute an obstacle to progress.[10]

The classical computational approach to cognition – the idea that cognition involves information processing similar in important respects to the way computers process information – is a good example of an anthropogenic approach, not least because it seems species-neutral. Cognition can be investigated in any organism – even a bacterium or a flowering plant – using a computational approach. Thus, it is not the physical system under study that makes an approach anthropogenic but the grounding assumption that guides the investigation. The computational approach is anthropogenic because it is grounded not merely in a human cognitive capacity but a highly sophisticated one.[11] Before machines were designed and built, 'computers' were human beings who computed, with pencil and paper, numbers used in statistics and look-up tables of diverse kinds. Computing machines thus were designed to carry out more rapidly and accurately specialised cognitive tasks that humans do.

The *biogenic* approach, by contrast, assumes that the best way to approach cognition is to recognise that it is, first and foremost, a biological function that assists an organism to make a living in an ever-changing environment and should be approached like other biological functions. The overarching theory of biology is evolutionary theory, so a degree of continuity among phyla – potentially a high degree – is assumed. The biogenic approach thus tackles cognitive problems from the ground up, as it were, beginning with the principles of biology. What are these principles? Biogenic explanations are currently clustered around three main frameworks for understanding living organisation: self-organising complex systems (SOCS), autopoiesis, and biosemiotics. Each framework derives its distinctive character from the aspect of biology it emphasises. SOCS approaches to cognition emphasise the *physics* of organisms as complex, dynamic, self-organising and thermodynamically open systems. Autopoietic approaches focus on the *biology* of biology, namely, the distinctive recursive, self-producing organisation of vitality. Biosemiotic approaches focus on the *intentionality* of biology, the ubiquity of sign action in living systems, from molecules to whole organisms. These frameworks yield a number of 'family traits' that necessarily constrain biogenic theorising. Certain moves cannot be made with a biogenic approach because of them. For example, one cannot pretend that physiology and evolution don't matter in theorising about cognition, a basic tenet of classic computational functionalism.

How does the anthropogenic/biogenic distinction concern ethics? Principally, as in cognitive science, most frameworks for making moral judgments are grounded in anthropogenic assumptions. This is true even of most frameworks that attempt to be biologically inclusive. The dominant assumption is that only human beings, and perhaps animals recognisably like us, act for reasons, which philosophers traditionally regard as meaningfully distinct from mere causes. Reasons, typically, are 'rational thoughts' (whatever those terms mean beyond common parlance), usually capable of being 'reflected' upon (ditto).

Moral arguments from sentience, too, often fall foul of the consciousness criterion, which requires the ability to reflect upon (not merely to have) an experience. For example, experiments demonstrating that trout feel pain when hooked by an angler[12] were challenged using precisely this argument, plus the absence in fish of that part of the human brain where subjective pain sensation is believed to be activated. In other words, despite displaying behaviour remarkably similar to that displayed by humans and other primates in pain, the fish couldn't be feeling pain because a) their brain structure wasn't sufficiently like that of human beings, and b) the fish didn't 'know' they were feeling pain.[13]

In sum, to the extent that systems of governance and law are based on ethical judgments, anthropogenic assumptions make it very difficult to make a case for including the interests of all living things and the inorganic matter and processes that sustain them, which collectively constitute the biosphere. A biogenic approach to ethics, by contrast, must take account of the biological realities that all organisms share. Even the most cursory look at those realities should demonstrate that normativity is intrinsic to living organisation. In the next section I will draw out three naturally normative facts derivable from the three different approaches to biology referred to above.

Making a Living is Intrinsically Normative

All organisms maintain themselves far from thermodynamic equilibrium by importing 'order' from their surroundings in the form of matter and energy, chemically transforming it to do work, and exporting 'disorder' in the form of waste products of various sorts. That they do so seemed to defy the Second Law of Thermodynamics until the crucial distinction was drawn between 'closed' and 'open' thermodynamic systems, leading the way to the science of dissipative structures and irreversible physical processes. What marks the difference between living systems and other kinds of self-organising complex physical systems that can maintain themselves far –from thermodynamic equilibrium – tornadoes and chemical baths, for example – is that (so far) only living systems are capable of acting to circumvent the consequences of resource depletion, by seeking resources elsewhere or by changing their own structure to survive stringent conditions.[14]

This is the first and most obvious naturally normative fact derivable from biology: *Every organism strives to persist; it is self-preserving.* Although technically, to use Foot's distinction, this is an 'attributive' norm rather than a 'prescriptive' norm – which would be of the form, *All things considered, every organism ought to strive to persist* – striving to persist is, in fact, what organisms normally do. An organism will do everything it can do to persist, unless it is defective in some way or has a conflicting higher-order goal. Every organism on Earth, from the simplest to the most complex, has a repertoire of responses – not just one response – to predictable environmental threats to its continued persistence. The organism may not know the difference between life and death, but its first choice, vouchsafed by its evolved physiology, is life. This norm is clearly not exceptionless. Other existential imperatives may trump the survival norm. Reproduction and altruism in social animals are two well-known examples. Singer's example of the suicidal person electing not to run when faced with a charging bull thus represents a 'defect' of natural normativity in Foot's sense. The person may have reasons for choosing to die, even rational ones, but she must overcome a collection of highly evolved embodied norms (impulses) to stand firm in the face of the charging animal.

In short, continuing to live is the primary value of – or act of 'valuing' by – all organisms. While Foot does not avail herself of his work in *Natural Goodness*, the pragmatist philosopher John Dewey shares with her, at least in part, the belief that 'evaluations of human will and action share a conceptual structure with evaluations of characteristics and operations of other living things, and can only be understood in these terms'.[15] Dewey, however, distinguished between two types of valuation in his moral philosophy: *valuing*, in the sense of prizing or liking, the basis of appetitive behaviour, and *evaluation*, which requires a modicum of thought or reflection.[16] Valuing is what all organisms do. What sorts of organisms are capable of evaluation is an open empirical question.

Organisms are not merely self-organising, far-from-thermodynamic equilibrium systems, they are also autopoietic systems.[17] That means they are continually producing the parts necessary for their functioning. Each part is produced by a network of components, which are themselves being continually produced by networks of components. Were a Boeing 747 an autopoietic system, the plane would be manufacturing its parts at the molecular level – down to the last rivet – and drawing in the necessary resources to carry this out, *while in flight*. This is a difficult balancing act, which is why organisms are constituted by a variety of control and regulatory mechanisms, including multiple kinds of feedback mechanism that maintain the system's steady state (homeostasis). All homeostatic processes operate within a relatively narrow range of values outside of which the organism's persistence is threatened. Too much of one variable (heat, cold, salt, atmospheric pressure, etc.) or too little of another (water, oxygen, food) means death (thermodynamic equilibrium). This is the second naturally normative fact derivable from biology: *Every organism exists within a relatively narrow range of values along many different parameters.* This fact intrinsically guides action for every normal, non-defective organism without a competing higher-order goal; it has prescriptive force.

In order to persist, organisms must establish causal relations with features of their surroundings that lead to exchanges of matter and energy. An organism is capable of interacting profitably with some, but not all, features of its environment as a result of its evolutionary and individual history of interaction with that environment. Thus, not every state of affairs is salient (significant) for every type of organism, or even every individual organism within a type. As a result of natural selection an organism's pattern of interaction with its surroundings generally tends to maximise advantage and minimise harm to the extent possible. Based on its history and current needs, an organism responds to different states of affairs according to an internal projection of value relative to its own persistence. Videos of a single bacterium joining a multispecies biofilm, a sessile community of bacteria, shows the cell testing and moving on several times before selecting what it perceives as a 'good' spot to settle down.

This is the third naturally normative fact derivable from biology: *All organisms have one or more mechanisms for assessing value (advantage/harm), namely, how the organism is faring in the immediate circumstance relative to some biological norm.* Organisms respond differentially to internal and external stimuli wholly or in part according to their assessments of (probable, relative) value – whether or not such assessment is accompanied by thought. How an organism interacts with its surroundings is significantly determined by the valence of the states of affairs with which it is confronted. Memory subserves this process. Differentiation among states of affairs involves the comparison of what is happening now relative to what was happening at some moment in the past. It is now well-established that all organisms, including bacteria and plants, can retain information about their surroundings for a non-zero period; they have forms of memory. Dewey saw that these mechanisms form the basis of hedonic value:

> Biological instincts and appetites exist not for the sake of furnishing pleasure, but as activities needed to maintain life – the life of the individual and the species. Their adequate fulfilment is attended with pleasure. Such is the undoubted biological fact. Now if the animal be gifted with memory and anticipation, this complicates the process, but does not change its nature.[18]

In short, most biological facts tend to have an attributive, if not a prescriptive, normative dimension simply because of the nature of living organisation. *Oughts* are implied by some biological facts, but these prescriptions are not exceptionless. That may be because living organisation is itself continually in flux, as is its surrounding medium, and natural selection doesn't plan for the future. Only variability – genetic, developmental, behavioural – is insurance against novel circumstances.

Taking the next step

I hope I have shown that there are norms in nature. Some empirical facts ineluctably guide action; they are normative in a prescriptivist as well as an attributivist sense. So what difference does that make? Over the past few decades, moral philosophers have argued (persuasively, I think) that there is nothing particularly privileged about 'we' defined in terms of kith and kin. If we wish to live a good life, those of us who enjoy a degree of affluence must recognise that we have a moral responsibility to assist the needy in distant lands.[19] We have a responsibility to other members of the human family, no matter how distant. The fact that many people feel this way is evident every time a catastrophic natural disaster strikes and aid pours forth from the unlikeliest of places. It would take a chapter much longer than this to argue that there is nothing particularly privileged about humankind. However, if *might makes right* is unstable or unattractive as an ethical foundation for human behaviour, it cannot be a good rule for human dealings with the rest of the natural world, such that the interests of this one species override those of all others.

On the other hand, all organisms are species-centric. Humans are no different than other animals for privileging their own kind in the actions they take. This tendency does not foreclose symbiotic mutualisms or individual animals acting in the interests of individuals of another species. Of all the animals of the Earth, human beings very probably are the only ones capable of understanding why species-centrism is fundamentally illogical and to choose to act in the interests of other organisms in a highly general way. The future of human life on this planet may well depend on activating this capacity.

Does this mean that, having been shown their deep kinship with other forms of life (and, moreover, their reliance upon them), human beings will suddenly change their behaviour? Of course not. We realise that other human beings are like us, but we do not treat them particularly well. The twentieth century, the bloodiest and most violent in all human history, is testament to that unfortunate fact. What it means, I think, is that there is at least a principled basis for devising an ecological ethics that could conceivably encompass the interests of all living things on the planet, which is critical if Earth Jurisprudence is to gain serious momentum as a movement.

Notes

1   See, for example, L.E. Johnson *A Morally Deep World: An Essay on Moral Significance and Environmental Ethics*, Cambridge University Press, Cambridge, 1991; N. Agar, *Life's Intrinsic Value: Science, Ethics and Nature*, Columbia University Press, New York, 2001.

2   See E.O. Wilson, *Sociobiology: The New Synthesis*, Harvard University Press, Cambridge, 1975; *Biophilia*, Harvard University Press, Cambridge, 1984 and *The Future of Life*, Vintage/Anchor, New York, 2003.

3   Perhaps because he did not argue for human ethics based on a more general biological nature, later Wilson (*Bibliophilia*) argued for an 'innate' human love for the natural world (biophilia) based on this shared biological heritage, which provided the foundation for his more recent passionate pleas for conservation and an end to anthropogenic species extinction (Wilson 2003, 2006).

4   P. Singer, *The Expanding Circle: Ethics and Sociobiology*, Clarendon Press, Oxford, 1981, pp. 72–73.

5   Ibid., p. 75.

6   Ibid., p. 79.

7   Of course, I didn't know then about the fierce resistance against 'biologising' the social sciences, including psychology, thanks in no small part to *Sociobiology*.

8   In the last decade it has become increasingly clear that simple organisms are more like us (less rigid and reflex-like) than previously thought, and we are more like them (more reflex-driven).

9   S. Rose, 'The Rise of Neurogenetic Determinism', in *Consciousness and Human Identity*, ed. J. Cornwell, Oxford University Press: Oxford and New York, 1998, pp. 86–100.

10   P. Lyon, 'The Biogenic Approach to Cognition,' 7 *Cognitive Processing*, 2006, pp. 11–29

11   Another reason computationalism is anthropogenic: For many years the Turing Test was the ultimate challenge for artificial intelligence and a central focus of the 'cognitive revolution' that began in the 1950s. To pass the Turing Test a machine had to fool a human being into believing she was communicating with another human being in a keyboard-based conversation.

12   L.U. Sneddon, V.A. Braithwaite, and M.J. Gentle, 'Do Fishes Have Nociception? Evidence for the Evolution of a Vertebrate Sensory System,' *Proceedings of the Royal Society of London: Biological Sciences (Series B)*, 2005, no. 270, pp. 1115–1121.

13   J.D. Rose, 'The Neurobehavioral Nature of Fishes and the Question of Awareness and Pain,' 10 *Reviews in Fisheries Science*, 2005, pp. 1–38.

14   P. Lyon, 'To Be Or Not To Be: Where is Self-preservation in Evolutionary Theory?' eds B. Calcott and K. Sterelny, *The Major Evolutionary Transitions Revisited*, MIT Press, Cambridge, 2011.

15   P. Foot, *Natural Goodness*, Clarendon Press, Oxford, 2001, p. 5.

16   J. Dewey, 'Valuing, Liking, and Thought', 20 *Journal of Philosophy*, 1923, pp. 617–622. See further E. Anderson, 'Dewey's Moral Philosophy,' *Stanford Encyclopedia of Philosophy*. First published 20 Jan 2005; revised 15 Feb 2010 (accessed 26/02/10). http://plato.stanford.edu/entries/dewey-moral/.

17   H.R. Maturana, and F.J. Varela, *Autopoiesis and Cognition: The Realization of the Living*, eds R.S. Cohen and M.W. Wartofsky, *Boston Studies in the Philosophy of Science,* D Reidel Publishing Company, Dordrecht, Boston, and London, 1980. See further P. Lyon, 'Autopoiesis and *Knowing:* Reflections on Maturana's biogenic explanation of cognition,' *Cybernetics & Human Knowing,* 2004, no. 11, pp. 21–46.

18   J. Dewey, 'Happiness and Conduct: The Good and Desire,' ed J.A. Boydston, *John Dewey, The Middle Works, 1899–1924*, Southern Illinois University Press, Carbondale 1908, pp. 241–260.

19   Singer, *Expanding Circle*; G. Cullity, *The Moral Demands of Affluence,* Oxford University Press, Oxford, 2004.

Elusive Lines and Environmental Rules[1]

Lawrence E. Johnson[2*]

> Why is his nature so ever hard to teach
> That though there is no fixed line between wrong and right,
> There are roughly zones whose laws must be obeyed?
> – Robert Frost, *A Further Range*

Ethics and law are about guiding our actions in the world and this requires drawing lines. We may draw our lines well or poorly but we must draw them if we are to proceed at all. This can be problematic. This much is true whether we are concerned solely with our human affairs amongst ourselves or with extending our imperatives or prohibitions to our interactions with the non-human world around us. This requires drawing. But of course this confronts us with the traditional and inevitable poser of *Where do you draw the line?* What is the rationale of putting a line just there rather than somewhere else? Can a line go there at all? Can it be enforced? Questions go on from there. They tend to become even more difficult when it comes to entities such as species or ecosystems that may have indeterminate boundaries or definitions. Their health or harm may be indeterminate and certainly ill-defined. When it comes to something like climate change, matters get no easier. I shall discuss some of the logic and difficulties of drawing lines in law or ethics. Law and ethics, complicating things, make somewhat differing demands on line-drawing. I shall offer no magic means for dealing with line-problems adequately across the board. There is none. Instead, I shall try to offer some ideas useful for engaging with such problems.

As we think about our world we do so by means some conceptual scheme with its distinctions, be they good, bad or mediocre. Though there are many and various ways of projecting our distinctions across the face of reality, some of which are better than others, reality does not come with our dotted lines already in place. No more does the earth's surface carry ready-made lines, be they lines of longitude or boundaries between legal jurisdictions. Even so, some maps are better than others by far. With maps or with concepts, some distinctions are easier to demark than others. The distinction between ice and water, for instance, is usually fairly sharp. But how many bands are there in a rainbow, how many colours? Answers vary from one individual or linguistic group to another, with some placing the transitions differently than do others. Yet certainly there are different colours. Just as certainly, there is a difference between the planet Jupiter and the rest of the universe that is no less real for Jupiter's lacking any exact boundary. I am told by astronomers that it does not have any real surface. From the outer atmosphere in, it just gets thicker toward the centre – becoming very thick indeed. We may say much the same thing about an ecosystem. The forest may be more or less in one place but its vital processes extend well beyond. Its health, for instance, may be adversely affected by salmon fishing off shore, reducing its input of nitrates.

It is a matter of cases how well the distinctions we make, and the conceptual schemes in which we employ them, work out in practice. It is a matter of what we are talking about and of our purposes and objectives in talking about it. As we develop them, our conceptual schemes may work well or poorly in serving our needs. Yet there will always be a touch of artificiality to them – and because of the sort of beings we are, and because of how we interact with the world, it is never entirely possible to avoid imputing to the world features that are projections of our own conceptual schemes. Some of what we project, such as lines of latitude or the bands in a rainbow may serve very useful functions. And some of what we project, as indeed, a distinction between a human and a non-human realm, may serve to mislead us.

It is not just *where* we draw our distinctions that is important. Important also is the degree of narrowness and rigidity with which our distinctions are interpreted. There are positives and negatives in the vagueness and indeterminacy that are such frequent features of our conceptual schemes. To be sure, everything else being equal, precision (when accurate) communicates more information than does imprecision. Moreover, imprecision and vagueness have come to have a very bad moral reputation – and one must grant that they are by no means entirely innocent of the charges that have been laid against them. We all know to our cost how insufficiently scrupulous people may take advantage of imprecise language to mislead people. Or to make shabby excuses for bad behaviour. But then again, we all know how language, very precise and narrowly interpreted, can be used to mislead people. Be sure to read the fine print. Hair-splitting can also be used for bad excuses. The fault is not with the degrees of precision afforded by language, but with our human use of and response to it. Those who deceive intentionally, or who intentionally leave open the possibility of so doing, are morally remiss. And we are imprudent if we do not take care to adequately understand what we accept from others.

To condemn rigidity in our conceptual schemes would be absurd. Our conceptual schemes would fall into a useless heap without an appropriate rigidity. Still, our conceptual schemes can never do our thinking for us, and sometimes we allow our use of them to distort our thinking. That can happen if we insist on interpreting and employing them *too* rigidly. Employing some flexibility is vital to much of our human communication. Yet an overly loose structure can lead us into sloppy thinking. We can be thankful that rigidity and flexibility can coexist within the same system. Without bones your arm would be as useless as it would be with no degree of flexibility at all.

Yet precision is not always possible and appropriate for human purposes in the world we humans face. Some things about which we might be concerned, perhaps such as moral rightness or life and death have considerably more vagueness and indeterminacy at their boundaries than do many other things. And sometimes we want to say something when there is no need for precision. ('John is quite tall. ' 'Exactly what do you mean by *tall*? 'Who cares? He is over two metres'.) And sometimes we want to stretch a term to make a point. 'Aye, there's the rub ', said Shakespeare's Hamlet. Not that *rub* can literally mean what it is made to mean there, but it was made to mean that most effectively. For that matter, *stretch* also springs from metaphor.

What I am saying here goes against a great deal of what we have been taught explicitly or by implication. It is widely presumed to be more rational, more useful, *better* if our thoughts can be shaped in terms that are defined with rigour and absolute precision. As intellectual heirs of the Enlightenment we are heirs of Descartes and his method of rational inquiry. This called upon us to think one step at a time, going logically from one clearly and distinctly perceived truth to the next, each shaped in precise terms. As it happens, Descartes's method works better in some applications than it does in others. It works quite well in mathematics. A principal feature of mathematics, comforting to some and infuriating to others, is that within its framework, questions that have answers at all have definite and demonstrable answers. Within that framework there are no grey areas or room for legitimate differences of opinion. Descartes, one of the finest mathematicians of all time, was seemingly trying to generalise to apply to all of reality that method which succeeded so well for him in mathematics.

Descartes's method works very well for dealing with those things that do (at least for relevant purposes) come in units with well defined properties and precise boundaries. It works well for account keeping, and for things of a mechanical nature, from Lego blocks to highly complex forms of structural engineering. It works well for quite a lot more than that. This is despite the fact that just about every thing in the real world, including Lego blocks, has boundaries that are at least *somewhat* indeterminate. And with properties that are not entirely self-contained. For some things, including some very important things, the fuzzinesses at the edges are substantially and crucially relevant on a practical level. Black-and-white thinking is frequently a too-crude tool for moral or biological thinking.

Another, related, shortcoming with Descartes's approach is that some things do not have an entirely self-contained identity. Rather, they have their identity in terms of a wider system. The plant is part of what it is to be a leaf. A discussion of life, or of most matters concerning it, is very much an enterprise in which thinking in terms of discrete well defined properties and units is quite problematic. Life tends not to come in that sort of unit, nor to be restricted to properties of that sort. An entity such as a species, an ecosystem, or a river system occurs and persists only in relation to its surroundings. Its surroundings and its interrelations with what is in its surroundings are part of its identity.

Line Drawing: Precision, Determinability and Accuracy

> Before I built a wall I'd ask to know
> What I was walling in or walling out,
> And to whom I was like to give offense.
> – (Robert Frost, 'Mending Wall ', *North of Boston*)

As well as posing conceptual difficulties, this business of drawing lines often involves us in practical difficulties of great magnitude and considerable urgency. In environmental affairs it is especially and often confrontingly true that moral and legal principles and their practical applications require specified limitations on our actions. We must decide what may or may not, shall or shall not, be done. This inevitably leads to controversies about where boundaries can best be drawn and what might be the consequences of drawing them in one way rather than another, and about whether they can properly be drawn at all. Where ought we to, where can we, draw the line? Some ways of making distinctions and drawing up the rules just do not work out well in practice. On top of that, there are often slippery slope arguments to the effect that were we to accept some rule or principle, or condone a certain course of action, then would be ultimately be forced to incur highly repugnant outcomes. If we permit/ban certain activities, what will it all lead to?

Frequently we are challenged to draw a line between some **A** and **B**. The inference is that if we cannot draw an adequate line between **A** and **B**, then causing/allowing **A** will have the effect, either logically or by some chain of human response, of bringing about **B**. Or, looking at it the other way, ruling out **A** may lead us to rule out **B**. Where-do-you-draw-the-line? posers are frequently put forward in connection with environmental management issues – water rights, fishing rights and the clearance of native vegetation being amongst the examples. If we are to outlaw the clearance of any native vegetation on any private land, what's to stop the government from banning all clearance on all private land? Is it possible to differentiate?

When one is challenged to draw a line between **A** and **B**, there characteristically will be at some apparent difference between them and some at-least-apparent gradation of cases between the two, challenging us to find a relevant and defensible way of distinguishing between them. Unless the evident difference is entirely illusory (which it can be), it is always possible to draw *some* sort of a line. We want more than just any old line. Yet what makes a good line good is not an easy matter to specify. Clearly, a good distinction is one that is useful for us with respect to whatever purposes we have in trying to draw the distinction. In order for them to be useful for us we need distinctions which do not create difficulties by putting things into unworkable categories or, more broadly, by being unfaithful to reality. Beyond that, our distinctions are better as they better help us to think or communicate about some matter of concern to us, and help us to achieve our goals in doing so. Insofar as we are concerned to draw distinctions for moral or procedural rules it is important that we have well chosen distinctions and coherent purposes.

A very important factor quite often involved in discussions involving whether suitable lines can be drawn is a call for *precision*. Arguments based on the supposed difficulty or impossibility of drawing a suitable line are at their best – most plausible and most useful – when, as well as an apparent difference between **A** and **B**, and a gradation between them, there is also some reasonable presumption that any acceptable distinction must be precise. Legal and other regulatory matters are commonly thought to require precision. After all, laws and rules have to be interpreted, followed, and enforced. So too, often, are moral principles held to require precision. It is very much preferable for us legally if we can have a clear and sharp distinction between legal and illegal, and it is useful for us morally to have clear and sharp distinctions between right and wrong. Our distinctions are supposed, by we who make them, to be about something – about apprehended dangers to the public welfare, or about the moral character of certain acts. Our distinctions are supposed not to be merely arbitrary, though it remains true that the exact location of the boundary might well be. Such distinctions as we can draw are able to track what they are about only to an approximation and somewhat arbitrarily.

To illustrate: common sense tells us that car headlights ought to be on when it is too dark to drive safely without them, and that traffic laws ought to require it. Public safety demands it. But, of course, just where do we draw the line? With no specific line, the defence would quite commonly be that 'it wasn't *really* dark, not as *I* understand the term'. We might draw a quite precise line in terms of *lumens,* a scientific measure of light intensity. There would be a degree of arbitrariness to any particular number of lumens we settled on, and no particular number of lumens would exactly correlate with any particular and measurable degree of road safety. The greatest problem in practice, though, is that precise lumen levels at particular times and places would be difficult to establish in retrospect, when trying to enforce the law, or in prospect when trying to obey it. A more workable way to draw the legal line would be to legislate that lights must be turned on when we are driving between certain times which are set with reference to sunset and sunrise. That would give us precise and mostly workable lines, ones that would vary sensibly with the changing daylight hours of the changing seasons. Yet even so, that would not take into account such factors as rain or cloud cover, or speed limits on particular roads.

As a general matter of utility (and of equal justice before the law), we do want our laws and other binding rules to be fairly precise, and drawn in such a way that their application in practice can be determined with workable ease. Of course *fairly precise* and *sufficient ease* are each rather rubbery terms – and where does one draw the line in applying them? It is better for us not to draw lines for them. Rather, for each rule, we should find the best fit we can between *precision, determinability,* and *accuracy* (fidelity to reality). It is important to remain alert to the difference between *precision* and *accuracy.* These are not the same thing, and they do not always go together. To say that I am 214.637 cm. tall would be quite precise. However, to say that I am about 177–178 cm, while less precise, would be far more accurate.

In general, it will not be possible to maximise all of these desirable qualities simultaneously. To specify conditions for the mandatory use of headlights in terms of time of day would be about as precise as specifying in terms of lumens, but would be far more determinable. Lumens more accurately, but less determinably, correspond with the requirements of road safety. To require that headlights be on when that would non-negligibly contribute to road safety would most accurately reflect what the law is there to do. But obviously such a requirement, without further stipulation, would be hopelessly imprecise and very difficult to determine. At best, a statement relating lighting to road safety might serve to clarify the intention of some more manageable rule.

Let us note some key line-drawing issues at stake here. Given that we cannot draw an entirely non-arbitrary line between **A** and **B**, what is that presumed to indicate? It might be that

1. There is no difference between **A** and **B**,
2. There is no important difference between **A** and **B**,
3. While there might be an important difference between **A** and **B**, there is no way to locate just where a transition takes place, or that
4. People being as they are, if we permit/require **A,** there is an unacceptably high risk that **B** will happen; or if we prohibit **A,** there is an unacceptably high risk that **B** will *not* happen.

Suppose the issue is whether to allow mining in the Arkaroola Wilderness Sanctuary, South Australia, which is, legally, a pastoral lease from the state. As the law is currently, mining companies can stake a claim and mine on any pastoral lease, subject to making due compensation to the lease holder. If it interferes with running sheep or cattle, then the economic loss to the grazier must be offset by the miner. However, though on a pastoral lease, Arkaroola is not operated as a pastoral lease. It offers splendid scenic wilderness experiences. And the existence-value of the area is significant to a vast number of people who do not have an economic stake in Arkaroola.

From one point of view, Arkaroola is just another pastoral lease on which to develop a mine. If it is permitted elsewhere it must be permitted there. If mining can be stopped there, then (horrors!) where can't it be stopped? From another point of view, Arkaroola is significantly different from other pastoral leases and embodies and highly important non-economic values. If mining can't be stopped there (horrors!), then (horrors!) where can it be stopped? So, if we are to permit/ prohibit mining at Arkaroola, how are we to draw the line so as to avoid (what we assume to be) disaster?

We must often draw the line somewhere, be it at a good somewhere, or at a bad or a mediocre somewhere, so we try to get the best fit we can. Yet if no line can be drawn which works better than having no line at all, we may be better off not trying to enact laws, even though we might well be convinced (and rightly) that there are important differences from one side to the other. Where no workable line is possible, wise legislators prefer to legislate as little as possible. Accordingly, the law generally prefers to have as little to do as possible with personal morality, social interactions, and domestic relationships.

People usually prefer their moral principles, as they do their statutory laws, to be clear as possible about where they do and do not apply, with clearly drawn lines. (That is, they usually do unless they wish to get away with something. Rightly, with moral principles as with laws, we tend to be suspicious of those who try to bend the rules so as to exempt their own actions.) Ethicists, though, I note with some chagrin, are rarely allowed the indulgence of non-specification commonly extended lawmakers. While we do not expect statutory law to guide us in all things – and we would deeply resent it if it did – it is often demanded of a set of moral principles that they resolve all matters that might be brought before them. Lines must be drawn through the greyness, resolving it into zones of determinate moral black and white. If a proposed moral principle does not draw the line, or does not draw it with comforting sharpness, a desire for moral clarity may lead us to want to replace it with one that does. However, an insistence on precision can be counter-productive. As we have noted, the most precise lines, and those that make it easiest to determine which side of the line one is on, are not necessarily the most accurate. A moral principle that draws precise lines and gives us definite answers may thereby give us easy or convenient answers with perhaps a comforting sense of knowing just where we are. Yet it may not give us the morally best answers, nor even give us answers that are adequately close to being the morally best answers.

Another indulgence rarely allowed to ethicists, though often to law-makers, is that of employing a degree of arbitrariness. If we need a speed limit, then the speed limit has to be *something,* so legislators enact one, even though we all know that a different one arguably as good could replace any specific limit. Legislators make legality. Ethicists, however, do not *make* morality. They give an account of it, but the difference between right and wrong is supposed not to be arbitrary, nor at their discretion. Fair is fair, and wrong is wrong. If we make the rules about environ-mental issues, or approve or disapprove particular courses of action, arbitrariness in our decisions is usually not well tolerated. Nonetheless, it is just true that there are some moral grey areas, and that sometimes we must cope with them and work around them. We may hope nonetheless that some line can be drawn through that greyness which at least has the virtue of ensuring that those matters which, one way or another, are not grey are put on their appropriate side of the boundary.

In sum, a *Where-do-you-draw-the-line?* challenge only has as much force as the presumption that a definitive line is requisite. For some purposes, an inability to adequately draw such a line between **A** and **B** does mean that we must take them to be equivalent (even if we know they are not). If some people of legal age are banned from drinking alcoholic beverages on the beach, all are. If the hoons and slobs cannot drink there, neither can we. If we can, they can. That is how the law works. Yet not everything works that way. Not everything can. We should not allow the asking of the question automatically to create a presumption that any acceptable line must necessarily have all three virtues of being precise, being accurate, and being fully determinable. In many situations in the real world, particularly in environment matters, there are very important differences which cannot be delimited by such lines, but with which we must cope nonetheless. We cannot leave them alone, because, like the natural world, they certainly will not leave us alone. I offer the following as points of reference in making our way through the difficulties of drawing lines and assessing slippery slopes possibly ensuing:

1. Things – in particular, courses of action – can be meaningfully different and successfully distinguished even when
   (a) no precise and non-arbitrary line can be drawn between them – *driving too fast and driving too slow*; and
   (b) they can both be subsumed under some description; – *apples and oranges are all fruit*
2. Lines that are the most precise, or the easiest to draw, may not be accurate, *as in the tallness example above.* [List continued below.]

Now imagine a parlour game: one person names two very different things and the other players try to state some features they have in common. For example, a gold brick and a mountain lake: neither is a living being, both are made of material heavier than wood, and both are usually attractive. A half-eaten orange and the action of flying a kite: each may feature in a pleasant spring outing. It will become clear that *any* two things will have something in common. The game soon becomes one of trying to add some humour to the description. I have heard an astronomer's joke to the effect that comets are like cats in that they have tails and do what they like.

One can go on to more elaborate forms of the game. Finding differences is just too easy, so consider a more complex game. In this game, one person names *three* different things, and the challenge is to find something that each pair of them have in common to the exclusion of the third. No matter how similar **A** and **B** might be, and how different **C** is, there will always be something which **B** and **C** have in common, but not **A**. Tweedledum, Tweedledee, and the Andromeda Galaxy: Tweedledee and the Andromeda Galaxy are the two further to the right.

Or whatever. There will always be something. The point of all this, of course, is that *any two things can be united by one description, and can be separated by another.* These are facts that can be put to use for many purposes – many of which are, to say the least, problematic (and possibly nefarious). In summary, this gives us two further points of reference:

3. Things can be meaningfully and importantly/relevantly similar even when
    (a) a precise line can non-arbitrarily be drawn between them; and
    (b) they can be subsumed under different descriptions.
4. Moreover, that things can be put under a common description does not necessarily mean that they are relevantly similar with respect to the particular issues under consideration. Most especially, it does not mean that causing/allowing one course of action will mean that we will, or must, or ought to, or cannot avoid allowing to happen some other course of action with which it has something in common.

Any proposed course of action **A**, can be described as being quite unprecedented, or as being only another instance of what has long been known and accepted. So too can any possible outcome **B**. Even a moral enormity on the scale of Auschwitz could be described as a social welfare initiative intended to alleviate unemployment. But do I dare to eat a peach? The time, place, circumstances and specific qualities of that particular peach unite to form a novel occurrence with unknown consequences. Our descriptions can be made to fit our inclinations. If we are opposed to course of action **A**, we can always find some heinous course of action **B** that has some features in common with **A**. We then argue that we must never allow any course of action that comes under the common description, lest we slide down the slippery slope to **B**. How persuasive the argument is will depend on our rhetorical skills of presentation and the plausibility of the assumption that the common description captures the essential features of **A** and **B**, with differences being too minor to arrest a likely progression from one to the other. This may indeed be the case. Or maybe it isn't. If we are in favour of some course of action **B** we can look for some course of action **A** that is generally approved and which can be put under some general description as **B**. Then *to be consistent we must* allow **B**. Those italicised words should set alarm bells ringing. Other parlour games would be to propose the most (seemingly) plausible slope to (seemingly) plausible bad consequences from an arbitrary starting point or to propose the most (seemingly) plausible reasons why, *to be consistent,* we must accept a given horrific conclusion.

Whether we are rule-makers or concerned citizens, mentally exploring slippery slopes may help us to better and more clearly determine our priorities. There might be more than one possible accommodation that could be developed between factual reality, our ways of doing things, and our evolving moral and non-moral values. Much of the art of lawyers, politicians, salespersons, and other persuaders lies in getting us to start our thinking along lines leading toward our structuring our evaluations in a way favourable to their cause. And much of their art lies in keeping us from thinking along dangerously rival lines. We are to start from the precedents and established principles from which *they* want us to start. Skilful persuaders know to, and know *how* to, get us to accept plausible starting points and then to lead us step by plausible step to their favoured conclusions.

There might be more than one possible accommodation that could be developed between factual reality, our ways of doing things, and our evolving moral and non-moral values. How we put it all together, which plausible fit we devise between theses factors, may well depend on what direction our thinking starts out on. I am reminded here of a phenomenon sometimes described in works of popular science, having to do with the behaviour of certain supersaturated solutions. Being supersaturated, they contain more of a dissolved substance than they can stably retain in solution. Once the process of crystallisation gets started, much of the dissolved substance will come out of solution, forming crystals. The process might start due to a jolt or to some small impurity, or it might start spontaneously. Some solutions are such that they can produce crystals of more than one form (composed of molecules of the same substance differently arrayed). Which sort of crystal one gets depends on which sort happens to start first. One might even get different sorts in different parts of the same container if crystallisation starts in different places independently. One way or another, though, something must eventually crystallise out.

I can offer no sure-fire way to navigate our way through to right answers. I can only say *be wary of the lines.*

> The lines of morality are not like the ideal lines of mathematics. They are broad and deep as well as long.
>
> Edmund Burke, *Reflections*

Note

This paper is adopted from a chapter on elusive lines and slippery slopes in a (now-unpublished) book of mine on bioethics. The book is currently in press with Cambridge University Press, with a title as yet to be determined.

Section 3: Customary Law

While unique in their application to Western law, the basic principles of Earth Jurisprudence are not new and can be found in the law of indigenous people throughout the world. This section will consider important examples of this through the Earth Jurisprudence network in Ethiopia and Kenya.

Anthropocentric and Ecocentric Versions of the Ethiopian Legal Regime

Melesse Damtie

We human beings consider ourselves as the centre of everything and that everything is created to meet human interest. With this mindset, we have exploited the Earth to the extent our technological capacity has allowed us. We have occupied or interfered with the majority of the Earth's surface, fragmented it and modified to our convenience and to the detriment of others. It is now time to challenge this mindset as it is not only bad for other beings, but also for the generators and perpetrators of this mindset, the human beings. This mindset is referred to as anthropocentrism. Nearly all aspects of human life, i.e. the economic, social and political aspects of most countries are now highly affected by this mindset. If we continue in this way, the Earth's life support system may not see the end of the present century, as many scientists assume. It is now time to shift from this human centred approach to an ecology centred approach as we humans are not master of but part of nature and we need to align with nature's laws.

Anthropocentric Versions of Ethiopian Laws

Influence by the Western Traditions

Most of the legal instruments at international as well as national levels are mainly dominated by rules and principles which favour humans by neglecting or by giving little consideration to all other creatures on Earth. It is normal that the legal documents and policy instruments of a nation would reflect the economic, social, political and psychological notions of the society. The latter in turn could also be greatly influenced by philosophers, thinkers, religious beliefs, cultural settings, etc. Therefore, the legal instruments, policy and strategic documents prevailing today are either directly or indirectly influenced by the thinking and perceptions of philosophers, moralists, thinkers, etc.

In our long history of interaction with nature we seem to have developed an attitude of dominating nature by giving it only economic values for human interests. These conceptions and attitudes have developed and taken shape through long and complex processes and ended up in human-centred values.

Alexander Gillespie identifies five main strands in the development of anthropocentric views.[1] Regarding the *first* factor, it was mainly based on the belief of the separation of the (immortal) soul from the (mortal) body.[2] The main actors in the development of this belief were Pythagoras and Plato. 'Plato believed that mind, acting on matter, is absolutely separate from it'.[3] In Descartes' proposition 'I think therefore I am', he perceived 'everything outside his own identity had a questionable existence ... that the 'outside' surroundings were not important to his material dependence'.[4] Today there are many technologists who believe that technology will fix everything for humans and even that nature will be replaced by scientific and technological innovations. Probably, Descartes was one of the major sources for such a belief. M.N. Pokrovsky also writes that: 'It is easy to foresee that in the future, when science and technique have attained to a perfection which we are as yet unable to visualise, nature will become soft wax in his [man's] hands which he will be able to cast into whatever form he chooses'.[5] For Descartes, 'nature consisted of only tangible qualities, like size and weight'.[6] According to this Cartesian view, the natural world has no other qualities except the physically quantifiable ones. Even in the anthropocentric standard, this is extreme, as beauties such as a scenic landscape are considered to be values of nature, even if they are not measured quantitatively. Galileo also revolutionised the 'natural philosophy from a verbal, qualitative to a mathematical one by his insistence that the book of nature was written in the language of mathematics (circles, squares, and triangles,)'[7] by making the scientific world focus on the mechanical and mathematical features of the natural world and ignoring its qualitative and intrinsic values.

The views of Descartes led to the belief that 'everything consists of insular understandable parts'.[8] This takes us to the *second* factor, that is, the individual nature of existence. That is, all things exist separately from each other and they have no or little influence over each other. This is the central idea of atomism. 'The importance of conceptual individualism predates Descartes by over fifteen hundred years. Pythagoras asserted that all things were numbers'.[9] 'Democritus and the Atomists expanded upon this theory by suggesting that not only were all things composed by numbers, but they were also isolated, individual units'.[10]

The scientific world was also influenced by this kind of 'individualistic' thinking. For instance, the 'Newtonian mathematical model, which was atomistic, captured the intellectual world ... and continued unabated until now'.[11] It was the spread of rationalist thinking that led Descartes to consider nature as a 'machine'. However, this ancient theory of atomism has been disproved by the latest research. For instance, quantum theory now proves that an atom is not irreducible and separated from the rest of matter. According to quantum theory; '... things are intertwined and interdependent to an unfathomable degree ... in the multiplicity of things there is unity. Matter is many things and one thing at the same time'.[12] Nothing is separated from another on Earth and all things are interconnected. The existence of one is necessary for the other. This interconnectedness invariably exists between living and nonliving, human and nonhuman. Failing to recognise this very important feature of the Earth will lead to loss of biodiversity, climate change, etc. and consequently to the collapse of the life support system of the Earth. That is why 'some contemporary theorists believe that the ideas developed by early modern philosophers, such as Descartes, contributed to a later careless attitude toward the environment'.[13]

The *third* element is about dichotomies between humanity and the rest of nature. This distinction is created due to the unique feature of humanity's rationality. It is true that we humans have unique capabilities like language, tool-making, etc. These unique features of ours 'have been put forward by many philosophers and thinkers like Aristotle, Aquinas, Locke, Rousseau, Kant, Hegel, etc'.[14] Uniqueness is not only a characteristic of humans. In nature, all things are unique. There is no sufficient reason to highlight the uniqueness of humans and ignore the rest of nature. Other organisms also have unique capabilities which we humans do not possess. Rather than arriving at ignorance, our rationality should have led to a more responsible and inclusive view of the world.

Religious grounds were also the basis used by some thinkers for the distinction between humans and the rest of nature. A number of significant thinkers have based the distinction between humanity and the rest of creatures 'on religious grounds in Western Christianity such as Augustine and Thomas Aquinas who have emphasised the dominion over Creation given to Adam and Eve as creatures made in the image of God'.[15] Thomas Aquinas also suggested that: 'It matters not how man behaves to animals, because God has subjected all things to man's power ...'[16] These conceptions have led humans to infer a different value for themselves, the value of great love for self. This is not in itself a problem. As Joseph Butler stated: 'the trouble with human beings is not really that they love themselves too much; they ought to love themselves more. The trouble is simply that they don't love others enough'.[17]

The *fourth* element in the creation of the anthropocentric position derives from the consideration that 'labour is the only valuable factor in production'.[18] This view was magnified by Marx as he stated: 'the purely natural material in which no human labour is objectivised ... has no value'.[19] This Marxist view was one of the dominant principles in socialist Ethiopia from 1974 to 1991. This view makes nature valueless so long as it is not modified or worked on by human labour. The idea of making human labour the main factor in value assessment was suggested by John Locke. Locke suggested that 'in its natural state, nature was almost worthless'.[20] This concept was taken by Adam Smith and developed further. Smith explained that: 'The real price of everything, what everything really costs to the man who wants to acquire it, is the toil and trouble of acquiring it ...'[21] In the Ethiopian context this labor theory of value has been applied, especially in relation to land.[22]

The *fifth* consideration is the belief of dominance and mastery over nature.

This conception goes back to Ancient Greece. For instance, Aristotle suggested 'nature ... has made all animals for the sake of man'.[23] He also continued in saying in his *Politics* that 'plants are created for the sake of animals, and the animals for the sake of men'.[24] This position is also attributed to Francis Bacon, 'whose importance and influence stretches from the Enlightenment to the present day'.[25] Fichte argued: 'I will be the Lord of Nature, and she shall be my servant. I will influence her according to the measure of my capacity, but she will have no influence on me'.[26] Francis Bacon from a similar viewpoint hoped that, 'humanity would subdue nature with all her children, to bind her to service, and to make her a slave'.[27]

These philosophers clearly shaped the thinking of mastery of humans over nature, not only in Western societies but also in the Eastern socialist camp. While Ethiopia was a socialist country, it was very common to shout slogans which stated 'We shall bring under our control, not only the reactionaries, but also nature!'

Religious teachings were also not immune from the thinking of the philosophers of ancient and medieval times. The influences of philosophers and thinkers on the religious teachings were not linear. They were so complex and in many instances indirect. It is general knowledge that Greek philosophy greatly influenced the Roman legal system. However it is less well known that 'the Roman Catholic Church has been influenced substantially by Roman legal theory'.[28] Individual thinkers' influence has occasionally been immense. For instance, 'Plato was a major influence on the Church, particularly through Augustine and his successors'.[29] One of the prominent Christian thinkers, Thomas Aquinas, 'applied Aristotelian Categories to theology, in an attempt to develop it into a logical system. It is widely believed that Aquinas based his theology on Aristotle and in the process developed an Aristotelian theology'.[30] These facts can show the link between the Christian Church and the ancient and medieval period philosophers. The Christian Church played a pivotal role in shaping Western culture, especially 'the medieval Western culture'.[31]

In the preceding discussions we have seen briefly how anthropocentric views have originated from propositions of thinkers and philosophers. Anthropocentric views were not limited to shaping societal mindset, rather they leaked into the major legal systems of the world and influenced nearly all countries. Even if Ethiopia was not colonised, it did not escape the Western influence and its laws, policies, and strategic documents clearly reflect this.

Nearly all of the Ethiopian laws, from the ancient *Fetha Nagast*[32] (The Law of Kings) to the present day proclamations are highly influenced by Western legal traditions. The following paragraphs will show this fact.

In his preface address to the English translation of the *Fetha Nagast* His Imperial Majesty Haile Selassie, Emperor of Ethiopia said:

Our people have always both administered and lived according to law. Our people were at first ruled by Mosaic Law, but after the advent of Christianity to Ethiopia they came later to be governed by Fetha Nagast – a work combining both spiritual and secular matters ... The Fetha Nagast has been venerated, supported and applied by both the government of Our Empire and by the Church ... By the province of the Almighty this bulwark of the law was preserved for Our people ...[33]

Moreover, the Emperor revealed that the Penal Code of Ethiopia was enacted on the basis of *Fetha Nagast*.[34] Ethiopian oral tradition traces the *Fetha Nagast's* origins back as far as the 318 sages of the Council of Nicaea, during the reign of the (Christian) Roman Emperor Constantine,[35] whereas writers such as Peter Sand locate the origin of the *Fetha Nagast* to the ancient Roman law.[36]

'The affinity between Ethiopian law and European law, especially Roman law, has often been pointed out, particularly by the draftsman of the Ethiopian Civil Code of 1960, Professor René David, who specifically alludes to Roman origins of the Ethiopian "Law of Kings", the ancient *Fetha Nagast*'.[37] René David was a French professor who drafted the 1960 Ethiopian Civil Code, which is still functional. Anyone can simply see that Roman (or European) legal principles and thinking were directly incorporated into the Ethiopian legal system. One of the ways in which this is illustrated is by legal rules in the 1960 Civil Code that are totally alien to the Ethiopian cultural setting.

As has been indicated above, even *Fetha Nagast* did not grow in the Ethiopian setting and it was introduced law. 'The non-indigenous origin of the *Fetha Nagast* is well reflected in its popular Amharic designation as *yabaher heg*, i.e., "the law overseas".'[38] As Ibn Al-'Assal himself identifies 'the spiritual parts of the *Fetha Nagast* were taken from the canons of the Coptic Church and its secular parts were derived from the "Canon of Kings", consisting of four books said to have been written at the Court of Emperor Constantine'.[39]

There is also evidence that shows the Islamic influence on the *Fetha Nagast*. As Peter Sand explains, 'the cultural environment that Ibn Al-'Assal lived and wrote the book viz. the Islamic civilisation under whose domination the Coptic community has existed since the 7th century' [40] was one of the evidences that *Fetha Nagast* was influenced by Islamic thoughts and principles. Even certain provisions of the *Fetha Nagast* were directly taken from 'Islamic law (more specifically, from the Malikite school), particularly in the areas of sales ...'[41]

In his foreword message, the translator of the *Fetha Nagast* explains about the 'Roman background of the "Law of Kings", especially the secular part of Ibn al-'Assal's work is pervaded by principles of Roman law'.[42] From this it is evident that Roman laws have found their way to the Ethiopian legal system. Moreover, there is clear evidence that the modern Ethiopian laws have been highly influenced by the *Fetha Nagast*. For instance, in the Imperial Preface of the 1930 (1923 E.C.) Penal Code, Emperor Haile Selassie noted that:

in order that the people may be able without difficulty to distinguish what is forbidden by law and what is not forbidden, and that by learning European practice they may attain to a high degree of knowledge, because the basis of our code of laws in many places fits in with the European code, we have, without changing the law which has been in the country up to now, harmonised the two and established this law in the year 1923 (1930).[43]

The Emperor made clear that we should not totally abolish the Ethiopian system. Instead, European laws would be incorporated into the Ethiopian system and the two systems would be harmonised. This gives us good evidence that Ethiopian legal thinking is greatly influenced by the Roman/European legal thinking. The Emperor continued to state in His Imperial Preface of the same code that: 'Our Lord has said in the Gospel that he who knows much shall be punished much but he who knows little shall be punished little'[44] showing also the religious source of the Ethiopian laws. The Ethiopian law was not only influenced by incorporation of the European laws but also by appointing European judges in the courts, especially in the higher courts. This was not only existed in practice but was also supported by the law itself.[45]

It was the 1930 Penal Code that had replaced the criminal provisions of the *Fetha Nagast*. Yet the *Fetha Nagast* was the starting point for the modern code.[46] 'The 1930 Penal Code drew its inspiration not only from *Fetha Nagast* but also from the European Penal Codes as adapted for use in Asia (mainly the Code of Siam and Indochina)'.[47]

Professor Jean Graven, the drafter of the 1957 Penal Code, confirmed that 'The Law of Kings, deeply rooted in the Old and New Testaments, the writings of the Fathers of the Church ... the Ethiopian law is faithful to its origins'.[48] Graven further expresses his views on this point and clearly shows the link that the Ethiopian law had established with the Code of Justinian or Constantine and the Council of Nicaea in 325.[49] Moreover, Philippe Graven, son of Jean Graven, stated that:

This is not to say that Anglo-Saxon law, which in any event could not readily serve as a model, has been disregarded altogether; quite the contrary, it has inspired the solutions adopted, e.g., with regard to juveniles ... However, its impact on the new Code [the 1957 Penal Code] is not comparable with that of European Codes, in particular, Italian, Swiss ... Penal Codes. The Swiss Code has been given special attention ... Certain provisions of the Ethiopian Code either are directly derived from the Swiss Code or incorporate views expressed by Swiss courts and legal writers.[50]

Moreover, Emperor Haile Selassie I, in his Imperial Preface to the 1960 Civil Code stated that:

In preparing the Civil Code, the Codification Commission convened by Us and whose work We have directed has constantly borne in mind the special requirements of Our Empire and of Our beloved subjects and has been inspired in its labours by the genius of Ethiopian legal traditions and institutions as revealed by the ancient and venerable *Fetha Nagast*.[51]

From this short historical survey we can see that the Ethiopian legal regime is totally overwhelmed by the European legal systems and thinking, which were the result of anthropocentric thinking. Even if Ethiopian laws have incorporated some traditional and customary practices, they have never incorporated the customary environmental protection practices and the traditional cosmological aspects.

Examples of Legal and other Instruments that Reflect Anthropocentric Views in Ethiopia

Although the degree varies, nearly all the laws, policies and strategic documents of Ethiopia are basically anthropocentric. The following are some of these laws and other documents.

Wildlife Development and Conservation Authority Establishment Proclamation No. 575/2008

The reason this law was enacted can be seen from its preamble. It can be read here that Ethiopian wildlife is needed to fulfil economic benefits of the country. Other values, even if not economic, are purely anthropocentric. Some of the preambular statements include:

- WHEREAS, Ethiopia possesses diverse, rare and endemic species of wildlife which are of great value to tourism, education and science;
- WHEREAS, it is necessary to undertake appropriate conservation and development of wildlife for its sustainable use;
- WHEREAS, by halting the ever growing wildlife threatening conditions and enable the country to obtain economic and social benefits from its wildlife resources ...

Article 5 of the Proclamation clearly indicates the objectives of the law by stating: 'The objectives of the Authority shall be to ensure the development, conservation, and sustainable utilisation of the country's wildlife resource'. Sustainable utilisation does not escape the anthropocentric nature of this objective. As Adler and Wilkinson clearly noted, 'to slaughter fish for our own consumption up to the limits of maximum sustainable yield so as to keep the fish "stock" replenished ...'[52] is an instrumental value for nature.

Similar provisions also exist in the forestry proclamation. The main idea here is not to criticise the existence of such provisos, but it is to state that statements in favour of intrinsic values of nature should also exist side by side with the instrumental values.

The Ethiopian Water Sector Strategy, 2001

This water sector strategy is an extreme example, even in anthropocentric aspects. It is designed to drain the country's wetlands and convert them into agricultural fields. Let us see some of its provisions in relation to wetlands:

- Reclaim existing wetlands, and prevent the formation of the new ones by using appropriate mechanisms;
- Develop preventative mechanisms to avoid formation of waterlogged areas;
- Develop guidelines as how to reclaim wetlands, and enforce these guidelines;
- Carry out appropriate drainage works on all wetlands.[53]

Draining wetlands and waterlogged areas is one of the strategies in the Ethiopian agricultural development plan. The main purpose of this plan is ensuring food security for Ethiopia by using the water resources of the country. This is not a bad plan in itself. The bad thing is totally sacrificing one of the most significant ecosystems, wetlands, for the purpose of ensuring food security for humans. As it is believed now, it is not possible to sustainably ensure food security by degrading the natural environment. This strategy totally ignored the ecological functions of wetlands in the life support system of the Earth.

The Ministry of Agriculture and Rural Development (MOARD) notifies that 66.6% of the total area of the country (i.e. 74.3 million ha from the total area of 111.5 million ha) is suitable for crop production.[54] If all this area is allocated only for food production of a few species, where should all the rest of species go? If we utilise this extent of the country exclusively for human food production and we also use up additional areas for our other purposes, such as building our cities and towns, grazing fields for our livestock, or dams for our hydroelectric or irrigation programs, it is obvious that we are pushing most of the species to extinction. This should raise ethical questions.

Examples of Legal and other Instruments that Reflect Ecocentric Views in Ethiopia

In Ethiopia nearly all legal, policy and strategic documents reflect anthropocentric views. However, there are some instruments which faintly reflect ecocentric views. The following will show these.

The FDRE Constitution

Article 92(2) and (4) provide respectively that: 'The design and implementation of programs and projects of development shall not damage or destroy the environment. Government and citizens shall have the duty to protect the environment'. These constitutional rules, which are provided in the policy objective of the Constitution, should be the basis for the environmental protection works in the country. According to these rules, development activities can be conducted; however, they should not damage the environment. It appears here that the constitutional environmental objective has acknowledged the intrinsic values of nature.

The Environmental Policy of Ethiopia

Section 2(3) (q) of this Policy provides that: 'Species and their variants have the right to continue existing, and are, or may be, useful now and/or for generations to come'. This provision started with an idea that appears to be ecocentric, however, immediately converted into anthropocentric views as it resorted to the human value of species and the environment in general. The Policy mixed up both views in a single sub-section. If species have the right to continue existing, their existence should not have been related to their use value to the present and/or future generations.

Examples of Traditional Cultures

In Ethiopia there are more than 80 different linguistic and cultural societies. Many of these societies have lived harmoniously with nature for millennia.

Oromo People

The Oromo are the most populous people in Ethiopia. They have many clans and these clans have different customary practices. Even so, many clans share related cosmological visions. 'The Oromo people have fostered belief systems and social norms that encouraged or even enforced limits to the exploitation of biological resources'.[55] For the Oromo people, *Waaqa* (God) is father and Earth is mother. According to this belief, *Waaqa* and Earth are not separable. The father gives rains, and the mother grows living creatures using the rain. They also believe that mountains are kings of the land. Kings are graceful because they are in their royal costumes, so is true for the mountains when they are in their costumes, which are forests. The Oromo people have special attachment with the fig tree and the Oromia (one of the 9 regional states of Ethiopia) Regional State has a fig tree at the centre of its flag. The Oromo people usually praise nature, especially Earth as their mother who brought up them by suckling her breasts. The following verses are taken from the songs of one clan of the Oromo, the Gumaro, in praise of the Earth:

> 'Oh Earth, mother of grasses,
> under you is water,
> on top of you is grain,
> we dig and eat on you,
> we raise cattle and lead them out
> to the pasture on you,
> you carry us on your back,
> Please, give us your peace!'
> Translation from Oromo language *(Afan Oromo)* by Addisu Legas

Southern Region

The Southern Region has the largest number of cultural and linguistic groups. This Region alone contains more than half of the country's cultural groups. The Sheko communities are among the people of the south-west Ethiopian rainforest. These people have a special belief system that led them to live harmoniously with nature. They have a belief in the guardian spirits, which are the protectors of everything in nature. As the result of this, they have developed the concept of sacred sites. For the Sheko people, mountains, forests, rivers, wetlands, rocks, etc. are sacred.[56] The sacred sites are accessible only upon permission by a spiritual/clan leader for justified reasons. It must be remembered here that even the spiritual/clan leaders have no exclusive power to pass decisions on such issues, they need to consult elders of the clan. If these rules are violated, the sinner may be condemned to punishment, which is usually given by sacrifice of a bull, goat, sheep, honey, grain, or others depending on the extent of the offence. This sacrifice is believed to purify the offender.[57]

Another fascinating activity of the Sheko communities is that they divide the forests for individuals to hang beehives and collect non-timber forest products (NTFP). A piece of land that is allocated (or a person can also inherit such land from ancestors) for an individual is known as *kobbo*. The management of *kobbo* is based on customary rules that identify the rights and obligations of the individuals who have their own forestland. Hence, *kobbo* is a portion of one's own 'forestland' mainly used for harvesting honey and other NTFP. The owner has customary right to hang beehives for honey production and to collect different types of NTFP. The owner can also use timbers from his *kobbo* for making traditional beehives. Only owners can extract wild coffee, house furniture, agricultural tools, non-timber construction materials including different types of climbers and spices, etc. for their own consumption and market sale. The holder of the forest who is allotted or has inherited a block is responsible for its management through traditionally known use and conservation rights. Other people are not allowed to use resources in *kobbo* for hanging beehives and extraction of other NTFP unless it is with the consent of the owner of a particular *kobbo*. Clan leaders enforce the customary rules of forest management, though to a lesser extent nowadays.

Traditionally, clan leaders control the holder of *kobbo,* who is responsible for illegal timber extraction and other damages brought within his boundary. The clan leader has the right to impose different forms of punishments on the illegal use of forest resources. Currently, both the government and the owners control forest and NTFP in *kobbo* area. Clan leaders still provide informal advices for promoting conservation of resources in *kobbo* area.[58]

Amhara Region

This is the second largest group in Ethiopia, within one of the most environmentally devastated regions of the country. But there are some islands of forests which could serve as a refugee centre for some species of animals and plants. One of these areas is the Zegie peninsula. The Zegie peninsula is an extension of a land-mass into Lake Tana, the biggest lake in Ethiopia. The traditional people of the peninsula, having seen the limited resources of their land, decided 500 years ago to totally eliminate cattle ranching as a means of livelihood.[59] From this practice we can see that the traditional communities of the peninsula had realised how ranching of cattle was unsustainable in their small area centuries ago. The people secure their livelihood by fishing, farming, beekeeping, poultry and other related activities.[60]

The other island of forest is the Denkoro forest, which is found in this region. The people who are living in and around the forest are led by a traditional chief known as *Kire Dagna*.[61] These people have a special respect and consideration for the Denkoro forest. The customary chiefs are in charge of giving permission to forest access, except for women and children who collect firewood for domestic purposes.[62] The people even once challenged Emperor Haile Selassie when he intended to award some of his war heroes (patriots) with pieces of land from the forest, and the Emperor reversed his decision after having seen how efficiently the people were administering the forest.[63]

Although the people follow Christianity (67 per cent) or Islam (33 per cent), still there are traces of customary practices and belief systems which lead these people to respect and live harmoniously with their natural environment.

Conclusion

The examples of customary practices mentioned in this work are just a few which are selected from various regions of Ethiopia. These and many other indigenous people and local communities in Ethiopia have clearly understood the rhythms of the Earth's systems and led sustainable lives for millennia. However, our legal system has not well addressed the knowledge systems and practices of these people and now our environmental situations are in a more aggravated state than ever before. These remaining knowledge systems are on the verge of extinction unless serious efforts are made to save them. We have lost many even before we have known and understood them. We need to learn from these traditional knowledge systems and from the ethical principles which have been developed by contemporary environmental ethics so that we can further develop our Earth Jurisprudence visions on an ecocentric basis. This is not fun; it appears now a matter of life or death – not only for the human species but for all living beings together within such a complex and interconnected natural ecosystem. This takes us to the paradigm shift in our way of thinking. That is, we need new type of jurisprudence, which conforms to the laws of nature.

Notes

1   See A. Gillespie, *International Environmental Law, Policy, and Ethics*, Oxford University Press, Oxford, pp. 5–12.

2   Ibid., p. 5

3   E. Hicks, *Traces of Greek Philosophy and Roman Law In The New Testament*, Society For Promoting Christian Knowledge, London, 1896, p. 23

4   Gillespie, *International Environmental Law*, p. 5.

5   J. Passmore (1974), *Man's Responsibility for Nature*, Gerald Duckworth & Co. Ltd, p. 25.

6   'Meditations' One, Two (especially ss. 303), 'Meditation Five', (Concerning the Essence of Material Things), and ss. 71, and 74 of 'Meditation Six', as cited in Gillespie, *International Environmental Law*, p. 6.

7   http://www.crystalinks.com/galileo.html, 07/09/09, See also J. Baird Callicott and Robert Frodeman (2009) (eds.) *Encyclopaedia of Environmental Ethics and Philosophy*, (Vol. 1) MacMillan Reference USA, a part of Gale, Cengage Learning, p. 87.

8   A. Gillespie (1997) cited above, p. 6. (See also ibid, note 18.)

9   Ibid.

10 See Marshall, P., *Nature's Web: An Exploration of Ecological Thinking* (London, Simon and Schuster, 1992), 6971, as cited in Alexander Gillespie, cited above, p. 6.

11 Koskenniemi, P., *From Alienation of Reason: A History of Positivist Thought* (UCL Press, London, 1969), 8 cited in Alexander Gillespie, cited above, p. 8

12 http://www.thebigview.com/spacetime/quantumtheory.html, 09/09/09

13 J. Baird Callicott and Robert Frodeman (2009) (eds.) *Encyclopedia of Environmental Ethics and Philosophy*, (Vol. 1) MacMillan Reference USA, a part of Gale, Cengage Learning, p. 362

14 Alexander Gillespie (1997) cited above, p. 10.

15 J. Baird Callicott and Robert Frodeman (2009) cited above, p. 149.

16 Aquinas, T., *Summa Contra Gentiles*, in the *English Dominican Fathers* (Burns and Oates, London, 1928), Vol. 1, Q 64.1 and 65. 3, as cited in Alexander Gillespie (1997), cited above, p. 74.

17 Mary Midgley, 'The End of Anthropocentrism?' in Robin Attfield and Andrew Belsey (eds.) (1994) *Philosophy and the Natural Environment*, Cambridge University Press, p. 103

18 Alexander Gillespie (1997) cited above, p. 11.

19 Karl Marx, *Capital* (Foreign Publishing, Moscow, 1981), Vol. 1, 2067, cited in Alexander Gillespie (1997) cited above, p. 11.

20 Locke, J., *Two Treatises of Civil Government* (Dent, London, 1936), ss. 37, 423, 304, 305, 314, 316, 308, as cited in Alexander Gillespie (1997) cited above, p. 12.

21 http://en.wikipedia.org/wiki/Labor_theory_of_value, 10/09/09

22 In Ethiopia, the wilderness is said to be '*Tef Meret*', which roughly m.eans the land that has no worth. The word '*Maqnat*' is used to indicate that it has been 'developed' by clearing it. Therefore, it is a common understanding that 'the worthless nature becomes valuable by using human labour'.

23 Aristotle, Politics (Everyman, New York, 1972), 10, cited in Alexander Gillespie (1997), p. 12

24 Aristotle, Politics, cited in John Passmore (1974) *Man's Responsibility for Nature*, Gerald Duckworth and Co. Ltd., p. 14.

25 Leiss, W., *The Domination of Nature* (Brazillier, New York, 1972), 4571; cited in Alexander Gillespie (1997) p. 12.

26 Fichte, J.G., *The Vocation of Man* (Routledge, London, 1946), 29, cited in Alexander Gillespie (1997) p. 13

27 Bacon, F., *Novum Organum* (1620), Book 1, XV; *Essays*; *The Wisdom of the Ancients and the New Atlantis* (Oldham Press, London, 1977), 134, 166, cited in Alexander Gillespie (1997) p. 13.

28 Gordon Arthur (2006) Law, Liberty and Church, Ashgate Publishing Company, p. 4.

29 Ibid, p. 59.

30 Ibid, p. 191.

31 Encyclopedia Britannica, Vol. 25 Encyclopedia Britannica, Inc. 1994 p. 551.

32 *Fetha Nagast* was the traditional source of law for Ethiopia's Coptic Christian community.

33 Paulos Tzadwa, (Translator, 1968) *Fetha Nagast: The Book of Kings,* HSIU, p.v.

34 Ibid, p. v

35 Ibid, p. xxxiv

36 See generally Peter H. Sand, 'Roman Origin of the 'Ethiopian Law of Kings' (*Fetha Nagast*)' *Journal of Ethiopian Law*, Vol. 11, 1980, pp. 71–81.

37 R. David, A Civil Code for Ethiopia: Considerations of the Codification of the Civil Law in African Countries 37 *Tulane Law Review* 187, 192 (1962–63)

38 S.D. Messing, The Highland-Plateau Amhara of Ethiopia (Diss. Univ. of Pennsylvania, Pittsburg 1957) 309, cited in Peter H. Sand JEL, Vol. 11 p. 74.

39 See Paulos Tzadua, cited above, preface.

40 Peter H. Sand, cited above, p. 79.

41 Peter H. Sand, cited above, p. 79.

42 *Fetha Nagast*, cited above, p.xvi

43 The 1930 Penal Code, Imperial Preface, paragraph 5

44 Ibid, paragraph 15. Note that the Emperor was citing a verse from the Bible Luke 12: 47.

45 See the preambular Proclamation to the 1930 Penal Code, Article 2 which states that: 'These tribunals shall be composed of judges appointed by Us [the Emperor] from time to time. Not less than six shall be British Judges proposed by the Deputy Chief Political Officer at our request.

46 *Fetha Nagast*, cited above, p. xxxiv.

47 Aberra Jembere, (2000), An Introduction to the Legal History of Ethiopia: 1434–1974, Transaction Publishers, New Brunswick, London, p. 11.

48 Steven Lowenstein, (1965) Materials for the Study of the Penal Law of Ethiopia, Faculty of Law, Haile Selassie I University, Addis Ababa, p. xv. [Professor J. Graven stated this in his foreword message to this book].

49 See Ibid, pp. xii-xvii.

50 Philippe Graven (1965), An Introduction to Ethiopian Penal Law, Haile Selassie I University, Addis Ababa, p. 2.

51 Civil Code of the Empire of Ethiopia, Proclamation No. 165 of 1960, pp. v-vi.

52 John Alder & David Wilkinson (1998) Environmental Law & Ethics, MACMILLAN

53 The Ethiopian Water Sector Strategy, 2001 Section 4.1

54 Agricultural Investment Potential of Ethiopia, MOARD, March 2009, p. 4 available at www.moard.gov.et

55 Workineh Kelbessa, The utility of ethical dialogue for marginalised voices in Africa – Discussion paper, iied, 2005, p. 3

56 Yilma Miressa, (2004) The Significance of Cultural Laws and Beliefs of the Sheko Community for Environmental Protection, senior thesis, (ECSC), unpublished, pp. 32–64

57 Ibid, p. 48

58 http://www.melca-ethiopia.org/kobbo.html, accessed on 10 October, 2009

59 Abebaw Adamu (2004) Indigenous Community, Environment and the Law: A Case Study in Amhara Region with Special Reference to Zgie Peninsula, Senior Thesis, Unpublished, ECSC, p. 32

60 Ibid. pp. 2–3

61 Jemal Kassaw (2005) Role of Society for the Existing Denkoro Natural Forest, Senior Thesis, Unpublished, ECSC, p. 37

62 Ibid, p. 12

63 Ibid, p. 14

Earth Jurisprudence in the African Context[1]

Ng'ang'a Thiong'o[2]

What is Earth Jurisprudence in the African context? What are our sources of law? What is the meaning of law within an African customary set-up?

In Africa, wilderness, or what you call 'wild law', is the great source of law, not written common law. In fact, our traditional law is oral and is passed from one generation to another orally, through music, art, dance, drumming, and through the 'do's and don'ts' of the community.

Reviving Customary Law in Africa

After the 1884 Berlin Conference, Africa began to be colonised. The policy of assimilation in West Africa was to work with indigenous institutions to integrate them into the dominant society. In East and Southern Africa it was direct colonisation. A completely new concept of administration was introduced with this colonisation. A completely new system of justice was introduced. Later the native tribunals were retained. These were the courts of the African people, but the difference was that the law being applied to them was English in order to suppress customary law.

In Africa we have a cosmovision of where there are no objects within the context of customary law. Everything is living. The sky is part of us, so is the Earth, air, water, the plants and the animals. We have to keep the balance between all these aspects for the community to survive. For example, one cannot eat all the food; some must be left for visitors and to feed the birds or other animals. When you cook, you don't cook exactly the amount of food you require, because there may be a guest passing by. So even when you build your house, you make some extra space for visitors and plant trees or leave space for an anthill, for example.

The community is not just a human community. It is a community with many other subjects. You have no reason to kill animals or destroy plants unnecessarily, because this will destroy the whole. Customary law is a source of law that contains all these principles we are talking about. These principles have stood the test of time, since time immemorial, and they have been transmitted from one generation to another.

So, what we are doing now is bringing back customary law, so that it can grow. One of the things we have been saying is that when you trans-locate a law

from one part of the world to another, you need to take into account local circumstances such as peoples' customs. In the African context, the traditional worldview
is Earth-centred. All species have rights and responsibilities and duties. For the
whole system to function you need each other. There is no species that is superior
to the other. We must learn to coexist. Human beings have duties and responsibilities too, in relation to their rights. No species has the right to wipe out another one.
In Africa, this is reflected in our dances, our music, languages, our stories and our
customs.

What Space is There for Recognising Customary Law and Earth Jurisprudence?

How does customary law fit with human rights norms? It is important to appreciate that the language of human rights is derived from 1948. This language is very
young and we are talking of human rights in relation to the use of state power.
However, in indigenous knowledge systems, human rights and responsibilities are
as old as the community.

So, the first important thing in customary law is that one must not oppress
weak members of society – people with disabilities, women, or children – nor must
one deplete other species. Secondly, one must not allow oneself to be oppressed.
One has a duty and a responsibility to make sure we are not oppressed. You don't
engage in the issue of community rights versus individual rights to question who
is superior to the other. There is only one circle, one path. The point is to maintain
the balance. When you want to use herbs, for instance, you don't use all the plants,
because you have a duty and a responsibility to transmit them to the next generation and to hand over a healthy territory.

The concepts of ownership, property, and property rights regime are alien to
customary law. We are guardians and trustees, both for ourselves and for others.
Human rights and related responsibilities are a core component of customary law,
but it encompasses a wider understanding of community.

Customary law is dynamic. It changes. It is important to see it as a way of life,
rather than hard, cold, legal norms imposed from elsewhere. I remember hearing
that our Adjudicator Act said that common law and the doctrines of equity will
guide our courts and that they should not be repugnant to justice and morality.
But whose morality are we talking about? We must be talking about the morality
of the whole community. Where we differ here is that we are defining 'community' to include other species as defined by Earth Jurisprudence. So, wilderness is
the source of law. Nature is the source of law. To us, in terms of the community,
human law must be consistent with and learn and draw from the law of Nature.

What space is there in the current Kenyan constitution for recognising this
natural law? It is important to thank The Gaia Foundation because of the interest
they have shown in the constitution-making process. Cormac Cullinan, author of
Wild Law, was given a brief to look at the Kenyan Draft Constitution, especially

the chapters on land, environment and culture, to see how they could be made Earth-centred. Cormac wrote a proposal, which we were able to push through the delegates at the Constitution Conference.

Kenya's constitution is the highest law in the land and the Constitutional Review process was about asserting sovereignty with customary law high up as a source of law. The hope was that it would be on a par with conventions on international law, and that customary law would be recognised as our way of life, the way we live. And indeed Earth Jurisprudence and Wild Law principles were drawn upon and accepted by the delegates. They formed part of the proposed new Constitution of Kenya.

Unfortunately, the Draft Constitution fell in the referendum. It was not approved due to other areas of contention. The chapter on the environment stated that anyone who owns a part of Kenya must treat living species with respect, that people must use their land according to the principles of Earth Jurisprudence. Everyone agreed on these principles and the chapters on environment, land, culture, human rights, and rights of people with disabilities, and women and children. The Draft Constitution was defeated because of other issues.

Some of the Challenges of Recognising Customary Law

The force behind customary law is that legitimisation comes from the community. People are beginning to realise that some of the solutions to their problems are found in customary law. People are getting cases out of the ordinary courts to go to the elders. Given a choice between going to the ordinary court or the elders' court, people are opting to go before the community, which for me is a great victory.

There is a case in one of the communities with which we work, in Kenya. It is similar to Andrew Kimbrell's ideas of the need for a *guardian ad litum*, to speak for Nature in court. This case relates to how we defend the rivers, mountains, lakes or waters. The local community in Kenya tells the story of a hyena that comes out of the forest at night to hunt for food. The farmer has left his goat out at night. The hyena sees the goat; he hunts it down, kills it and eats it. The farmer, enraged by the loss of his livestock, tracks down the hyena and takes his revenge. However, now the family of the hyena want compensation for their loss. They want restorative justice. The character of our customary justice system is non-retributive, non-punitive. It restores. It heals. So, this case being presented by the community will soon go before the local jury to decide upon. It will be an interesting case to follow. This system of justice is not about technicalities. You want to go to the substance of the matter.

There is also the story of a pastoralist, a herdsman, who would travel large distances with his animals to graze. Over a period of time some of this land became privatised. On his travels one day, this pastoralist was arrested for trespassing. He did not know why, since he had covered these lands for years without any previous problems. Later, before the judge, he was found guilty for entering private lands without permission and he was told he would be sentenced to jail for one month. Upon hearing his sentence, he looked at the judge quizzically, and said, 'I'm sorry Sir, I'm afraid I will not be available to serve this sentence, since I'll be busy looking after my animals.' This form of jurisprudence pronounced by the judge was completely alien to the pastoralist. It was not from the customary source of law, which had not been recognised when his territory had been privatised.

Community Dialogues and Working with the Elders

How can we restore customary law so it sits next to common law? First of all, we need to dig to find the concepts and the consciousness of communities to get it right. We can learn from the existing remaining indigenous communities, those that have held their knowledge for generations and have not been too severely influenced by other knowledge systems.

Our ancestors and our elders are a core component of the work we are doing. If the elders aren't deemed as important in this process, then we cannot communicate directly to the ancestors. Ancestors communicate directly through elders. Elders are the true knowledge-holders. It is important to clarify that elders are not only male. Eldership is an important concept. So how do our ancestors speak to us? They speak to us through memory. Memory is inspiring the process. The primacy of Nature inspires. This work helps us to develop our skills to listen and learn from Nature, to observe and see. The work is done in a community environment. Community is very, very important. There is a primacy of elders in this whole system through the transmission of knowledge. They emphasise the importance of the way you live, and being respectful to every other being in the ecosystem.

In the five sites where we [Porini Trust] work in Kenya the aim is to protect what is remaining and reclaim what is lost. To do this you must first enter into a community dialogue. This is a dialogue that takes place within the community – sometimes accompanied by mapping of the community and what we call 'social-cultural profiles'. This is where the community talks. They discuss who they are, where they came from, what have they been doing, how they have survived colonialism, how they survived the gun, and all of those other invading forces like the mining and logging merchants. The story is told in their own language and in their own way. We [Porini] just facilitate that dialogue to take place. The community itself starts to see what has worked and what has not worked through this reflection.

At one of these sites, Giitune Forest, merchants had completely destroyed an indigenous forest. They had harvested all the trees. For many years, we tried working with the local administration, but this did not work. We tried working with religion, with the Church. We appealed to people there to stop the destruction, but that did not work either. We tried working through Kenya's legal system. We went to court. We threatened to sue them, but that did not work either. It was only when the elders decided to do a cleansing ceremony, where they decreed that anyone desecrating the forest would be doomed, that things changed. The forest has returned. It has regenerated.

Another site where we work is with a community who are the traditional custodians of a mountain called Karima. The entire hill had been desecrated. Alien tree species were introduced when the local tea factory planted Eucalyptus trees to dry the tea and tea leaves. Eucalyptus poisoned all the indigenous tree species. We started working with the community there by initiating a social cultural profile. The people there know their story. They know what has happened over this period of time. Their story has been passed orally from one generation to the next. We tracked their history back to 1650. We also tracked the history of the hill and its physical features. We identified 26 streams that flow down the hill. We noted the hill's animals and plant species. We also started working with Kenya's environmental court, which decreed that the hill is a sacred site. Many ceremonies and cultural practices took place there. We now need to remove the exotic plants that are destroying it, replant the indigenous plants that were cleared away, and return the hill to its former status, so that it can be revived as a sacred site where people can worship and carry out their ceremonies.

The communities where Porini works have been able to withstand a tsunami of colonialism. They have managed to keep their territory and their indigenous wisdom from being totally destroyed. The indigenous women still retain their knowledge of rainmaking. They can predict the length of the rains. This prediction will tell them which seeds to sow. This traditional knowledge has withstood pressure from over a hundred years of corporate colonialism. Our challenge now is to protect these communities and to find ways of working with Kenya's laws and Constitution.

An Effective Response to the Devastation of Kenya's Forests and Environment

Kenya is suffering from desertification and loss of forest cover. Our forest cover has been decimated to just 8 per cent of Kenya's total land area. Planting indigenous trees must be a priority here in Kenya. We must also support local communities to protect the few remaining indigenous forests.

Traditional societies have coexisted with Nature for generations. Everything they do is inspired by Nature. For example, they can tell the time by looking at the sun. They can tell when it is going to rain by the strength of the winds and the direction from which they are blowing. They can tell what time of the year it is by observing the vegetation. They can tell what the weather will be like for the next three to four weeks by looking at animal dung.

I came across a law from 1903, which required that every landowner must have 10 per cent of his or her land covered by forest. The law recognised that a certain amount of forest cover was needed in Kenya in order for it to survive. But this law is no longer valid because of the many logging interests.

However, we have been able to re-create some spaces within our dominant legal framework. We have good laws enacted by parliament, such as the *Environmental Coordination and Management Act 1999*, a very progressive law that contains principles of Earth Jurisprudence; and we have the *Forest Act 2005*, another progressive law that allows forest-adjacent communities to be responsible for their forests and recognises indigenous knowledge in this local governance of the forests. Those to me are important victories.

We have also been able to recognise community lore in some cases as law. For example, there are elders' courts called *Njuri Ncheke*. Here the elders can impose sentences and punishments. Its power is not derived from Kenya's law, common law, or from the court or the judge. The power is derived from the community, which recognises that whoever destroys the forest will be cursed. [The curse] is likely to affect the perpetrator's family, whether in this generation or the next. Future generations will have to pay [the perpetrator's] debt. If animals are going to die and our plants disappear, then our ancestors are going to get annoyed. If we allow this to happen, then our spirits will call us and punish us, because of what we are doing now.

Africa – A Sleeping Giant?

Africa has often been called a 'sleeping giant' due to the potential to exploit its remaining 'resources'. But how do we feel about this? After we have succeeded in decolonising our minds, we need to think about a new African leadership. The current leaders only believe in economic growth and economic development. This, they believe, supersedes everything else. The mantra of economic growth seems to justify and legitimise messing up the environment and violating human rights. Our war in Africa is the struggle against poverty. It is the fight against HIV/AIDS. As long as we are fighting these, nothing else seems to count. Global warming and environmental degradation are being ignored. Instead, we increasingly open our markets to achieve competitiveness in the international market. China can dump whatever it likes inside our shores. In the name of economic growth and development, mines can be opened up and our forests destroyed.

I believe Africa is a 'sleeping giant' in many other ways. We have very rich traditions, deep-rooted cultures, great philosophers and sports people, and great women. Our leaders are commoditising this greatness. We need to develop another form of African leadership across the continent. This leadership will have values other than money, profit, and economic globalisation. It will be rooted in something other than the doctrines of the London School of Economics. However, creating this new vision will be a challenge too. Even our lawyers are trained to think of land as a commodity and available for purchase to the highest bidder.

It is important to remember that we also have ancestral lands that cannot be sold. You cannot mortgage ancestral lands because the graves of your ancestors and your sacred sites are located there. I am happy that sacred sites are now being protected because these sites link our past and future to our ancestry. I believe that, yes, Africa *is* a giant, but in a way that is very different to how our current leaders think of us as a giant. They see our potential grandeur solely in terms of the market place.

Latin America and Asia are also giants. Not because they are 'tiger economies', or because they have big steel structures or big hydroelectric dams. They are giants because of what they have contributed to the world through their ideas, their diverse cultures and their resilience.

The Sources of Inspiration for Earth Jurisprudence

Let me end by reminding you what our great Elder and ancestor, Father Thomas Berry, has said about Earth Jurisprudence – Nature and indigenous peoples are the two sources of inspiration. We need to open our hearts and begin to learn the language of Nature, and it is important to appreciate that indigenous communities hold knowledge millions of years old. We can learn to listen to both the communities and to Nature.

Our journey in Kenya wholly affirms what Thomas Berry has said.

Notes

1   Extract from an Interview with Ng'ang'a Thiong'o: Thoughts on Earth Jurisprudence and the work of Porini Trust in Kenya, The Gaia Foundation, 24th September 2007.

2   In Memory of Ng'ang'a Thiong'o, a leading light in evolving Earth Jurisprudence in Africa, who died in February 2010 after an illness bravely borne. Thiong'o was co-founder and Legal Policy and Advocacy Officer for Porini Trust, Kenya. A respected barefoot lawyer and activist, he was Chair of Kenya's Release of Political Prisoners campaign and a member of Kenya's High Court. As a former Legal Advisor to the Green Belt Movement he worked alongside Nobel Peace Prize winner Prof. Wangari Maathai during her most difficult times. He was also a leading light on Earth Jurisprudence thinking and practice in Africa, and a popular participant at international Earth Jurisprudence events. Sadly, with many activities on the go and a young family to support, Ng'ang'a Thiong'o died in February 2010.

Part Three

Earth Jurisprudence in Practice

Section 1: Ecocentric Law

Ecocentric law is a term used to describe law that is Earth-centred and recognises that human beings exist as one part of a broad Earth community. As the title suggests, Earth Jurisprudence is one part of this broad movement. This section will first consider the extent to which existing law, from around the world, embodies ecocentric principles. Following this it will investigate the importance of sustainability in modern law and consider how the principles of Earth Jurisprudence can influence dispute resolution processes.

Where the Wild Things Are: Finding the Wild in Law

Nicole Rogers

My purpose in this article is to consider what is wild in existing law and I shall start in a somewhat unconventional fashion, with a children's book. Maurice Sendak's, *Where the Wild Things Are* is a prize-winning classic tale which has always been popular and which has, no doubt, become even more well-known with the 2009 release of the movie with the same title.

A cursory search of legal databases suggests that, thus far, environmental law scholars have not utilised this story in developing critiques of existing environmental laws, although the *title* of the book has been appropriated by some authors.[1] On the other hand, a leading figure in the law and literature movement, Professor Desmond Manderson, has argued that this text constitutes young children as legal subjects and instils in them an understanding of the 'meaning, function and interpretation of law'.[2] If Manderson's emphasis on the importance of such stories as a mechanism for social control is justified, then surely it is worthwhile to consider how such texts construct our relationship with the wild (as well as our relationship with law). In the first part of this article, I shall use this text as a starting point in identifying some pivotal assumptions about the nature of the wild and the nature of human interaction with the wild, and then consider whether such assumptions also underpin Western legal constructs of the wild.

Max's Journey and Wilderness Law

Max is a little boy in a white wolf suit who is called 'wild thing' by his exasperated mother and sent to bed without his supper after making mischief. He finds himself, to his clear delight, on a fantastical journey into the wild.

> That very night in Max's room a forest grew and grew – and grew until his ceiling hung with vines and the walls became the world all around and an ocean tumbled by with a private boat for Max and he sailed off through night and day and in and out of weeks – and almost over a year to where the wild things are.[3]

The wild things are described as 'terrible'; Max, however, is not intimidated. He tames them with a 'magic trick' which resembles hypnosis and then finds himself king of the wild things.

Max calls for a 'wild rumpus' and the wild things (including Max) bay and howl at the moon, swing from trees, dance and generally cavort. But Max puts an end to all this fun. He sends the wild things supperless to bed and sits inside his regal tent looking disconsolate. He is lonely and homesick and is lured home by the smell of 'good things to eat'.

The most striking aspect of Max's relationship with the wild things is that, very quickly, he becomes their king. He tames them and then sternly metes out the same undeserved punishment which he had himself received from his mother while an 'outlaw' and 'barbarian'.[4] The notion that we can control and subdue 'wild things' and that, furthermore, we as humans are at the pinnacle of an illusory hierarchy of living beings is hardly a novel one; it can be found in Genesis and, of course, forms part of our legal relationship with wild things. In national parks legislation[5] in all Australian states and in the wilderness-specific legislation in New South Wales and South Australia,[6] we find the assumption that we can and should manage wild places, even if they are to be managed with minimal interference with natural processes. Furthermore, the assumption that we can and should manage wild species underlies the different Acts that deal with wildlife, including threatened species.[7] Ownership and control are basic tenets in our legal interactions with the 'wild'.

The characteristics of the wild in Sendak's book also have a wider significance. The wilderness inhabited by the wild things is separate from civilised or inhabited spaces. In the fragmented legislative regimes set up by each State, wilderness is invariably identified as separate and apart from everyday human lived-in environments, as Other. Sendak's wilderness is not only physicially separate but geographically remote. Max sails over an ocean for nearly a year in a journey reminiscent of that undertaken by the white Europeans who colonised the Pacific islands. We find a similar emphasis on remoteness in some wilderness legislation.[8]

Yet this concept of wilderness as Other is problematic, as writers such as William Cronon have pointed out. Cronon critiques the preoccupation with wilderness preservation at the expense of the transformed environments in which people tend to congregate and where they live and work;[9] he writes that 'wilderness tends to privilege some parts of nature at the expense of others'.[10] He advocates celebrating *wildness* rather than wilderness[11] and I shall return to this point in a little while.

Another problematic feature of the wild in Sendak's text is that it is devoid of humans, at least until Max arrives. The Eurocentric vision of wilderness as uninhabited is one which has been challenged by indigenous people. Marcia Langton, for instance, has argued that 'the valorisation of 'wilderness' has accompanied an amnesia of the fate of indigenous people'.[12] The legislative criteria and management principles for wilderness align more closely with the 'European fantasy'[13] of wilderness as terra nullius than with indigenous ideas. Such criteria and management principles emphasise that wilderness is not compatible with sustained inappropriate human activities or with modern technology.[14]

Sendak's text also contains some commonplace assumptions about the function of wilderness as a space for recreation and play, and these assumptions again are to be found in wilderness legislation. Richard White points out that 'nature has become an arena for human play and leisure'[15] 'a paradise where we leave work behind'[16] and indeed Max, having established his supremacy over the wild things, joins them in the 'wild rumpus'. The association of wilderness with recreation, play and leisure is apparent in wilderness legislation, which states that wilderness areas must provide opportunities for both solitude and appropriate self-reliant recreation.[17] Yet again, the association between wilderness and play is problematic. The pressure to manage national parks as 'playgrounds rather than preserves'[18] reduces the conservation value of such areas and reinforces the artificial separation between wilderness and cultivated environments that we envisage as work environments. To conceive of nature and everyday work as quite separate is, according to White, to engage in 'self-deception';[19] this is, however, one of the messages in Sendak's book. Max's wild play at the beginning of the story is unsuited to his suburban home and Max finds refuge in a faraway wild land where his playful antics are condoned and encouraged.

A final salutary lesson which can be derived from Sendak's book is that wilderness and the wild are vulnerable to commodification in our modern Western world. Sendak's book itself markets and exploits the wild but even more so, the movie has and will create countless opportunities for profiting from 'wild' things and 'wild' places. We are all more or less aware of the extent to which nature as a commodity or as a 'consumable experience' can be brought and sold in the marketplace.[20] For instance, the legislative emphasis on tourism in national parks[21] represents a commodification of nature.[22] The use of special legislation by the New South Wales government[23] to facilitate the making of a feature film in a wilderness area lends further support to the argument that the law will and does promote commodification of the wild. A legislative framework for the commodification of wild things can also be found in biobanking provisions,[24] which enable developers to destroy habitat in areas earmarked for development if they protect off-site biodiversity. Proponents of biobanking describe this as 'harness[ing] the potential of the market' to achieve sustainability.[25] In fact, biodiversity offsetting fosters the illusion that we can buy and sell that which we do not truly own: the wild.

Wilderness law and its related offshoots, including national parks legislation and threatened species legislation, may not be the best place to find the seeds of Cormac Cullinan's wild law. Our current legislative regimes reinforce the notion of wilderness as pristine and separate from human lived-in environments. Somehow, we can reconcile our 'business as usual' approach with environmental priorities by protecting relatively small remote areas while carrying out environmentally destructive activities elsewhere. The challenge for wild law is to break down this distinction between the wild and the human environments. Yet there is little in existing environmental law to assist us.

Eco-Pragmatism and Wild Law

Environmental law generally does not resemble wild law and has been likened, instead, to development law.[26] How, then, do we find the seeds of wild law in a legal system generally dedicated to the protection of property rights? I am suggesting that wild law necessitates a lateral creative approach, an appropriation of existing legislation with all its flaws for purposes which are radically different from those for which it was designed; this is what Keith Hirokawa calls eco-pragmatism.[27] Legislators and the powerful lobby groups which dictate government policy may not have had wild intentions when they drafted such legislation. This does not, however, prevent the development of wild law from such unpromising raw materials.

Think, for instance, of damaged wilderness – almost an oxymoron if one associates wilderness with the pristine and inviolate. However, damaged wilderness can still become a haven for wild things. William Cronon notes that the one of the most toxic waste dumps in the United States, the Rocky Mountain Arsenal, has become, in the absence of humans, the habitat for large quantities of endangered species.[28] Probably the most compelling example of damaged wilderness is, however, Chernobyl. Mary Mycio admits that she envisaged the land around Chernobyl, which has been labelled by the Russian bureaucracy the Zone of Alienation, as a 'dead zone'.[29] Yet when she visited the area years after the nuclear accident, she found 'a unique, new ecosystem'.[30] In Chernobyl we find 'the paradox of man-made contamination';[31] a contaminated wilderness that has become a sanctuary for wild animals because humans can no longer live there.

I think that the paradox of damaged or unnatural wilderness, which provides a counterbalance to what Cronon has described as the extremism of the wilderness myth,[32] is also a useful metaphor when considering how a legal system which is intrinsically hostile to the wild can nevertheless be used for wild purposes. Keith Hirokawa has embraced the concept of eco-pragmatism as 'a promising negotiating strategy';[33] eco-pragmatism involves 'working within the established legal framework and using existing legal vocabulary'.[34] This strategy is effective, he believes, because of the way in which the common law evolves, as Ronald Dworkin's chain novel. Each author contributes to another chapter but must retain existing themes.[35] According to Hirokawa, it is the incremental changes which matter[36] and which remain far more influential than radical critique.[37]

I am suggesting that the seeds of wild law are to be found not in existing wilderness and threatened species legislation but in these small-scale incremental changes, as lawyers and judges re-interpret existing legislative provisions from a wild or playful[38] perspective. I can think of many examples of such re-interpretations, including my own argument that the constitutional provision which requires the payment of just terms compensation in the event of Commonwealth acquisition of property could be used by organic and/or conventional farmers to argue that the *Gene Techology Act 2000* (Cth) is invalid.[39] It might seem counter-intuitive to develop wild law through constitutional law and arguments based on property rights; at the time, I expressed some reservations about such a strategy.[40] Nevertheless, Carol Rose has written, somewhat tantalisingly for those of us in search for the wild in law, that 'even in the most tame, the most human-centred realm of property, one often catches a glimpse of wildness'.[41]

When I was developing my argument, I was influenced by the work of English historian, E.P. Thompson, who pointed out that eighteenth century legislation designed by the English ruling classes to protect their own property interests could also be used by members of the working class to protect their common property interests. Thompson observed that the ruling classes became 'prisoners of their own rhetoric';[42] it was 'inherent in the very nature of the medium which they had selected for their self-defence that it could not be reserved for the exclusive use only of their own class'.[43]

Of course, engaging with our anthropocentric, property-focused existing laws can be a dispiriting exercise for wild lawyers. Laurence Tribe, for example, has expressed concern that the process of translating a 'felt obligation' into the 'terminology of human self-interest'[44] may not only ultimately disempower the enthusiastic activist but might also serve 'to legitimate a system of discourse which so structures human thought and feeling so as to erode, over the long run, the very sense of obligation which provided the initial impetus for his own protective efforts'.[45] Wild lawyers must resist the cynicism and disenchantment which can be the end result of 'shoehorn[ing] their world views and values into an alien set of concepts and laws'.[46] Wild lawyers must be stubborn and tenacious; they need to focus on 'the big picture'.[47]

In the remainder of this article, I want to look at the eco-pragmatic characteristics of current climate change litigation and the ramifications of climate change for wild law.

Eco-Pragmatism at Work: Climate Change Litigation

Climate change is the formidable challenge which makes a mockery of our attempts to lock up and preserve the wild in carefully protected reserves: wilderness areas are as vulnerable to the devastating effects of climate change as are more degraded landscapes. I want to briefly look at contemporary climate change litigation because we can find in this litigation possibly the most compelling contemporary examples of eco-pragmatism at work. In Australia, there is no comprehensive national or state legislation on climate change. Thus it is fascinating to watch the evolution of Australian climate change litigation as scholars, lawyers and judges interpret existing legislative provisions in both creative and lateral ways. International climate change litigation is equally audacious in its re-reading of existing legal categories and authorities.

In Australia, although climate change lawsuits have not always been successful, cases like the first *Walker* case,[48] the *Anvil Hill* case,[49] and the *Northcape Properties* case[50] have imposed requirements on decision-makers to take climate change impacts into account in planning and development control decision-making.[51] However, climate change litigation is by no means confined to such case law. Brian Preston has acknowledged the enormous diversity of actions both nationally and globally,[52] as litigants draw on negligence, nuisance, conspiracy law, misrepresentation, trade practices law, constitutional law, international human rights law and international law principles. For instance, in the United States, industrial polluters have been and are being sued on the basis of negligence or nuisance.[53]

Such creative climate change litigation constitutes the sowing of the seeds of wild law.[54] It raises public awareness and thus, according to Judge Preston, provides a 'catalyst for executive action'.[55] It also highlights the opportunities in existing legislation, both national and international, for creative interpretation and ecocentric argument.

Conclusion

Max's encounter with the wild, as a place where he has an enjoyable romp with wild things before returning home to a hot dinner, can be seen as the archetypal Western experience of the wild. At the end of the book, Max is willingly subsumed into an orderly, rule-bound suburban existence from which all disorderly, unruly, wild elements have been expunged. Max, subdued, chastened and, according to Manderson, ready to become a rule-abiding legal subject, is no role model for a wild lawyer.

The task of the wild lawyer is to constantly interrogate the wild/civilised dichotomy which is exemplified in this children's book, to re-examine all legal rules and conventions with wide open, wild eyes. The wild does not only reside in remote and pristine places, and in unlikely monsters. It can be found and cultivated everywhere, even in the most conservative legal doctrines. Ultimately I would contend that it is through the ongoing subversive and playful re-interpretation of such doctrines, rather than by way of the legislation which purports to protect the wild, that wild law may one day take seed and flourish.

Notes

1   See, for instance, H.M. Babcock, 'Should *Lucas v. South Carolina Coastal Council* Protect Where the Wild Things Are? Of Beavers, Bob-o-Links, and Other Things that Go Bump in the Night', *Iowa University Iowa Law Review*, vol. 85, 2000, p. 849 and R. Meltz, 'Where the Wild Things Are: The *Endangered Species Act* and Private Property', *Environmental Law*, vol. 24, 1994, p. 369.

2   D. Manderson, 'From Hunger to Love. Myths of the Source, Interpretation, and Constitution of Law in Children's Literature', *Law and Literature*, vol. 15, 2003, p. 93.

3   M. Sendak, *Where the Wild Things Are*, HarperCollins, London, 1963.

4   Manderson, *From Hunger to Love*, p. 106.

5   See, for instance, *National Parks and Wildlife Act 1974* (NSW); *National Parks Act 1975*
    (Vic); *Nature Conservation Act 1992* (Qld); *National Parks and Reserves Management
    Act 2002* (Tas); *Reserves (National Parks and Conservation Parks) Act 2004* (WA);
    National Parks and Wildlife Act 1972 (South Australia); and *Territory Parks and Wildlife
    Conservation Act 2001* (NT).

6   *Wilderness Act 1987* (NSW); *Wilderness Protection Act 1992* (SA).

7   See, for instance, *Threatened Species Conservation Act 1995* (NSW); *Flora and Fauna
    Guarantee Act 1988* (Vic); *Nature Conservation Act 1992* (Qld); *Threatened Species
    Protection Act 1995* (Tas); *Wildlife Conservation Act 1950* (WA); *National Parks and
    Wildlife Act 1972* (South Australia); and *Territory Parks and Wildlife Conservation Act
    2001* (NT).

8   See, for instance, the Schedule of the *Nature Conservation Act 1992* (Qld) which identi-
    fies wilderness as an area that is, inter alia, remote at its core from points of mecha-
    nised access and other evidence of society.

9   W. Cronon, 'The Trouble With Wilderness' in *Uncommon Ground. Rethinking the
    Human Place in Nature*, ed W. Cronon, W.W. Norton and Company Inc., New York
    and London, 1996, p. 85.

10 ibid p. 86.

11 ibid p. 89.

12 M. Langton, 'What Do We Mean By Wilderness? Wilderness and *Terra Nullius* in
    Australian Art', *The Sydney Papers,* vol. 11, Summer 1996, p. 11 at p. 19.

13 Langton, 'What Do We Mean by Wilderness?', p. 30.

14 See, for instance, the *Wilderness Act 1987* (NSW), which identifies wilderness as areas
    which have 'not been substantially modified by humans and their works' (s. 6). Other
    legislation stresses the absence of modern technology and exotic organisms (*Wilderness
    Protection Act 1992* (SA) s. 3(2)) or the lack of disturbance by modern society (Schedule,
    Nature Conservation Act 1992 (Qu)).

15 R. White, "Are You An Environmentalist Or Do You Work For a Living?': Work and
    Nature' in *Uncommon Ground. Rethinking the Human Place in Nature*, ed W. Cronon,
    W.W. Norton and Company Inc., New York and London, 1996, p. 173.

16 ibid.

17 See, for instance, *Wilderness Act 1987* (NSW) ss. 6(1)(c), 9(c), *National Parks Act 1975*
    (Vic) s. 17A (3)(a) and *Nature Conservation Act 1992* (Qu) s. 24(c).

18 R. Buckley, 'The Weakness of Wilderness Protection Policy', *Environmental Policy and
    Law,* vol.30, no. 4, 2000, p. 196 at p. 197.

19 White, 'Are you an environmentalist or do you work for a living?', p. 185.

20 William Cronon, 'Introduction: In Search of Nature' in *Uncommon Ground.
    Rethinking the Human Place in Nature*, ed W. Cronon, W.W. Norton and Company
    Inc., New York and London, 1996, p. 46.

21 For instance, in the *National Parks and Wildlife Act 1974* (NSW), both conserva-
    tion and the 'fostering [of] public appreciation, understanding and enjoyment
    of nature and cultural heritage and their conservation' are primary objects of the
    legislation; the latter object is arguably best achieved through tourism in national

parks. Furthermore, changes to the legislation recommended by the NSW Taskforce on Tourism and National Parks and accepted by the government further promote tourism objectives; in fact, according to the National Parks Association, tourism has now become the major focus. See J. Huxley, 'Eco-tourism No Walk in the Park', *The Sydney Morning Herald* (Sydney), 23 June 2009; available at <http://www.smh.com.au/opinion/ecotourism-no-walk-in-the-park-20090622>.

22 See Langton, 'What Do We Mean by Wilderness?', p. 24.

23 The *Filming Approval Act 2004* was enacted by the New South Wales government after the Land and Environment Court had held that intensive filming of a commercial feature film in the Grose Wilderness in the Blue Mountains National Park was outside the scope of the objects and purposes of the *National Parks and Wildlife Act 1974*, as well as inconsistent with the management objectives of declared wilderness areas under the *Wilderness Act 1987*. See *Blue Mountains Conservation Society Inc v Director-General of National Parks and Wildlife and Ors* [2004] NSWLEC 196 (29 April 2004).

24 See, for instance, *Threatened Species Conservation Amendment (Biodiversity Banking) Act 2006* (NSW).

25 P. Curnow and L. Fitz-Gerald, 'Biobanking in New South Wales: Legal Issues in the Design and Implementation of a Biodiversity Offsets and Banking Scheme', *Environmental and Planning Law Journal*, vol. 23, 2006, p. 298.

26 G.W.G. Leanne, 'Environmental Law's Liberal Roots: (Not) A Green Paradigm' in *Green Paradigms and the Law*, ed N. Rogers, Southern Cross University Press, Lismore, 1998, p. 21.

27 K. Hirokawa, 'Some Pragmatic Observations About Radical Critique in Environmental Law', *Stanford Environmental Law Journal*, vol. 21, 2002, p. 225. The term was originally used by Daniel Farber in D. Farber, *Eco-pragmatism. Making Sensible Environmental Decisions in an Uncertain World*, University of Chicago Press, Chicago, 1999.

28 Cronon 'Introduction', pp. 27–8.

29 M. Mycio, *Wormwood Forest: a Natural History of Chernobyl*, Joseph Henry Press, Washington, 2005, p. 1.

30 ibid p. 2.

31 ibid.

32 This extremism is evident in the argument that 'if nature dies because we enter it, then the only way to save nature is to kill ourselves'; see Cronon, 'The trouble with wilderness', p. 83.

33 Hirokawira, 'Some Pragmatic Observations', p. 259.

34 ibid p. 265.

35 ibid pp. 260 – 1.

36 ibid pp. 276 – 7.

37 ibid pp. 227 -8.

38 By playful, I am referring to, as Maria Lugones puts it, 'a particular metaphysical attitude that does not expect the world to be neatly packaged': M. Lugones, 'Playfulness,

'world'-travelling and loving perception', *Hypatia,* vol. 2, 1987, p. 3 at p. 16. Playfulness suggests a delight in ambiguity and uncertainty and encompasses 'play[ing] with the frames of play' itself: see B. Sutton-Smith, *The Ambiguity of* Play, Harvard University Press, Cambridge Massachusetts and London, 1997, p. 148.

39   N. Rogers, 'Seeds, Weeds and Greed: An Analysis of the *Gene Technology Act* 2000 (Cth), its effects on property rights, and the legal and policy dimensions of a constitutional challenge', *Macquarie Law Journal,* vol. 2, 2002, p. 1.

40   ibid, p. 28.

41   C.M. Rose, 'Given-ness and Gift: Property and the Quest for Environmental Ethics', (1994) 24 *Environmental Law,* vol. 24, 1994, p. 1 at p. 30.

42   E.P. Thompson, *Whigs and Hunters. The Origins of the Black Act,* Pantheon Books, New York,

1975, p. 263.

43   ibid p. 264.

44   L.H. Tribe, 'Ways Not to Think About Plastic Trees: New Foundations for Environmental Law', *Yale Law Journal,* vol. 83, 1974, p. 1315 at p. 1330.

45   ibid at p. 1331.

46   This quotation is from an article by Naomi Roht-Arriza, who was writing about the predicament of indigenous people who are confronting the appropriation or theft of their traditional knowledge; see N. Roht-Arrioza, 'Of Seeds and Shamans: The Appropriation of the Scientific and Technical Knowledge of Indigenous and Local Communities', *Michigan Journal of International Law,* vol. 17, 1996, p. 919 at pp. 956 – 7.

47   E. Rivers, 'How to Become a Wild Lawyer', *Environmental Law and Management,* vol. 18, 2006, p. 28 at p. 28.

48   *Walker v the Minister for Planning and Ors* [2007] NSWLEC 741. This decision was, however, overturned on appeal.

49   *Gray v Minister for Planning and Ors* (2006) 152 LGERA 258.

50   *Northcape Properties Pty Ltd v District Council of Yorke Peninsula* [2008] SASC 57.

51   See also *Re Australian Conservation Foundation v Latrobe City Council* (2004) 140 LGERA 100; *Charles Howard Pty Ltd v Redland Shire Council* [2007] QCA 200.

52   B. Preston, 'Climate change litigation', *Environmental and Planning Law Journal,* vol. 26, 2009, p. 169.

53   For instance, the native Inupiat village of Kivalina, which will have to be relocated as a consequence of climate change erosion, is currently seeking damages from nine oil companies, fourteen power companies and a coal company, arguing public nuisance and conspiracy; ibid at p. 171.

54   Cormac Cullinan has used the metaphor of sowing the seeds of wild law in 'Sowing Wild Law', *Environmental Law and Management,* vol. 19, 2007, p. 71.

55   Preston, 'Climate Change Litigation', p. 189.

Wild Law: Is There Any Evidence of Earth Jurisprudence in Existing Law?

Begonia Filgueira & Ian Mason

In 2008 we set ourselves the task of testing the hypothesis that there was existing legislation already consistent with Wild Law principles.[1] The result was the report on which this paper is based.[2] So far no one had established criteria for creating wild law or for testing whether a law was valid following Earth Jurisprudential principles, thus this was our first challenge.[3]

After much discussion, research and creative thinking it was decided that wild laws would have the following three main characteristics: Wild laws would centre governance on the Earth, they would provide mutually enhancing relations to all members of the Earth Community and they would be based on community ecological governance. As we needed to be more specific to pinpoint down exactly what each of these indicators meant from a legal perspective, we decided on the following three indicators and a number of sub-indicators:

1. Earth-Centred Governance:
    *   Respect for the intrinsic value of Earth and all its members/components;
    *   Dominant rationale behind the law environmental protection;
    *   Governance informed by the laws of nature;
    *   Respect for the 3 key Earth Rights of an Earth Community member.
2. Mutually Enhancing Relations to promote the well-being of the whole Earth Community:
    *   Recognition of interconnectedness between members/components of the Earth Community;
    *   Reciprocity;
    *   Conflict resolution mechanism for interests/rights of humans and those of non-human members for the wellbeing of the whole Earth Community (procedural and substantive);
    *   Restorative mechanism/process to (re)establish mutually enhancing relations for the wellbeing of the whole Earth Community;
    *   Adaptive mechanism/process in light of evolving challenges to pursue mutually enhancing relations.

3. Community Ecological Governance (CEG):
    *   Participation of all members of the Earth Community in ecological governance;
    *   Legal recognition of 3 key rights of public participation:
        *   Access to information
        *   Public participation in decision-making
        *   Right to access to justice;
    *   Respect of other key issues of CEG.

Methodology

The next step was to choose the laws we would analyse by giving them scores for their wildness when tested against the indicators. A team of researchers was recruited and allocated to a number of regions across the globe. Regional supervisors were largely chosen on the basis that the authors had contacts with people with experience in Earth Jurisprudence in the relevant jurisdictions. Best examples of wild law (cases, statutes, regulations and constitutional instruments) from various parts of the world were then recommended by the regional supervisors. We analysed laws from the European Union, South Africa and Ethiopia, India, New Zealand, the United States, Ecuador and Colombia.

The report evaluated laws applying to mountains, forests and endangered species. Initially laws on sacred sites were also to be included but given time, resources and the fact that we did not find any laws specifically on sacred sites sadly meant that these were left out. This paper does not reproduce all of the results of the report. It focuses on drilling down on the indicators to get a new and closer understanding of Wild Law and on reviewing practical examples of the indicators in some of the laws analysed in the report.

Earth Jurisprudence and the law

Although the term Earth Jurisprudence was not coined until Thomas Berry published the *Great Work*[4] in 1999, in his 1972 essay, *Should Trees Have Standing?*[5], Professor Christopher Stone was already thinking along the same lines: could a non-human entity such as a tree be given legal personality and legal rights as corporations are?

This argument was also raised, but rejected, in the U.S. Supreme Court's *Sierra Club v Morton*[6] where Justice William Douglas dissented:

> The critical question of 'standing' would be simplified and also put neatly in focus if we fashioned a federal rule that allowed environmental issues to be litigated before federal agencies or federal courts in the name of the inanimate object about to be despoiled, defaced, or invaded by roads and bulldozers and where injury is the subject of public outrage ... This suit would therefore be more properly labelled as Mineral King v. Morton.

Inanimate objects are sometimes parties in litigation. A ship has a legal personality, a fiction found useful for maritime purposes. The corporation sole – a creature of ecclesiastical law – is an acceptable adversary and large fortunes ride on its cases. The ordinary corporation is a 'person' for purposes of the adjudicatory processes, whether it represents proprietary, spiritual, aesthetic, or charitable causes.

Thomas Berry called for a new *rights-based* relationship between the human and the Earth Community. A relationship where the world did not exist only for humans to covet and use, where a sustainable future could be achieved but only if the inherent rights of the natural world were given legal status.[7] To coincide with the 2002 Earth Summit, Cormac Cullinan developed these ideas in his book, *Wild Law*.[8] As a lawyer he wanted to formalise Earth Jurisprudence in a way that could be incorporated into modern legal theory. Laws based on Earth Jurisprudence would be wild laws – wild not because they were irrational or out of control, but wild because they derive from nature free from human interference. Cormac Cullinan argued that:

> [I]n order to change completely the purpose of our governance systems, we must develop coherent new theories or philosophies of governance ('Earth Jurisprudence') to supplant the old. The Earth Jurisprudence is needed to guide the re-alignment of human governance systems with the fundamental principles of how the universe functions … Giving effect to Earth Jurisprudence and bringing about systemic changes in human governance systems will also require the conscious fostering of wild law.[9]

The essential concept of Earth Jurisprudence is that the universe itself is the source of law because it is where all activity takes place. We are also not alone but are part of nature, where there is an intimate connection between every being and the universe, a connection that determines time scales, life spans, seasons and temperature ranges and provides all of the elements on which all creatures, animate and inanimate, depend and from which they are formed Thus human laws which are aimed at living in harmony with the universe should be designed to correspond with the universe or universal laws. Laws constructed in this way would create the basis for a mutually enhancing relationship between humans and all other members of the Earth Community.

There are two ways of studying our relationship with the Earth. One is the more conventional, objective, scientific approach of measuring, observing, verifying and recording based on Western philosophy and methodology. The other is a more intimate and sometimes intuitive experiential mode of connecting with the natural world and understanding it from within. This later mode of learning is more common among indigenous peoples whose own life is much closer to the natural world and whose laws appear more as lore and custom than as formulated regulations.

Earth Jurisprudence aims to draw on the best of both methods to forge a new understanding of law, constitutionality and lawfulness that is conducive to establishing and maintaining a naturally mutually-enhancing relationship between the human and all other species. Earth Jurisprudence therefore draws its principles primarily from experience of that relationship, adapting the methodology of formal constitutions and law-making to comply with those principles.

This assumes that the natural relationship between the human presence and the natural environment has been disturbed. While this is not the place to consider the evidence for that, the fact is that existing legal systems have been unable to prevent or mitigate widespread loss of biodiversity; environmental pollution; de-forestation; climate change; and the whole related range of human degradation of the planet. Legal systems commonly treat the Earth as a *resource* and value it only as such when in fact it is the organism that sustains all forms of life. For present purposes we define 'Earth Jurisprudence' as: *the philosophy of law and regulation that gives formal recognition to the reciprocal relationship between humans and the rest of nature.*

Justification for granting nature rights

As lawyers we tend to focus on legal instruments and as Earth Jurisprudence clearly supports granting rights to nature, the legal reason for granting rights to nature was paramount in our mind. The reason for granting rights to nature is to protect nature, but Earth Jurisprudence argues that there is a deeper, more philosophical basis for granting these rights. The reason is that nature deserves to be valued for itself, because it is and exists, not because of the value we place on it as a resource. Humans are not granted better or more rights because they contribute more or less to society. They are granted rights in recognition of their existence, equality and for their protection. The Declaration of Human Rights was borne out of the crimes against humanity during the Second World War. The Earth's balance is now in peril, in need of protection, and here is where rights can give a greater equilibrium to the relationship between man and nature.

Given the differences between humans and nature, granting rights to nature cannot just be an extension of what we understand to be human rights. The concept is more akin to the neighbour principle[10] and the duty of care well known in common law systems. The question is who is the neighbour of the human race? Earth Jurisprudence argues that we are all interdependent and interconnected, thus our neighbours include not only our human neighbours but everything that is around us. We are part of an Earth Community made up of humans, animals, plants, insects, water, bacteria and all that surrounds us.

Another reason for granting rights to nature is because it also would protect humans. Our interconnectedness with nature means that we depend on nature; if nature is in peril so is our survival.

The indicators in theory and in practice

Earth-centred governance in theory

The greatest difference between modern jurisprudence and Earth Jurisprudence is the place where it is centred. Modern jurisprudence is anthropocentric, the assumption being that all laws are made entirely for the benefit of human beings. This attitude is exemplified by concepts such as human rights, public benefit and private ownership, which do not take into account the other-than-human. Even environmental protection is frequently for the purpose of enabling some human scheme to continue and is not for the protection of nature for its own sake.

Laws that centre their governance on the Earth would derive their respect for the Earth from the Earth's own value, not for what the Earth can do for humans.[11] Their dominant rationale should be protection of the environment, and thus maintaining ecosystems in balance rather than tipping the balance in favour of humans.[12] Further, these laws must be informed by the laws of nature, specifically on ecological criteria including life cycles, diversity and ecological limits. Finally this Earth-centred law must respect the Earth Rights of all the members of the Earth Community, and in defining these Earth Rights we take Thomas Berry's lead:

> the right to be, the right to habitat, and the right to fulfil its role in the ever-renewing process of the Earth Community.[13]

The right to be would be akin to the human right to life, e.g. freedom from disturbance during reproductive and migratory cycles. The right to habitat gives a right to basic conditions of wellbeing, such as a prohibition to pollute a river as it destroys a habitat. Thirdly the right to fulfil its role in the Earth Community would include a right to contribute to the cycle of life, such as creating the right conditions for bees to pollinate.

Earth-centred governance in practice

The Indian case *K.M. Chinnappa v. Union of India*[14] provides a really good example of earth-centred governance. It dealt with a dispute over the extension of a mining lease granted to Kudremukh Iron Ore Co. Ltd (Iron Ore Co) who had been carrying out open cast iron ore mining in the Kudremukh National Park for the last 30 years. This park was now a hotspot for biodiversity due to the harm caused by mining waste. The extension of the lease was not granted by the relevant State Committee and the Court affirmed the validity of the Committee's decision.

Chinnappa endorsed the World Charter for Nature's[15] statement that 'mankind is part of nature and life depends on uninterrupted functioning of natural systems' and the Supreme Court fully recognised that humans are part of a greater community. The Court went so far as to say that by 'destroying nature man is committing matricide'[16] thus declaring that man comes from nature, is part of nature and has a duty of care to protect the environment.

New Zealands' National Parks Act[17] is also well earth-centred. It is founded upon the principle that National Parks will be protected in perpetuity for their 'intrinsic worth' in the national interest.[18] However this purpose is also to be balanced with the public's use and enjoyment of the park. Some parts of parks may be set aside for conservation only[19] and other parts may be set aside for public amenities.[20] Governance is informed by the laws of nature to the extent that the law mandates that parks be preserved as far as possible 'in their natural state'[21] and be administered for their soil, water and forest conservation value[22].

The Ecuadorian Constitution goes the furthest by granting rights to nature by stating that:

> Communities and ecosystems possess the unalienable right to exist, flourish and evolve within Ecuador. Those rights shall be self-executing, and it shall be the duty and right of all Ecuadorian governments, communities, and individuals to enforce those rights.

On the other hand *The South African National Environmental Management Act 107 of 1998* is primarily aimed at the management of the environment in the human interest without even considering the value of nature per se:

> Environmental management must place people and their needs at the forefront of its concern, and serve their physical, psychological, developmental, cultural and social interests equitably.[23]

In India, the main purpose of the *Plant Varieties and Farmers Rights Act 2001* is to protect the rights of farmers and their intellectual property rights over local plants and seeds. In terms of Earth-centred governance this Act scored very poorly as environmental protection features only in the conservation of plant varieties likely to become extinct.[24] The laws of nature are only considered in so far as plants are seen as distinct from each other for registration purposes[25] and the acknowledgment that overexploitation by commercial agriculture can potentially destroy plant varieties.

Mutually enhancing relations in theory

As humans are part of the Earth and dependant on nature we have a responsibility to ensure that we have a sympathetic arrangement with the natural process, an arrangement that allows both humans and other members of the Earth Community to thrive. In order for this to occur laws would have to recognise the interconnectedness and reciprocity of all things on Earth. Humans would need not only to recognise in laws that they are dependent on the planet to survive but that they are interdependent. James Lovelock ponders on our interdependency in his book *The Revenge of Gaia* when asking why we pee:

urea is waste for us and wasting it loses valuable water and energy. But if we and other animals did not pee and breathed out nitrogen instead, there might be fewer plants and later we would be hungry. How on Earth did we evolve to be so altruistic and have such enlightened self interest? Perhaps there is wisdom in the workings of Gaia and the way she interprets the selfish gene.[26]

Wild law's conflict resolution mechanisms would need to recognise the need for restorative justice rather than focus exclusively on punishment and retribution. These mutually enhancing laws would have to be flexible and have an in-built adaptive mechanism to provide for change when relationships are not longer mutually enhancing.

Mutually enhancing relations in practice

Again *Chinnappa* presents a good example of the law supporting mutually enhancing relationships. The court did not only require the mining operations to cease but supported a number of measures which are key elements of restorative justice. The Iron Ore Co was to make funds available for the government agencies to implement a plan of rehabilitation, reclamation and restoration of the mined area prepared by the appropriate government agency. Further, it was required to pay yearly monetary compensation to be used for research, monitoring and strengthening the protection of the Kudremukh National Park and other protected areas of the Kudremukh State and to wind up and transfer all the buildings and other infrastructure to the Forest Department of the State of Karnataka at book value[27].

The South African Constitution[28] on the other hand, although recognising the effect of the environment on human health and well-being,[29] contains no corresponding recognition of the effect of human activities on the environment and other members of the Earth community let alone any reciprocal obligations or enforceable rights. There could be an opening in s. 38[30] for arguing Earth Jurisprudence principles under the umbrella of *public interest* but the possibilities are limited by the obvious view that *persons* empowered to approach the court are human or juridical persons and do not include non-human members of the Earth community (although they do include non-human members of the human community such as corporations and associations). There is no requirement to establish or maintain mutually enhancing relations and there is no mechanism for adapting to changing conditions or challenges.

Community ecological governance in theory

In order for laws to be wild law they would have to be formulated after having listened to nature, after giving nature a voice, rather than only be based on an understanding of the effect a law will have on human possessions or human life. Indigenous people who live closer to the land are often better at this process than urban dwellers. However, many laws today are made in offices far away from the area that they affect, by those who do not understand local realities.

Ecological community law requires not only interest-led consultation but also local consultation and consultation with nature and its laws. Wild laws are impregnated with local lore and its understanding of nature. Consequently laws must respect traditional knowledge, cultural heritage, benefit sharing, human rights, nature rights, self-determination and community land rights. These laws also necessarily recognise a right to access to information, public participation in decision-making and a right to access to justice.

Community ecological governance in practice

The South African National Environmental Management Act[31] provides a good example of community ecological governance as it requires participation in the decision-making process of *all interested and affected parties in environmental governance*[32] with particular regard being given to traditional (indigenous) communities, women and youth, a provision which may give scope for the interests of nature as well. There is extensive provision for access to information[33] and protection for whistle-blowers that act in good faith.[34] Equitable access to the environment for all is required[35] and traditional knowledge is to be taken into consideration in the decision-making process[36].

New Zealand's *National Parks Act* and its inclusion of the Treaty of Waitangi[37] principles – which include active protection of Maori interests and values such as traditional knowledge and practices and interests in or associations with land – makes it score high in respect of community ecological governance.

In the Indian *Plant Varieties and Farmers Rights Act 2001*[38] there is no mention of the Earth Community. Access to information is limited to requiring the publication of applications to register a variety of plants and for opening the register to public inspection.[39] Access to justice is concerned with unsuccessful applications for registering plants.

The EU's *Habitat Directive*[40] is the result of a top-down decision-making process between the Council (Member States), the Commission and the European Parliament, far removed from those it directly affects and thus is not seen as a good example of community ecological governance. There is no provision for indigenous communities[41] or non-human communities to participate in the designation process, and very little for the wider public anywhere in the Directive. Even when it may be appropriate, there is no requirement to heed public opinion once obtained. The public has no rights of access to information other than to see the six-yearly reports of national governments. There are no direct means by which the public can engage in enforcing or giving effect to the Directive and little or no recognition of the significance of traditional knowledge, customary practices and co-management.

Conclusion

Although there were many elements of wild law in the laws we reviewed, we did not find a coherent wild law approach in existing regulation. There were many instances of recognition of the interdependence of humans and the Earth but in very few instances was the Earth valued as anything else than a resource. Laws were clearly anthropocentric and listening to the needs of nature or granting rights to nature (except in the case of the Ecuadorian Constitution) was non-existent. Protection was granted in extreme specific cases where species faced extinction and thus interfered with human rights and needs. Public participation in decision-making was limited and not underlined by a true respect for nature. Indigenous communities were seen as minorities who needed protection and whose views were, in few cases, taken into account.

However, we did find positive elements of wild law in 17 of the 24 laws analysed and thus suggest that moving on to a deeper Earth Jurisprudence may not be impossible. The main hurdle appeared to be the conflict between human interests and those of nature, which in some way negates much of the acknowledgment of interdependency that we found. But this is precisely where we need a shift in philosophy. The relationship between man and the Earth needs to be seen not only as one of conflict, but as a relationship that is mutually enhancing for all parties concerned. The difficulty here is that from an age of abundance where mutual respect may have been easier to achieve, climate change is now making the Earth become more hostile to human life and the balance may be tipping in the opposite direction. In itself this provides humans with sufficient motivation to rebalance this relationship, but will the world order want to change? Following the Copenhagen negotiations on climate change it could be argued that for most governments development and trade are more important than harmony and survival.

Practical suggestions

This paper and the report on which it is based do not have all the answers. Still, we wanted to come up with some practical suggestions that in our view would make laws wilder:

- Look at the Earth as a living being and stop using the word 'resource' when we speak about nature, as it implies that we value the Earth for its economic value only.
- Redefine public interest to include the interests of the other members of the Earth community.
- Redefine sustainability, as development[42] can no longer be part of its definition.
- Promote the enjoyment of nature – it should be the right of every human to have access to nature; the more we know about nature the more we will value it for itself – schools can have a large role to play.

- The right to protect and respect the environment should be part of the right to life which most constitutions recognise, thus acknowledging the interdependency of human life and the rest of nature. This may be more palatable to governments than granting rights to all members of the Earth community.
- Support the call for a binding Universal Declaration of the Rights of Nature. Greater balance between humans and the other members of the Earth community can be achieved by granting rights to all members of the Earth community. Where there is a dispute someone will be able to represent those other members and a balance can be achieved on an individual basis in a court of law.
- Support the call for Nature's Rights Acts at national level akin to the Human Rights Act in the UK.
- Educate judges, lawyers and environmental professionals about the need to promote the interests of nature, environmental challenges we face and ancient societies' relationship with the Earth, and how to integrate these elements into the decisions they make in their professional capacity. They will then be better able to make a judgment when trying to balance interests and they may engage emotionally with the subject.
- Promote the use of intuition as a valuable resource – we are part of nature and have something embedded within us that may allow us to make the right choice.

Finally, since the report was written in 2008 there has been major progress in the Earth Jurisprudence debate. We now have a Mother Earth Day[43] on 22 April when we can all formally engage with the Earth with respect and at an emotional and interdependent level. We also soon may understand governments' position on granting rights to the Earth at an international level. The Bolivian government tabled a draft resolution[44] at the UN titled 'Harmony with Mother Earth' which asked of Member States, relevant United Nations and other organisations to present their views to the Secretary-General on a *possible declaration of ethical principles and values for living in harmony with Mother Earth*.[45] The resolution was passed and we expect a report to be presented to the UN's General Assembly 65th session.

Earth Jurisprudence is not a fanciful philosophy, it is penetrating more and more minds and we would encourage all readers to support and progress its understanding and dissemination.

Notes

1   This discussion was started at the fourth Wild Law UK Conference.
2   Begonia Filgueira, Ian Mason et al, *Wild Law: Is There Any Evidence of Earth Jurisprudence in Existing Law and Practice*, UKELA and the Gaia Foundation, 2008.

3 Thanks go to all the researchers and supervisors for their invaluable contribution to the report, with special thanks to Professor Lynda Warren for supervising the project, Carine Nadal of the Gaia Foundation and Peter Kellett of UKELA for their work on the indicators, and Mary Munson of the Center for Earth Jurisprudence for her work on the USA sections.

4 T. Berry. *The Great Work: Our Way into the Future*, Bell Tower, New York, 1999.

5 C.D. Stone. *Should Trees have Standing? And other Essays on Law, Morals, and the Environment*, Oceana Publications Inc, New York, 1996.

6 Sierra Club v Morton 405 US 727

7 Berry, *The Great Work*, p. 61.

8 C. Cullinan, *Wild Law: A Manifesto for Earth Justice,* Green Books, Devon, 2002.

9 Ibid., p. 30.

10 *Donoghue v Stevenson* [1932] AC 562

11 There is inspiration in the World Charter for this as it declares that every form of life is unique irrespective of its value to humans.

12 The report at p. 5 states that this indicator is consistent *'with maintaining the natural environment as far as possible in its natural state'*. However the authors have now revised this to mean maintaining an equilibrium or living in harmony with the natural environment as it is not clear what 'a 'natural state' is,' and because humans as part of nature will have a measure of impact on Earth.

13 T. Berry, *Evening Thoughts*, Sierra Club Books, San Francisco, 2006, p. 10.

14 *K.M. Chinnappa v. Union of India*, WP 202/1995 (30.10.2002) AIR 2003 SC 724, 2003.

15 The World Charter for Nature adopted by the United Nations General Assembly on 29 October 1982.

16 *K.M. Chinnappa v. Union of India,* para. 1.

17 National Parks Act (New Zealand) no 66 Public Act, 17 December 1980.

18 Section 4, National Parks Act (New Zealand).

19 Sections 12, 13 and 14, National Parks Act (New Zealand).

20 Section 15, National Parks Act (New Zealand).

21 Section 4(2) (a), National Parks Act (New Zealand).

22 Section 4(2)(d)

23 Section 2(2) – – This prompted our researcher to comment that the title of the Act should really be the National Human Management Act!

24 Section 29.

25 Preamble of the Protection of Plant Varieties and Farmer's Rights Act 2001.

26 J. Lovelock, *The Revenge of Gaia*, Penguin Books, New York, 2007, p. 25.

27 *K.M. Chinnappa v. Union of India,* para. 7.

28 South African Constitution, Act 108 of 1996.

29 Section 24 (a), South African Constitution.

30 Section 38 – Enforcement of Rights, South African Constitution.

31 Act 107 of 1998.

32 Section 2(4) (f), The South African National Environmental Management Act.

33 Section 31, The South African National Environmental Management Act.

34 Section 31(4), The South African National Environmental Management Act.

35 Section 2(4)(d), The South African National Environmental Management Act.

36 Section 2(4)(g), The South African National Environmental Management Act.

37 The Treaty of Waitangi is the founding document of New Zealand and is the record of the agreement made between the British Crown and the Maori chiefs or rangatira. Under the Treaty the Maoris agreed to cede sovereignty to the British Crown which was given exclusive rights to purchase any lands that the Maoris wished to sell. In return, the Maoris were given full rights of ownership of land, forests, fisheries etc. The Maoris were also granted the full rights and protection of British citizens. For further information see www.nzhistory.net.nz.

38 Act 53 of 2001.

39 Sections 21 and 84, Plant Varieties and Farmers Rights Act.

40 EC Council Directive 92/43/EEC of 21 May 1992 on the conservation of natural habitats and of wild flora and fauna, OJ L 206 22.7.92, p. 7.

41 The authors do query what indigenous community means when applied to Europe.

42 At least unless the word 'development' is carefully defined so as not to imply never ending economic growth.

43 UN GA/10823, 22 April 2009.

44 This draft resolution was co-sponsored by Algeria, Benin, Belarus, Bosnia and Herzegovina, Cape Verde, Cuba, Dominica, Ecuador, Eritrea, Georgia, Guatemala, Honduras, Mauritius, Nepal, Nicaragua, Paraguay, Saint Vincent and the Grenadines, Saint Lucia, Seychelles and Venezuela.

45 UN GA/10823, 22 April 2009.

From Reductionist Environmental Law to Sustainability Law

Klaus Bosselmann

This chapter explores some of the deeper characteristics of the failure of law in general and environmental law in particular. The thesis is that this failure can be overcome by incrementally incorporating sustainability into the interpretation of existing laws and the design of new laws.

To summarise the argument, environmental law is hampered by a reductionist approach to its subject, i.e. the environment or more precisely, the human − nature relationship. This relationship is misconceived through the domination of certain philosophical and cultural traditions in European history. As a consequence, modern legislation to protect the natural environment has developed in a compartmentalised, fragmented, economist and anthropocentric manner. For environmental legislation to become effective, therefore, broader coverage and better enforcement are not enough. The inherent design flaw is the absence of a fundamental rule to not harm the integrity of ecosystems. Such a rule requires the acceptance of sustainability as an overarching legal principle.

The Flawed Design of Environmental Law and Policy

Practitioners, politicians and lawmakers are not normally concerned with the philosophical foundations of environmental law. For the most part, they see environmental issues as factual items to be dealt with in a practical, 'rational' manner. Statutes and cases may raise some deeper questions of interpretation and application, but those questions, it is felt, can and must be resolved with legal argument, not ethical reasoning.

The same is true for legal education. Law students are hardly exposed to environmental and philosophical concepts. They can achieve their degrees without any training in environmental law, let alone training in legal reasoning within a discernible ethical context.

On the other hand, students of environmental law are increasingly confronted with ethical and philosophical questions. Many of today's textbooks of environmental law contain introductory chapters on theoretical foundations including ethics.[1]

The question might be asked why lawyers should be concerned with the deeper dimensions of legal constructs when there a more competent experts like philosophers, anthropologists, theologians and so on. But are they? Arguably, the issues involved here are still at the periphery of academic training and scholarship. The very fact that there are environmental philosophers, environmental economists, environmental lawyers and other 'specialists' is a manifestation of reductionism in academia.[2] Environmental lawyers have nowhere else to go but to recognise their subject as interdisciplinary. That such recognition and expertise hardly exists also explains, to a certain degree, the flawed design of environmental laws.

The size, severity and urgency of the global ecological crisis makes a deeper examination of the prevailing values and worldviews necessary and worthwhile. We need to address such questions as: What is the value of nature to humanity and how does it relate to our health, culture and social well-being?; How did we come to this state of mind and affairs, and whose interests are being served by maintaining the status quo?; Why do most people in modern society fail to relate personally or collectively with this unfolding human and environmental tragedy? This is not the place to attempt some answers,[3] but a main reason why current policies and laws are not getting to the bottom of the crisis is the dominance of economic rationality.[4]

Economic rationality has shaped both the capitalist world and the socialist world. Neither capitalist USA nor communist China have ever developed a philosophy of long-term sustainable economies, and this should not surprise us. Economic rationality has its roots in the project of Modernity or, as German philosopher Rudolf Bahro termed it, the 'European cosmology'.[5] This can be summarised in a few characteristics, namely dualism (of humans and nature), anthropocentrism, materialism, atomism, greed (individualism gone mad) and economism (the myth of no boundaries and limitless opportunities).

These six characteristics, *DAMAGE* for short[6], have resulted in what I call environmental reductionism.[7] It consists of a compartmentalised, fragmented and anthropocentric idea of the environment.

To get an idea of the compartmentalisation of the environment, we only have to consider how the media neatly separate environmental issues from the 'rest', in particular financial news and economic growth stories. This is mirrored by government structures with their specific portfolios.

While compartmentalisation refers to the boxing-off of the environment from other policy areas, fragmentation refers to the environment itself. The issue here is that lawmakers do not think of the environment as a whole, but more as a generic word for natural resources (an economic term). Most laws have natural resources as their subject, not the natural world. As a result of this fragmented view, environmental security is only partially noticed and only in competition with economic objectives. This is ecological nonsense of course. Imagine a child protection law that says: 'Do not beat your child too often and too much'. Environmental law does just that: 'Do not pollute the environment too often and too much'. The flawed thinking behind such reductionism is the assumption that the environmental crisis can be solved within the current economic, political and legal system and without challenging underlying values.

Finally, the anthropocentric-utilitarian approach to the environment. Most legal definitions emphasise the utility value of the environment. Typically, this includes natural resources and amenity values such as recreation and beauty. The health of ecosystems is of little concern.

Unravelling the Problem

During the 1980s, a new concept emerged that supposed to shift the environment from the periphery into the centre of public decision-making. The concept was called sustainable development. The challenge here was and is to define what exactly needs to be sustained.

Sustainable development is, of course, nowhere defined. At present, neither international law nor domestic laws prescribe a way to resolve conflicting priorities. Rather sustainable development is accomplished when everything is taken care of – the environment is protected, the economy is developed and social equity is achieved. Such view assumes that the three objectives are not really in conflict with each other and can be achieved simultaneously. Sustainable development resists definition and avoids the hard questions, which is precisely why it has become so popular among governments.

What is missing is an environmental absolute. Law and society are not principally opposed to absolutes. Think, for example, of criminal law protecting life, health, well-being, integrity, freedom, property. They are absolutes not to be compromised. You get punished for harming people or property, but not for harming the environment. Specific laws may prohibit incidents like felling a tree, killing an animal without reason,[8] building a house without resource consent, or uncontrolled discharge of waste, wastewater or chemicals. But these are exceptions from the individual right to use natural resources. In their accumulation, these user rights amount to large-scale destruction of the global environment. There is no general environmental rule to qualify individual entitlements.

It is, however, quite possible to fashion a general rule, one that draws a line in the sand and sets a bottom-line limitation. This rule would apply throughout the system of law and governance and is not confined to a single act.

The Principle of Sustainability

A recent book of mine[9] describes such a rule and finds it in the principle of sustainability. If we trace the historical and philosophical foundations of sustainability from their beginnings and relate them to the corresponding developments of legal theory and practice,[10] we can see an ever-increasing gap between individual entitlement and responsibility for the commons. This can best be observed in the way the Lockean idea of property rights emerged and how the modern concept of private property isolated itself from any common property responsibilities.[11]

On the other hand, there is a wealth of sustainability wisdom in the history of all cultures. European culture is no exception. The experience of pre-industrial Europe, in particular, is worth noting. By the mid 1800s Europe's forests had all but gone. Deforestation had reached a degree that threatened the entire economy of Europe.[12] This opened up two possibilities for the future. One was to look for a new energy source to fuel the economy, the other to look for an alternative economy. Of course, coal replaced wood and fired up the industrial revolution. But the alternative was available too, i.e. the rediscovery of sustainability.

Forest management scholars in Germany proclaimed the wisdom of replacing every tree felled with planting a new one. They cited the medieval land use system (*'Allmende'*) as the mother of sustainable economies. The *Allmende* system recognised public ownership of the land to guide any form of private land use. That way the substance of the land could be protected from overuse, thereby preserved for future generations. In 1714, this effect was termed *'Nachhaltigkeit'* by German accountant and administrator Hans Carl von Carlowitz.[13] The term and concept eventually dominated forest economic theory and were exported, for example, to the French Forest Academy where, in 1837, its director Adolphe Parade translated it to 'soutenir' (showing its Latin roots: 'sustinere' = to keep, preserve, sustain). From there it reached the English translation of sustainability. By the mid 1800s the notion 'living from the yield, not from the substance' was widespread among forest academies and indeed science faculties throughout Europe. It was state-of-the-art knowledge!

The fact that the industrial revolution ignored this knowledge does not render it useless, obviously. It only meant that the idea of sustainability did not fit the all-pervasive idea of progress. Essentially, this has not changed until today – except for the fact that the case for sustainability has never been stronger.

The modern chapter of the sustainability discourse began with the Report of the UN Commission for Environment and Development[14] (*Brundtland Report*) that created the composite term 'sustainable development', but did so – or should have done so – on the basis of a well-established history of the sustainability concept. The famous Brundtland definition[15] is, of course, incomplete. It leaves the question open what the needs of future generations might be and consequently what may have to be passed on. It is fair to assume that the Brundtland Commission called for a fundamental duty to keep the basic options open for future generations. The only way to keep these options open, however, is to sustain the ecological basis of development. The Commission was quite clear about this. Already the inaugural meeting of the Commission in October 1984 set out the objective 'to build a future which is more prosperous, more just, and more secure because it rests on policies and practices that serve to expand and sustain the ecological basis of development'.[16] In many passages the report emphasised that we are borrowing the environmental capital from future generations and that economic growth must be constrained to preserve the Earth's ecological integrity.[17] In the literature this interpretation is often referred to as the 'strong' approach to sustainable development.[18]

The alternative so-called 'weak' approach, favoured by governments and corporations, considers ecological sustainability as one concern next to social and economic 'sustainability'. To consider the ecological, economic and social dimensions of development simultaneously is important, as mentioned, but this in itself does not make development 'sustainable'. The sustainability of ecological systems must be the bottom-line, yardstick and benchmark. History, science and ethics all point to the same, rather simple idea: any form of development must respect ecological boundaries to avoid decline or collapse. This idea is so prevalent that we may consider it common knowledge of humanity. That modern society has ignored this common knowledge cannot change its truth. Sustainability certainly deserves a status similar to other guiding ideas: freedom, equity and justice.

Looking over 20 years of international law and policy, there has been increased recognition of the sustainability principle, but equally an increased gap between soft law development and hard law. Soft law represents a consensus of the international community of states that is considered legally relevant, although not binding. It must not be forgotten though that international environmental law has emerged through the pressures of the environmental movement, to which states only responded. This gives the institutions and groups of civil society special importance. Like human rights, environmental rights are being pushed by civil society, rather than governments and the same is true for the overarching principle of sustainability.

It is not surprising, therefore, that the essentials of ecological sustainability have only ever been expressed in soft law documents. Here are just two samples:

> Mankind is a part of nature and life depends on the uninterrupted functioning of natural systems which ensure the supply of energy and nutrients,
>
>   Civilisation is rooted in nature, which has shaped human culture and influenced all artistic and scientific achievement, and living in harmony with nature gives man the best opportunities for the development of his creativity, and for rest and recreation, ...[19]

> We reaffirm our common fundamental values, including freedom, equality, solidarity, tolerance, respect for all human rights, respect for nature and shared responsibility, are essential to international relations. ...[20]

Soft law references to 'uninterrupted functioning of natural systems' and 'respect for nature' aim for overcoming anthropocentric reductions. They had, however, little impact on international or national decision-making. States have never acted upon its own soft law promises and, for example, designed free trade regimes within the functioning of natural systems. However, we can also view soft law as a spanner in the system of global governance. Sooner or later, it will have an impact on the international and national laws.

Compare, for instance, the ecocentric approach of the 2000 Earth Charter[21] with the anthropocentric approach of the 1992 Rio Declaration (Principle 1: 'Humans are at the centre of concerns for sustainable development'). The Earth Charter is a declaration of fundamental principles for building a just, sustainable, and peaceful global society for the 21st century.[22] Created by global civil society and endorsed by thousands of organisations and institutions, the Charter is not only a call to action, but a motivating force inspiring change the world over. It has been endorsed by a number of states, by UNESCO and by the IUCN [23], the world's oldest and largest environmental organisation with a membership of 1200 NGOs and 88 states.[24]

Ecological integrity as central reference

The Earth Charter defines the principle of sustainability as preserving the integrity of ecological systems. Ecological integrity is a central category not just of the Earth Charter[25], but for a wide range of disciplines of environmental study including science, public health, philosophy, anthropology and law.[26] The meaning of ecological integrity is well defined in the literature, but also in some existing laws. The legal concept of integrity has its origin in the 1972 US Clean Water Act. It has then widely been used in North America, for example, in the Great Lakes Water Quality Agreement[27] between Canada and the United States. The Preamble reads:

> The purpose of the Parties is to restore and maintain the chemical, physical and biological integrity of the waters of the Great Lake Basin Ecosystem ... where the latter is defined as ... the interacting components of air, land, water and living organisms including humans ...

Like the Great Lake Basin, each ecosystem has certain characteristics. Ecosystems exist in infinite variation. Like snowflakes, no two systems are identical. But again like snowflakes, all ecosystems have a number of characteristics in common, for example they all:

1. Contain living and non-living elements;
2. Have a measurable degree of diversity (species, genes, chemicals etc);
3. Have some degree of resilience (defined as the system's ability to maintain relationships between system elements in the presence of disturbances);
4. Have a one-way net flow of energy (from outside to inside);
5. Have a carrying capacity for particular kinds of organisms;
6. Exist in a state of non-equilibrium (i.e. they change through time);
7. Exist in a state where changes are irreversible (ecosystems do not return to a previous state, but evolve to a new form).

Taken together, we can see these characteristics as the integrity of an ecosystem. In biological and ecological sciences a common and tangible concept is, therefore, ecological integrity.[28]

An example of the use of this concept is the Canada National Parks Act 2000:

Definitions

'ecological integrity' means ... a condition that is determined to be characteristic of its natural region and likely to persist, including abiotic components and the composition and abundance of native species and biological communities, rates of change and supporting processes.

Sec. 8(2)

Maintenance or restoration of ecological integrity, through the protection of natural resources and natural processes, shall be the first priority of the Minister when considering all aspects of the management of parks.

A 'first priority'! What would be required now is quite simple. First, Canada should expand its sustainability approach beyond the confined space of national parks to include the entire country. Second, other countries should follow the Canadian example. A blueprint for the design of such law is simple and could read[29]:

'Sustainability Act'

Part I – Purpose and Definitions

Purpose

1. The purpose of this Act is to achieve social and economic development within the boundaries of ecological sustainability.

Meaning of ecological sustainability

2. 'Ecological sustainability' means the absence of damage to ecological integrity caused by human impact in an ecosystem of any size within which the impact is found.

Definitions

3 In this Act,

(...)

'Ecological integrity' means the ability of an ecosystem to recover from disturbance and re-establish its stability, diversity and resilience.

Sustainability has to consider the three dimensions of ecology, economy and society. A corresponding law, therefore, should follow three design criteria:

     Effectiveness for the ecological dimension;

     Efficiency for the economic dimension; and

     Equity for the social dimension.

At present, environmental law is ineffective and inefficient. There is also a growing body of evidence that environmental law fails to prevent inequity. When we think of food prices, for example, the issue of environmental justice becomes ever more urgent. But environmental justice cannot be isolated from ecological justice as the broader concept embracing humans and nature alike.[30]

A law based on sustainability will further incorporate legal norms such as the precautionary principle, the polluter pays principle, the principle of common but differentiated responsibility and the concept of intragenerational, intergenerational and interspecies equity. Some of these are accepted principles in law, others are only evolving. In their significance and mutual reinforcement they are most poignantly expressed in the Earth Charter. The Earth Charter should serve, therefore, as the overall ethical and legal framework for achieving a sustainable society.[31]

Conclusion

Environmental law has followed the Western idea of progress. This idea is based on the modern secular myth that humans are in control and above nature. We now 'know' that nature's boundaries are ignored at our peril, but this knowledge has not yet guided us. The underlying myth is too strong.

Any myth reflects our search for meaning. Our thoughts and aspirations seek some symbolic language through which we can talk about, and participate in, what we otherwise cannot see, touch, or taste. What is our goal, our meaning, our purpose as human beings? These are the questions a myth can answer.

As the greatest thinkers, artists and spiritual leaders have maintained,[32] only a new myth can inspire the cultural change that we need, a new myth that celebrates live in all its bounty and variety. Earth Jurisprudence is a beautiful expression of this new myth and we lawyers have all reason to be guided by it.

Notes

1   This is true, at least, for text books in Germany and New Zealand, however, less so in the UK or USA. At global level, virtually all texts in the area of international environmental law include chapters or paragraphs with deeper theoretical reflections.

2   See further below

3   For an attempt see, K. Bosselmann, *Im Namen der Natur* [*In the Name of Nature*: *the concept of eco law*] Scherz: Munich, Germany, 1992; *When Two Worlds Collide: Society and Ecology*; RSVP: Auckland, NZ, 1995.

4   K. Bosselmann, *The Principle of Sustainability: Transforming Law and Governance*, Ashgate, Aldershot, UK, 2008; K. Bosselmann, 'Earth Democracy: Institutionalizing Ecological Integrity and Sustainability' in *Democracy, Ecological Integrity and International Law*, eds R. Engel, L, Westra, and K. Bosselmann, Cambridge Scholars Publication, Cambridge, UK, 2010, pp. 319–33.

5   See, R. Bahro, *Avoiding Social and Ecological Disaster: The Politics of World Transformation: An Inquiry into the Foundations of Spiritual and Ecological Politics*, eds P. Jenkins, trans D Clarke, Gateway: Bath, UK, 1994, p. 219.

6   Composed of the first letter of each.

7   K. Bosselmann, 'Losing the Forest for the Trees: Environmental Reductionism in the Law' in *Law for Sustainability*, Special Issue of online journal *Sustainability* eds J. Dernbach and J. Mintz 2010 http://www.mdpi.com/journal/sustainability/special_issues/env-laws.

8   Although in most jurisdictions, this would be permissible so long as the animal is killed in a 'humane' manner without suffering that would constitute 'cruelty'. No justification *per se* is required.

9   K. Bosselmann, *The Principle of Sustainability*.

10   Ibid., pp. 13–25.

11   K. Bosselmann, 'Property Rights and Sustainability: Can They Be Reconciled?' in *Property Rights and Sustainability: The Evolution of Property Rights to Meet Ecological Challenges*, eds D. Grinlinton and P. Taylor, Martinus Nijhoff, Leiden Boston, (forthcoming) 2011.

12   See, J. Radkau, *Natur und Macht Eine Weltgeschichte der Umwelt*, Beck: Munich, Germany, 2000, p. 245. Recently published in English as *Nature and Power: A Global History of the Environment*, Cambridge University Press, New York, 2008.

13   H.C. von Carlowitz, *Sylvicultura Oeconomica, Anweisung zur Wilden Baum Zucht [Forest Economy or Guide to Tree Cultivation Conforming with Nature]* TU Bergakademie Freiburg und Akademische Buchhandlung: Freiburg, Germany, 2000 [1713]. See also, U. Grober, Tiefe Wurzeln: Eine Kleine Begriffesgeschichte von 'Sustainable Development' – Nachhaltigkeit'. *Natur und Kultur* (2002) *3:1*, pp. 116–28; U. Grober, *Deep Roots: A Conceptual History of Sustainable Development (Nachhaltigkeit)*; Wissenschaftzentrum Berlin für Sozialforschung: Berlin, Germany, 2007.

14   *Our Common Future*, Oxford University Press, Oxford, UK, 1987.

15   'Sustainable development is development that meets the needs of the present generation without compromising the ability of future generations to meet their own needs'. Ibid., p. 8.

16   Ibid., p. 356.

17   For example: 'We borrow environmental capital from future generations with no intention or prospect of repaying. They may damn us for our spendthrift ways, but they can never collect on our debt to them. We act as we do because we can get away with it: future generations do not vote; they have no political or financial power; they cannot challenge our decisions'. Ibid. 'From One Earth to One World', at para 25.

18   K. Bosselmann, 'Strong and Weak Sustainable Development: Making the Difference in the Design of Law', *South African Journal of Environmental Law and Policy,* vol. 13, 2007, pp. 14–23.

19   United Nations, World Charter for Nature A/RES/37/7 22 ILM 455 (1983), (1982).

20   United Nations General Assembly, 60/1. 2005 World Summit Outcome, *in* A/Res/60/1 (2005). p. 5.

21   See, for example, Earth Charter Principle 1: Respect Earth and life in all its diversity.
     a. Recognise that all beings are interdependent and every form of life has value regardless of its worth to human beings.

b. Affirm faith in the inherent dignity of all human beings and in the intellectual, artistic, ethical, and spiritual potential of humanity.

22 See www.earthcharter.org; K. Bosselmann, 'Earth Charter', in *Max Planck Encyclopedia of Public International Law*, eds R. Wolfrum, Oxford University Press, 2009, http://www.mpepil.com/

23 www.iucn.org.

24 Ibid.

25 Part I ('Respect and care for the community of life') and Part II ('Ecological Integrity') represent the first component, the remaining Parts III ('Social and Economic Justice') and IV ('Democracy, Nonviolence and Peace') the second component of the Charter's framework for global governance.

26 Global Ecological Integrity Group www.globalecointegrity.net, a network of over 250 academics working in this field.

27 Signed in 1978, ratified in1988.

28 The ethical foundations go back to the work of environmental philosopher Laura Westra; they have considerably influenced the interdisciplinary dialogue on measuring ecosystem health. See the books by members of the Global Ecological Integrity Group www.globalecointegrity.net

29 For a similar example see B. Pardy, 'In Search of the Holy Grail of Environmental Law: A Rule to Solve the Problem', *McGill International Journal of Sustainable Development Law and Policy*, vol. 1, 2005, p. 29.

30 K. Bosselmann, *The Principle of Sustainability* pp. 79–109.

31 K. Bosselmann, 'In Search for Global Law: The Significance of the Earth Charter', *Worldviews: Environment, Culture, Religion*, vol. 8, no.1, 2004, pp. 62–75; K. Bosselmann, and P. Taylor, 'The Significance of the Earth Charter in International Law', in *Toward a Sustainable World: The Earth Charter in Action,* eds P. Blaze Corcoran, The Hague, Kluwer International, 2005, pp. 171–173.

32 And also economists, bureaucrats and politicians know in their bones.

Nature in Court: Conflict Resolution in the Ecozoic Age

Liz Rivers

The Challenge

If nature has rights that are recognised by our legal systems, and those rights come into conflict with the rights of humans, how is that clash to be resolved? The apparent impracticality of bringing plants, animals and ecosystems into court is often cited as an obvious reason why wild law is naive and unworkable. Indeed, when the owner of a tree in the USA sued a negligent driver for damage to the tree, the Michigan court dismissed the claim with the following judgment:

> We thought that we would never see
> A suit to compensate a tree.
> A suit whose claim is prest
> Upon a mangled tree's behest.
> A tree whose battered trunk was prest
> Against a Chevy's crumpled chest.
> A tree that may forever bear
> A lasting need for tender care.
> Flora lovers though we three
> We must uphold the court's decree.[1]

The court clearly thought the notion of a tree having rights enforceable in court was ridiculous. Given the characteristics of our existing court system with its adversarial litigation process, I agree that it is difficult to see how nature (or more accurately the 'more than human' world) could be represented within that system.

The challenge is to show that there are systems for reconciling the rights of all the different elements of the planetary system in meaningful ways, that respect the rights of the individual elements as well as the integrity of the whole.

The Answer

Just as we need to change our jurisprudence and our ideas of who and what is entitled to protection before the law, we also need to change our systems for resolving conflicts of rights between different parts of the system.

To do this I will start by contrasting two different worldviews, then look at the underlying assumptions of our current court-based litigation system, which is not fit for purpose as it perpetuates old ways of thinking and relating. I will contrast the current system to new practices and methodologies that are emerging around the world at grassroots level, which I believe could provide an answer to this challenge.

The Old Paradigm: 'Empire'

Writer David Korten uses the term 'Empire'[2] to describe the current dominant paradigm of our age, which we are seeking to evolve beyond. It is also described as the 'Industrial Growth Society' and is characterised by beliefs such as: it is possible to have unlimited economic growth; the Earth is a resource for the use of humans and corporations; the so-called 'free market' can solve all our problems; the way to be happy is to consume more; and the dangerous illusion that humanity can exist independently of the biosphere ('the illusion of independence' – Cullinan[3]). We have been literally brainwashed with these ideas, such that is hard to recognise them as ideology rather than reality.

When we live in a world where such assumptions go largely unnoticed and unchallenged, this leads to an atmosphere of scarcity, competitiveness, overconsumption and above all – fear. A world in which even the comparatively wealthy feel chronically insecure most of the time.

This in turn leads to a legal system where the law recognises no higher power than itself, only humans and corporations have rights, and nature is protected only to the extent that it has utility for humans.

Our Current Litigation System[4]

I was trained and practiced in the English legal system, from which the Common Law came, so I will speak from this perspective. This is the system which has now been adopted by most of the English-speaking world.

Litigation in the Common Law system can fairly be described as mired in a Newtonian worldview (mechanical, reductionist, cause-and-effect thinking). Undoubtedly the rule of law was a significant improvement over 'might is right' as a method of resolving differences. However, it has the characteristics of a system in decline, i.e. it is rigid, adversarial and not able to adapt to the evolutionary leap that we need to make. It is essentially about fault-finding and blame, only concerned with the past, and only takes into account information that it considers to be 'objective'. It does not value feelings or relationships.

It is hard to see how this process can be adapted to respond to the interests of the natural world in any meaningful way. As Cormac Cullinan says, we need a 'Copernican shift' of our systems,[5] and I believe that this applies to our conflict resolution systems as well as our jurisprudence and primary laws.

Some attempts have been made to adapt the Common Law system to recognise the natural world in a different way. Andrew Kimbrell, a US Public Interest lawyer has identified the Guardian ad Litem process, the Public Trust Doctrine and Citizen Suit provisions as possible mechanisms to extend our existing system.[6] These are valuable as ways of reforming our current system but I do not think provide a complete answer to the challenge.

The New Paradigm: 'Earth Community'

If our current system is not fit for purpose, what could we replace it with? Before looking at specific options I will go back to the concept of worldviews. Earlier I described the dominant worldview as 'Empire'. I would like to contrast this with the emerging worldview of 'Earth Community'.[7] Thomas Berry describes this eloquently when he says 'We are a communion of subjects, not a dominion over objects'. This worldview recognises that humans are only one part of a wider system, with which we are interdependent. It knows that nature has inherent value, irrespective of its utility to humans and corporations. When embraced fully, this approach leads to the enhancement of creativity, spontaneity, community, wildness, sufficiency in material things and abundance in the non-material. As Arnold Toynbee said 'Nature is going to compel posterity to revert to a stable state on the material plane and turn to the realm of the spirit for satisfying man's hunger for infinity'.

This leads to a very different type of jurisprudence, where the Universe is seen as the primary lawgiver, human law must operate within environmental limits and rights are seen as inherent in existence, rather than something granted by human society.

From this it flows that all members of the Earth Community enjoy the following rights:
- To be
- To habitat
- To participate in the unfolding Universe story[8]

Assuming we get to the point where such rights are recognised by our legal systems, how do we then resolve differences between different parts of the system (e.g. human-nature or nature-nature)? At this point I realise that our current language is an obstacle in itself. We talk about 'conflict resolution', 'dispute resolution' and 'dispute management'. The language is transactional and reductionist, it betrays a Newtonian worldview of separate entities clashing, with the only outcomes being win, lose and compromise, and assuming that at the end the disputants go their separate ways. If we apply such language to an ecosystem it sounds nonsensical, just as the idea of animals litigating in court sounds nonsensical to the conventional worldview. In an ecosystem there are many discrete entities that have their own qualities and yet they all come together and find a way to coexist that not only honours their individual qualities but creates a whole that is greater than the sum of all the parts. The system is in a constant dynamic state of change, with the different parts of the system adjusting to each other all the time – this is a *process* rather than an *event*.

The concept of biomimicry has become very popular in design and engineering.[9] This is the idea that we can solve technical design problems by looking at how nature approaches these challenges. We need an equivalent biomimicry in conflict resolution – to learn from nature how it works with differences for the good of the whole, whilst protecting the integrity of the parts.

The Growth of Mediation: A New Paradigm Process

Although litigation remains the dominant formal method of resolving legal disputes across the developed (or the more accurately termed 'ecological debtor') world, a quiet revolution has been underway for the last 25 years. In an exquisitely Taoist fashion, the excesses of the US litigation system in the 1980s gave rise to the mediation movement, an evolutionary step in how we resolve our differences. Unlike litigation it is flexible, forward-looking, enables the participants to create their own outcomes rather than suffering an imposed outcome, and it seeks to uncover and meet underlying interests and needs rather than simply arbitrating between positions. It is much better suited to addressing complexity, and it values feelings and relationships.

Mediation has been spreading around the industrialised world since the mid 1980s (most indigenous cultures have had their own form of it for centuries). Although mediation is sometimes presented merely as a pragmatic alternative to adversarial processes, at its heart it is fundamentally different and represents an evolution in consciousness. Litigation tends to bring out the worst in people whereas mediation supports the expression of the best – it rehumanises, encourages listening, empathy, flexibility, learning and honouring of emotions. It recognises and enhances our interconnectedness.

Mediation sits as part of a wider movement which includes processes such as consensus building, stakeholder dialogue, conciliation and a range of other processes that share the same underlying assumptions and values. These are the foundations on which we can build our new processes. However, these processes are firmly rooted in the human dimension and so need to develop further. How can current mediation processes evolve to respond to a legal system that recognises the inherent worth of the whole of creation? What qualities would such processes need?

Jung and the Four Dimensions[10]

Jung's four functions can assist us here. This is a model which has been in use since Platonic Greece. They are:

Thinking (Air Element) – the ability to use our rational minds to analyse, categorise and conceptualise. The thinking mind has brought us great technological advances which have transformed our lives. However, as Einstein warned: 'We should take care not to make the intellect our god. It has, of course, powerful muscles, but no personality. It cannot lead, it can only serve'.

Feeling (Water Element) – the water element allows us access to our emotions, to empathy, emotional intelligence, rapport and compassion. Research shows that most decisions are made on the emotional level and then retrospectively justified with rational justifications. Certainly emotions play a far larger part in our actions than we give them credit for. The feeling function is also where our sense of values reside – the part of us that knows what is right and what is wrong.

Sensing (Earth Element) – this represents the wisdom of our bodies. In order to feel fully alive and to communicate a sense of presence to others we need to be grounded in our bodies, breathing freely and alive to our senses. When we are in this state we have access to different ways of knowing. There is a Polynesian saying: 'All knowledge is only rumour until it is in the muscle'.

Intuiting (Fire Element) – this represents our passionate, inspired dimension. Where we get our vision and inspiration from, the sense of purpose that makes life worth living. 'I have a dream' is its rallying cry.

We live in a culture where Thinking dominates and the other ways of knowing have been devalued for millennia. This certainly shows up in mainstream culture and is reflected in our legal system, where a very narrow band of information is deemed to be legally 'relevant' to the resolution of a dispute. This dominant view has been challenged in recent years by concepts of emotional intelligence,[11] spiritual intelligence[12] and somatic intelligence.[13] However, it has yet to permeate our legal system, which remains stubbornly situated in a nineteenth century view and atmosphere.

In my view, we need to embrace methods of dialogue and communication which honour all four dimensions of intelligence. This will be profoundly challenging to those who are rooted in a purely intellectual approach, and will attract ridicule as not being 'objective' or rational. It reminds me of the early days of quantum physics, when the scientists were confounded by the paradoxes they were discovering at the sub-atomic levels. It seemed nonsensical and ridiculous to the Newtonian worldview, and yet that is what their experiments were showing them. We need to enter the realm of quantum conflict resolution – a magical world where conventional assumptions are turned on their head and alchemical transformations are possible.

Environmental Constellations – A Case Study

One methodology which meets these criteria is Environmental Constellations. This is a therapeutic process based on the work of Bert Hellinger which has been adapted for family, organisational and environmental issues. In the words of Environmental Constellations practitioner Zita Cox[14]:

> Environmental Constellations are a wonderfully versatile and creative tool. They allow us to observe our place in nature and our systemic relationship to other living beings. They assist us with new ways of thinking about and finding resolutions to

the problems we have created, such as climate change, pollution, alienation and accelerating species extinction etc; They are an aid to empathy and shared understanding in all situations.

Constellations facilitate 'joined up' thinking by mapping the environmental system in front of our eyes. They open up creative and imaginative possibilities, accessing the unconscious mind to work with us and for us.

We are all interconnected parts of family, organisation, community and ecosystem. The difficulties we struggle with are often best resolved when we work systemically. Constellations are an innovative way of gaining insight into the systems we are a part of. A constellation draws on emotional intelligence and intuition as well as logical thought. It provides a way of seeing below the surface to understand what the real issues are and what can actually be done to improve things.

How they work

To set up a constellation in a group of people the person exploring the issue is invited to choose individuals from the group to act as representatives for the various elements or parts of the system they wish to understand. The issue holder positions the representatives in the room as feels right to them, at this point a pattern becomes visible and the nature of the relationship between the different elements begins to be seen. The facilitator works with the dynamics that emerge, leading to further information, unexpected insights, ideas, strategies, a deeper understanding and sharing of knowledge and experience and, where appropriate, resolution.

One of the important tenets of the method is the belief that a system is by its nature inclusive, each person or element has their rightful place. If we attempt to exclude any element, the system becomes unbalanced, problems emerge and people and situations become stuck. The information revealed by the constellation allows the system to rebalance and the flow of energy to resume, or it may simply give us a new and unexpected way to understand a situation'.

At the 2007 UKELA Annual Wild Law conference, we held a mock trial in which the lawyers were invited to argue wild law principles. The particular legal problem chosen was that of palm oil plantations in the Borneo rain forest, where deforestation was affecting the orang-utan population.

We invited Zita Cox to attend the conference and to run a constellation on the same topic. For this process we chose as our starting point the question of how a lawyer representing the orang-utan could 'take instructions' from their client (i.e. find out what they wanted). One of the groups took the role of lawyer and another two were chosen to be the orang-utans. They were placed in relationship to each other. Next the remaining elements of the system were put in place. Representatives were chosen to play all the other elements: the rainforest, the indigenous people who lived there, the logger who wished to cut down the rainforest in order to plant a palm oil plantation and the palm oil plant. An interaction then started between the various different parts of the system, careful facilitated by the practitioner. I remember particularly vividly an interaction between the logger and the orang-utan. The logger ceased to be the all-powerful conqueror, laying waste to all before him in the name of his god-given right to maximise corporate profit. His youthfulness and immaturity in this system of ancient wisdom he had entered became very apparent. The orang-utan said to him with innate authority: 'you are so young', and with that I gained a visceral felt sense of the integrity of this ecosystem and the need for any incomer to tread lightly and respectfully. The logger's composure and his sense of entitlement were challenged, yet not in an aggressive way, more like a sense of Gandhian 'soul force' or 'satyagraha'. I gained a direct, felt experience of the system, its constituent parts and the relationships between them.

Although this event took place over two years ago I can still recall its power very clearly and have a strong sense of the atmosphere that was created. Cormac Cullinan, who took the role of the lawyer, described the experience as follows:

Learning how to engage empathically with the Earth Community as whole is essential if we are to act as responsive and responsible members of that community. My participation in Environmental Constellations guided by Zita Cox gave me an intriguing experience of what this might feel like. This is a fascinating technique for slipping past the self-imposed limits of 'logical' thinking into the communion of participation.

Other Possible Processes
I will briefly touch on a number of other processes which I think have value in this area.

For dialogue with the natural world

Council of All Beings[15]
This is a process developed by Deep Ecology pioneers Joanna Macy and John Seed to address the deepening sense of alienation from the natural world that many feel. Through a series of dynamic experiential exercises, movement work, reflection, visualisations, and time spent in nature participants explore their concern and love for our planet, rediscovering their 'deep ecology' – interconnectedness with the myriad species and landscapes of the Earth.

Shamanic Rituals
Liz Hosken of the Gaia Foundation talks about how the shaman in indigenous cultures has the role of maintaining the balance between the human group and the ecosystem in which they live. The shaman learns to read the signs from nature and will conduct rituals to restore the balance when things get out of kilter.

For Dialogue Between Humans About Worldviews

Applications of Systemic Family Therapy to Socially Divisive Issues

The Public Conversations Project is an organisation based in the USA which draws on systems theory and systemic family therapy in order to create powerful dialogue about socially divisive issues.[16] They specialise in topics where the difference is rooted in people's deepest values and worldviews, and which are therefore not capable of easy resolution. The aim is not for participants to reach agreement but rather to increase their understanding of one another.

Spiral Dynamics[17]

This is a psychological theory that holds that individuals, groups and societies evolve through different worldviews, with different levels of complexity. In order to influence effectively, it is essential to diagnose the paradigm in which people are operating and speak to them in terms that will make sense to them.

Some would argue that purely consensual processes as I have described above will never be sufficient in themselves, that there needs to be an element of coercion before those in power will come to the table. As Barack Obama said 'Power never concedes voluntarily'. I agree that there needs to be some muscle in our formal processes in order to bring powerful parties, who might otherwise ignore the rights of nature, to the negotiating table or dialogue process. However, we should see these as transitional steps on the way to a world where all parts of the Earth Community recognise that either everyone wins or nobody wins; that to be winning a war against nature is to find yourself on the losing side (to paraphrase E.F. Schumacher).

In this table I summarise the two worldviews described above and the methods for resolving differences which flow from them:

	Old Paradigm ('Empire')	New Paradigm ('Earth Community')
Underlying worldview	Unlimited growth Earth is a resource 'free market' Illusion of independence Myth of superiority	'We are a communion of subjects not a dominion over objects' (Thomas Berry) We are part of a larger system with which we are interdependent and need to start behaving appropriately in relation to it
Values, attitudes	Scarcity Fear Competitiveness Materialism Mechanistic	Abundance Creativity Spontaneity Wildness Community
Jurisprudence (how we conceptualise law)	Law recognises no higher power than itself Nature is protected to the extent it has utility for humans (anthropocentric)	Nature/universe is the primary law giver (human law must operate within environmental limits) Rights are inherent to existence Nature has inherent value, irrespective of human utility (geocentric)

Legal system – who has rights?	Only: Humans Corporations Law is anything that humans say it is	*All* members of the Earth Community, including Animals Plants Habitats (e.g. rivers, mountains) Fundamental rights: To be To habitat To participate in the unfolding Universe story
Conflict resolution methods	Litigation Rigid Adversarial About blame Win/lose Denies relevance of non-rational 'subjective' data Linear, reductionist, transactional, analytical Does not value feelings and relationships	**Mediation and dialogue** Flexible Consensual Forward looking Aspires to win/win Recognises interests and needs Can encompass complexity Values feelings and relationships *But:* *Has not yet started to think about how to meaningfully engage with the natural world*
What next?	Guardian ad litem concept Public trust doctrine Citizen suit provisions	Specifically for working with natural world: Environmental Constellations Council of All Beings Shamanic Practices For working with worldviews: Systems theory and family systemic therapy Spiral Dynamics

Rights-based thinking – is it part of the problem?

My experience when presenting these ideas is that people often respond very positively to the underlying philosophy of Earth Jurisprudence, but feel jarred when the concept of rights for nature is mentioned. Somehow it does not seem to fit with the new paradigm. I believe that the concept of 'rights' as we currently use it belongs to the old paradigm (Empire) and at best they should be viewed as a stepping-stone towards the new paradigm (Earth Community). Rights tend to emphasise separateness and encourage competitiveness, rather than emphasising interconnectedness and synthesis. If we really felt our interconnectedness in our gut, we would have little need for rights.

If we take the metaphor of a family, in a healthy functioning family there is no need for any member to invoke their rights. It is only in very dysfunctional families that this is the case. Relationships with the 'family' of the Earth Community have become very dysfunctional, with the upstart newcomer (humans) acting like an out-of-control teenager. Mediators encourage people to look beyond rights to thinking about interests and needs instead. Maybe we need the concept of rights in the short term to restore some semblance of balance, and when that has been achieved we can shift all members of humanity to a more sophisticated approach of looking at interests and needs.

The journey of the modern age has been to recognise the humanity in *all* humanity, and to extend recognition and respect beyond white wealthy males to women, children, people of colour and workers. In each case it was necessary for those groups to invoke their rights before they were treated with respect. So, maybe we need a concept of rights for nature as a transitional step towards widening our circle of compassion. As Einstein said:

A human being is a part of a whole, called by us 'universe', a part limited in time and space. He experiences himself, his thoughts and feelings as something separated from the rest ... a kind of optical delusion of his consciousness. This delusion is a kind of prison for us, restricting us to our personal desires and to affection for a few persons nearest to us. Our task must be to free ourselves from this prison by widening our circle of compassion to embrace all living creatures and the whole of nature in its beauty.

Notes

1   C. Cullinan, *Wild Law: A Manifesto For Earth Justice*, Green Books, Devon, 2002, pp. 106–107.

2   D.C. Korten, *The Great Turning: From Empire to Earth Community*, Berrett-Koehler Publishers, 2006, p. 54.

3   Cullinan, *Wild Law*, pp. 37–50.

4   I include arbitration in this term

5   Cullinan, *Wild Law*, pp. 44, 89–90.

6   For more detail see Andrew Kimbrell: 'Halting the Global Meltdown: Can Environmental Law Play a Role' *Environmental Law and Management* vol. 20, 2008, p. 1.

7   Korten, *Great Turning*, pp. 17, 37–38.

8   Thomas Berry, *Evening Thoughts: Reflecting on Earth as Sacred Community*, Sierra Club Books, San Francisco, 2006, pp. 110–111, 149–150.

9   Janine Benyus, *Biomimicry: Innovation Inspired by Nature*, Harper Perennial, New York, 2002.

10 I am indebted to Nicholas Janni for his clear explanation of this model www.olivier-mythodrama.com 'Elements of Leadership'. He also talks of a fifth element of Ether.

11 Daniel Goleman, *Emotional Intelligence*, Bantam Dell, New York, 1995.

12 Danah Zohar, *Spiritual Intelligence: The Ultimate Intelligence*, Bloomsbury Publishing PLC, London, 2001.

13 Stanley Keleman, *Love: A Somatic View*, Center Press, Berkley, 1994.

14 www.environmentalconstellations.com

15 www.joannamacy.net

16 www.publicconversationsproject.org

17 www.spiraldynamics.org

Section 2: The Rights of Nature

Fundamental to Earth Jurisprudence is the notion that human beings exist as one equal part of the Earth. Because we are not elevated above nature and because everything exists as part of a mutually dependent web of relationships, each component of nature has equal value. Following from this, environmental philosophers have argued that nature ought to have rights recognised and protected in law. This section will consider first the philosophical underpinnings of this argument and then note examples where the rights of nature have been recognised in law.

Rights of the Earth: We Need a New Legal Framework Which Recognises the Rights of *All* Living Beings

Thomas Berry

Our present legal system throughout the world is supporting the devastation of the nature rather than protecting it. In America the situation became critical in the last decades of the nineteenth century. At that time we moved from an organic, ever-renewing, land-based economy to an extractive, non-renewing, industrial economy. Now, this economy, supported by the political power and legal framework of the Western nations, has taken possession of much of the planet.

The force of industrial economy has invaded every aspect of human life including its political, legal, educational and religious functioning. So extensive is this economy that we must now speak of ourselves as living in an industrial civilisation. The difficulty is that the industrial process is so destructive of the natural world that we begin to envisage a terminal destiny of those life forms on which humans most depend.

We are told by biologists that we are extinguishing the living species of the natural world at a rate unequalled since the last great extinction, some 65 million years ago. Lester Brown of the Worldwatch Institute tells us that we are 'losing the war to save the Earth'. Vandana Shiva tells us, 'Humanity seems to be in a free fall towards total disaster'. The International Union of Concerned Scientists tells us that human disturbance of the natural life systems of Earth is putting human life itself into serious peril.

Yet those in control of our industrial civilisation continue to insist that the well-being of humans can be achieved only through the industrial processes carried out by globalising institutions such as the World Bank, the International Monetary Fund, the World Trade Organisation, and the multitude of transnational, multi-national and global corporations that have been actively engaged in extending the industrial way of life throughout the planet.

Since the continuation of our present industrial processes depends directly on the legal system, we must reconsider our present legal system in its deepest foundations. Critics of the present situation in America have consistently found that our existing legal structures cannot protect the natural world. The Federal Judiciary has so frequently ruled against the regulations of the Environmental Protection

Agency that a person might conclude that the environmental and ecological move-ments in America are themselves being declared unconstitutional.

It seems that such considerations of the basic function of law have never been a concern in recent times. The founders of the American Constitution were so concerned with their escape from the authoritarian controls of European monar-chies that their main concern was to establish a range of personal rights, especially rights to own and use property without restriction by government.

This protection for personal rights was later extended to industrial corpora-tions, with no recognition of the inherent rights of the natural world. Indeed, the basic purpose of government and of the entire legal system in America has been to assist and even to subsidise the industrial corporations in their exploitation of nature. The well-being of the corporations came to be identified legally with the well-being of the people. Morton Horowitz, Professor of History of Law at Harvard University, indicates in his book *The Transformation of American Law: 1780–1860*: 'As political and economic power shifted to merchant and entrepre-neurial groups in the post-revolutionary period, they began to forge an alliance with the legal profession to advance their own interests through a transformation of the legal system'.

In the resulting commercial – industrial society the corporations provide the jobs needed by people who no longer have their own life support in any direct rela-tion with the land. Most people now depend for survival on having a job within commercial corporations. We now live and breathe more in the world of industrial production and consumption than in the natural world.

Not only the industrial empires of the nineteenth and twentieth centuries but also the entire industrial civilisation achieved their juridic foundations within this context. For these reasons, any effort to diminish the devastating consequences of the industrial age might begin with discussion of this question of rights, their origin, their distinction, their role, and especially their function in human – Earth relations.

The primary supposition of this article is that the interdependence of every mode of being on every other mode of being requires humans to recognise that every living being has rights that are derived from existence itself. Otherwise the ordered structure and functioning of the entire planet are endangered.

This interdependence is immediately evident. We cannot have healthy humans on a sick planet. We cannot have a viable human economy by devastating the Earth economy. If the grass does not grow, we die. If the climate is altered, we are in deep trouble. We cannot survive if the conditions of life itself are not protected. What humans do affects every other being. If we cut the rainforest, then the land turns into desert and the entire planetary process is disturbed.

The well-being of each member of the Earth community is dependent on the well-being of the Earth itself. Within this context, then, I would make the following set of proposals expressed in terms of rights, which should be recognised in national constitutions and in courts of law:

1. The natural world on the planet Earth has rights, which come with existence. These rights come from the same source from which humans receive their rights, from the universe that brought them into being.

2. Every component of the Earth community has three rights: the right to be, the right to habitat, and the right to fulfil its role in the ever-renewing processes of the Earth community.

3. All rights are specific and limited. Rivers have river rights. Birds have bird rights. Insects have insect rights. Humans have human rights. Difference in rights is qualitative, not quantitative. The rights of an insect would be of no value to a tree or a fish.

4. Human rights do not cancel out the rights of other modes of being to exist in their natural state. Human property rights are not absolute. Property rights are simply a special relationship between a particular human 'owner' and a particular piece of 'property' so that both might fulfil their rules in the great community of existence.

5. Since species exist only in the form of individuals, rights refer to individuals and to those natural groupings of individuals into flocks, herds, packs, not simply in a general way to species.

6. These rights are based on the intrinsic relations that the various components of Earth have to each other. The planet Earth is a single community whose members are bound together with interdependent relationships. No living being nourishes itself. Each component of the Earth community is dependent on every other member of the community for the nourishment and assistance it needs for its own survival. This mutual nourishment, which includes predator-prey relationships, is integral with the role that each component of the Earth has within the comprehensive community of existence.

It is our responsibility to make these six principles the foundation of the new legal system all over the world. The time has come when human laws and Earth laws must be brought together.

Such is the challenge before us – in every aspect of our lives. Such is the basis of the proposals that I have presented here as basic to our human survival and the survival of that great community of living beings in the florescence that they once knew. Humans will always make significant demands on the surrounding world. Yet a more mutually beneficial situation must be found. The legal profession needs to cease its subservience to the industrial corporations to fulfil its larger responsibilities for the integral survival of the Earth in the fullness of its grandeur.

Acknowledgement

Special thanks to *Resurgence* magazine for allowing this piece to be republished. The original article appeared in Issue 214, September/October 2002. For more articles from Resurgence visit http://www.resurgence.org/.

If Nature Had Rights What Would We Need to Give Up?

Cormac Cullinan

It was the sudden rush of the goats' bodies against the side of the *boma* that woke him. Picking up a spear and stick, the Kenyan farmer slipped out into the warm night and crept toward the pen. All he could see was the spotted, sloping hind-quarters of the animal trying to force itself between the poles to get at the goats – but it was enough. He drove his spear deep into the hyena.

The elders who gathered under the meeting tree to deliberate on the matter were clearly unhappy with the farmer's explanation. A man appointed by the traditional court to represent the interests of the hyena had testified that his careful examination of the body had revealed that the deceased was a female who was still suckling pups. He argued that given the prevailing drought and the hyena's need to nourish her young, her behaviour in attempting to scavenge food from human settlements was reasonable and that it was wrong to have killed her. The elders then cross-examined the farmer carefully. Did he appreciate, they asked, that such killings were contrary to customary law? Had he considered the hyena's situation and whether or not she had caused harm? Could he not have simply driven her away? Eventually the elders ordered the man's clan to pay compensation for the harm done by driving more than one hundred of their goats (a fortune in that community) into the bush, where they could be eaten by the hyenas and other wild carnivores.

The story, told to me by a Kenyan friend, illustrates African customary law's concern with restorative justice rather than retribution. Wrongdoing is seen as a symptom of a breakdown in relationships within the wider community, and the elders seek to restore the damaged relationship rather than focusing on identifying and punishing the wrongdoer.

The verdict of a traditional African court regarding hyenacide may seem of mere anthropological interest to contemporary Americans. In most of today's legal systems, decisions that harm ecological communities have to be challenged prima-rily on the basis of whether or not the correct procedures have been followed. Yet consider how much greater the prospects of survival would be for most of life on Earth if mechanisms existed for imposing collective responsibility and liability on human communities and for restoring damaged relations with the larger natural community. Imagine if we had elders with a deep understanding of the lore of the wild who spoke for the Earth as well as for humans. If we did, how might they order us to compensate for, say, the anticipated destruction of the entire Arctic ecosystem because of global climate change, to restore relations with the polar bears and other people and creatures who depend on that ecosystem? How many polluting power plants and vehicles would it be fair to sacrifice to make amends?

'So what would a radically different law-driven consciousness look like?' The question was posed over three decades ago by a University of Southern California law professor as his lecture drew to a close. 'One in which Nature had rights,' he continued. 'Yes, rivers, lakes, trees ... How could such a posture *in law* affect a community's view of *itself*?' Professor Christopher Stone may as well have announced that he was an alien life form. Rivers and trees are objects, not subjects, in the eyes of the law and are by definition incapable of holding rights. His speculations created an uproar.

Stone stepped away from that lecture a little dazed by the response from the class but determined to back up his argument. He realised that for Nature to have rights the law would have to be changed so that, first, a suit could be brought in the name of an aspect of nature, such as a river; second, a polluter could be held liable for harming a river; and third, judgments could be made that would benefit a river. Stone quickly identified a pending appeal to the United States Supreme Court against a decision of the Ninth Circuit that raised these issues. The Ninth Circuit Court of Appeals had found that the Sierra Club Legal Defense Fund was not 'aggrieved' or 'adversely affected' by the proposed development of the Mineral King Valley in the Sierra Nevada Mountains by Walt Disney Enterprises, Inc. This decision meant that the Sierra Club did not have 'standing' so the court didn't need to consider the merits of the matter. Clearly, if the Mineral King Valley itself had been recognised as having rights, it would have been an adversely affected party and would have had the necessary standing.

Fortuitously, Supreme Court Justice William O. Douglas was writing a preface to the next edition of the *Southern California Law Review*. Stone's seminal 'Should Trees Have Standing? Toward Legal Rights for Natural Objects' ('Trees') was hurriedly squeezed into the journal and read by Justice Douglas before the Court issued its judgment. In 'Trees,' Stone argued that courts should grant legal standing to guardians to represent the rights of Nature, in much the same way as guardians are appointed to represent the rights of infants. In order to do so, the law would have to recognise that Nature was not just a conglomeration of objects that could be owned, but was a subject that itself had legal rights and the standing to be represented in the courts to enforce those rights. The article eventually formed the basis for a famous dissenting judgment by Justice Douglas in the 1972 case of *Sierra Club v. Morton* in which he expressed the opinion that 'contemporary public concern for protecting nature's ecological equilibrium should lead to the conferral of standing upon environmental objects to sue for their own preservation'.

Perhaps one of the most important things about 'Trees' is that it ventured beyond the accepted boundaries of law as we know it and argued that the conceptual framework for law in the United States (and by analogy, elsewhere) required further evolution and expansion. Stone began by addressing the initial reaction that such ideas are outlandish. Throughout legal history, as he pointed out, each extension of legal rights had previously been unthinkable. The emancipation of slaves and the extension of civil rights to African Americans, women, and children were once rejected as absurd or dangerous by authorities. The Founding Fathers, after all, were hardly conscious of the hypocrisy inherent in proclaiming the inalienable rights of all men while simultaneously denying basic rights to children, women, and to African and Native Americans.

'Trees' has since become a classic for students of environmental law, but after three decades its impact on law in the United States has been limited. After it was written, the courts made it somewhat easier for citizens to litigate on behalf of other species and the environment by expanding the powers and responsibilities of authorities to act as trustees of areas used by the public (e.g., navigable waters, beaches, and parks). Unfortunately, these gains have been followed in more recent years by judicial attempts to restrict the legal standing of environmental groups. Damages for harm to the environment are now recoverable in some cases and are sometimes applied for the benefit of the environment. However, these changes fall far short of what Stone advocated for in 'Trees'. The courts still have not recognised that Nature has directly enforceable rights.

Communities have always used laws to express the ideals to which they aspire and to regulate how power is exercised. Law is also a social tool that is usually shaped and wielded most effectively by the powerful. Consequently, law tends to entrench a society's fundamental idea of itself and of how the world works. So, for example, even when American society began to regard slavery as morally abhorrent, it was not able to peaceably end the practice because the fundamental concept that slaves were property had been hard-wired into the legal system. The abolition of slavery required not only that the enfranchised recognise that slaves were entitled to the same rights as other humans, but also a political effort to change the laws that denied those rights. It took both the Civil War and the Thirteenth Amendment to outlaw slavery. The Thirteenth Amendment, in turn, played a role in changing American society's idea of what was acceptable, thereby providing the bedrock for the subsequent civil rights movement.

In the eyes of American law today, most of the community of life on Earth remains mere property, natural 'resources' to be exploited, bought, and sold just as slaves were. This means that environmentalists are seldom seen as activists fighting to uphold fundamental rights, but rather as criminals who infringe upon the property rights of others. It also means that actions that damage the ecosystems and the natural processes on which life depends, such as Earth's climate, are poorly regulated. Climate change is an obvious and dramatic symptom of the failure of human government to regulate human behaviour in a manner that takes account of the fact that human welfare is directly dependent on the health of our planet and cannot be achieved at its expense.

In the scientific world there has been more progress. It's been almost forty years since James Lovelock first proposed the 'Gaia hypothesis': a theory that Earth regulates itself in a manner that keeps the composition of the atmosphere and average temperatures within a range conducive to life. Derided or dismissed by most people at the time, the Gaia hypothesis is now accepted by many as scientific theory. In 2001, more than a thousand scientists signed a declaration that begins 'The Earth is a self-regulating system made up from all life, including humans, and from the oceans, the atmosphere and the surface rocks,' a statement that would have been unthinkable for most scientists when 'Trees' was written.

The acceptance of Lovelock's hypothesis can be understood as part of a drift in the scientific world away from a mechanistic understanding of the universe toward the realisation that no aspect of nature can be understood without looking at it within the context of the systems of which it forms a part. Unfortunately, this insight has been slow to penetrate the world of law and politics.

But what if we were to imagine a society in which our purpose was to act as good citizens of the Earth as a whole? What might a governance system look like if it were established to protect the rights of all members of a particular biological community, instead of only humans? Cicero pointed out that each of our rights and freedoms must be limited in order that others may be free. It is far past time that we should consider limiting the rights of humans so they cannot unjustifiably prevent nonhuman members of a community from playing their part. Any legal system designed to give effect to modern scientific understandings (or, indeed, to many cultures' ancient understandings) of how the universe functions would have to prohibit humans from driving other species to extinction or deliberately destroying the functioning of major ecosystems. In the absence of such regulatory mechanisms, an oppressive and self-destructive regime will inevitably emerge. As indeed it has.

In particular, we should examine the fact that, in the eyes of the law, corporations are considered people and entitled to civil rights. We often forget that corporations are only a few centuries old and have been continually evolving since their inception. Imagine what could be done if we changed the fiduciary responsibilities of directors to include obligations not only to profitability but also to the whole natural world, and if we imposed collective personal liability on corporate managers and stockholders to restore any damage that they cause to natural communities. Imagine if landowners who abused and degraded land lost the right to use it. In an Earth-centred community, all institutions through which humans act collectively would be designed to require behaviour that is socially responsible from the perspective of the whole community.

A society whose concern is to maintain the integrity or wholeness of the Earth must also refine its ideas about what is 'right' and 'wrong'. We may find it more useful to condone or disapprove of human conduct by considering the extent to which an action increases or decreases the health of the whole community and the quality or intimacy of the relationships between its members. As Aldo Leopold's famous land ethic states, 'a thing is right when it tends to preserve the integrity, stability, and beauty of the biotic community. It is wrong when it tends otherwise'. From this perspective, individual and collective human rights must be contextualised within, and balanced against, the rights of the other members and communities of Earth.

On 19 September 2006, the Tamaqua Borough of Schuylkill County, Pennsylvania, passed a sewage sludge ordinance that recognises natural communities and ecosystems within the borough as legal persons for the purposes of enforcing civil rights. It also strips corporations that engage in the land application of sludge of their rights to be treated as 'persons' and consequently of their civil rights. One of its effects is that the borough or any of its residents may file a lawsuit on behalf of an ecosystem to recover compensatory and punitive damages for any harm done by the land application of sewage sludge. Damages recovered in this way must be paid to the borough and used to restore those ecosystems and natural communities.

According to Thomas Linzey, the lawyer from the Community Environmental Legal Defense Fund who assisted Tamaqua Borough, this ordinance marks the first time in the history of municipalities in the United States that something like this has happened. Coming after more than 150 years of judicially sanctioned expansion of the legal powers of corporations in the U.S., this ordinance is more than extraordinary – it is revolutionary. In a world where the corporation is king and all forms of life other than humans are objects in the eyes of the law, this is a small community's Boston tea party.

In Africa, nongovernmental organisations in eleven countries are also asserting local community rights in order to promote the conservation of biodiversity and sustainable development. Members of the African Biodiversity Network (ABN) have coined the term 'cultural biodiversity' to emphasise that knowledge and practices that support biodiversity are embedded in cultural tradition. The ABN works with rural communities and schools to recover and spread traditional knowledge and practices. This is part of a wider effort to build local communities, protect the environment by encouraging those communities to value, retain, and build on traditional African cosmologies, and to govern themselves as part of a wider Earth community.

These small examples, emerging shoots of what might be termed 'Earth democracy,' are pressing upward despite the odds. It may well be that Earth-centred legal systems will have to grow organically out of human-scale communities, and communities of communities, that understand that they must function as integrated parts of wider natural communities. In the face of climate change and other enormous environmental challenges, our future as a species depends on those people who are creating the legal and political spaces within which our connection to the rest of our community here on Earth is recognised. The day will come when the failure of our laws to recognise the right of a river to flow, to prohibit acts that destabilise Earth's climate, or to impose a duty to respect the intrinsic value and right to exist of all life will be as reprehensible as allowing people to be bought and sold. We will only flourish by changing these systems and claiming our identity, as well as assuming our responsibilities, as members of the Earth community.

Acknowledgement

This article was originally published in *Orion* magazine, January/February, 2008. For further information on *Orion*, please visit, http://www.orionmagazine.org/. Thank you to Cormac Cullinan for permission to republish.

Earth Day Revisited:
Building a Body of Earth Law
for the Next Forty Years

Linda Sheehan

In a poster for the first Earth Day in 1970, Pogo trumpeted: 'We have met the enemy, and he is us'.[1] Earth Day evolved from the ashes of the Cuyahoga River fiercely burning in Ohio, from bewildered marine birds sagging under the weight of the massive Santa Barbara oil spill, and from the spreading silence beneath the surface of the dying Great Lakes. With a hard look at our faults and the best of intentions, we adopted new environmental legal systems that we thought would rein in our most destructive tendencies. Fuelled with major influxes of funding, and catalysed by key lawsuits by public interest groups, these new legal systems slowly picked off the most obvious insults, such as raw sewage discharges, free-flowing industrial wastewater, and ubiquitous toxic waste dumping.

Four decades of experience later, we again face large-scale environmental revolt, now on a global scale. Earth-wide climate change, shrinking rivers and aquifers, accelerating extinction rates, and other, potentially irreversible transformations raise searching questions. Why, after decades of what we thought were 'cure-all' environmental laws, have we failed to protect the basic structure and integrity of the Earth's ecosystems? What must we do to set a legal path for the next forty years that will reverse this trend? This essay begins to answer these questions, drawing from examples arising from California's ongoing battles over water.

Earth Day Hits 40, Sits in Self-Reflection

Despite some focused success stories, studies abound showing that our current system of environmental laws has largely failed to protect ecosystems. Known rates of extinction among mammals and birds have climbed far higher in recent years than the average rates through geological time.[2] Waterways have been particularly hard-hit; where examined, the status of inland water faunas has been considerably worse than originally suspected.[3] Climate change will exacerbate these impacts, with already-stressed water sources predicted to 'shrink under even the most conservative climate change scenario,'[4] resulting in untold costs.[5]

Lack of funding, political backtracking, understaffing, weak enforcement and other hurdles have prevented existing environmental laws from achieving their intended potential. Their segregation by environmental impact, which fails to consider that water, air, land, and wildlife are connected, further reins in their effectiveness. Our current, single-stressor laws simply did not envision, for example, that we could fundamentally change the structure of entire ecosystems with air pollution, particularly on a global scale.

However, even full implementation of existing environmental laws will fail to ensure a thriving planet, because these systems assume at their core that that the natural world is something to control. Regardless of the level of benevolent intent, this fundamentally flawed foundation has built up a legal house of cards that now quivers dangerously. Ecosystems are registering objections, whether we choose to acknowledge them or not.

The environment is objecting locally, as legally permitted pollution discharges and water diversions slowly, increasingly compromise the health of rivers, fish, and communities' water supplies. It objects regionally, with the rapid death spiral of the San Francisco Bay-Delta Estuary, the largest on the west coast of North America, after decades of damming and pumping Estuary waters and pouring back toxic irrigation runoff and municipal wastes. We also see the Earth objecting globally, through climate change-related droughts, fires, floods and collapsing glaciers.

If our laws allow this level of destruction, we need better laws. We need a legal system that flows from a core recognition that ecosystems and their inhabitants have inherent rights to struggle to be healthy, thrive, and evolve. Our current environmental laws focus on the rights and needs of people to thrive and evolve, but pay relatively little attention to the same rights on the part of the natural world. They assume that the environment will be protected if humans take from it a little less, and a little less quickly. But this simply slows, never stops, the downward slide. Since we are inextricably intertwined with our environment, this trend does not bode well for humans either.

The U.S. Endangered Species Act does operate from a basic premise that species have a right to exist, independent of their direct benefit to people. However, it comes with a 'God Squad'[6] loophole that is drawing increasing attention as ecosystems and species decline further. Moreover, the basic premise of the Act, which allows intervention only when species are poised to vanish, fails to support the actions needed to prevent – rather than simply react to – declines.

Like the rest of our legal system, the Endangered Species Act presumes that we can act apart from the environment and that we can control and manipulate it as we choose, to maintain a (false) sense of security in our lives. These dangerously outmoded operational assumptions ignore the fact that we are bound to this Earth. To thrive, we must accept this reality, and the corresponding inherent rights of all of the Earth's ecosystems and inhabitants to have a fighting chance to thrive and evolve as well.

How do we implement this recognition in our legal system so that it effectively ensures the well-being of our shared planet? Effective laws allow and constrain behaviour to promote desired relationships and implement each party's respective rights. The two-way relationship piece is key: without mutual respect for the rights being protected, laws will skew their benefits to one side or the other. The currently marginalised rights of the natural world accordingly must be incorporated into the core assumptions that undergird our laws. The goal is to develop a legal system based on equal, fundamental rights and responsibilities – one that reflects and implements the mutually enhancing relationships that exist among humans and the natural world.

To date, we have sought the relatively easy way out, setting up legal systems that allow us to continue widely-accepted human behaviours that degrade our environment and ultimately ourselves. We will institute meaningful change only when we release our tightly held, but inherently flawed, assumptions about our ability to 'manage' a second-class status Earth. This is a tall order. However, digging into the origins of those assumptions will help us to begin to envision a body of 'Earth law' that incorporates modern science and our growing ethical awareness of our responsibilities to current and future Earth generations.

As we reflect on the fortieth anniversary of Earth Day, our challenge – and our duty to future generations – is to develop this new legal system and incorporate it into other aspects of our laws, policies, and governance systems. This evolution will become more urgent and correspondingly more difficult as we increasingly 'hit the wall' in terms of essential ingredients for life, such as water, food and energy. Rather than reacting from fear, though, we have the ability right now to plan with wisdom. Lessons from past will shed light on our missteps, and help us illuminate a new path.

Archeological Dig Uncovers Long-Lost Assumptions: Inquiry Commences

Our current legal system was built from an operative – and faulty – assumption that water, land, forests, air and wildlife are in essence 'resources', or 'wealth' to be extracted, manipulated and controlled for human benefit. At the turn of the twentieth century, the first Chief of the United States Forest Service, Gifford Pinchot, firmly established the growing nation's 'conservation ethic' as one that would myopically focus on 'controlled' use of these natural 'resources'. Pinchot described the ethic as 'the art of producing from the forest whatever it can yield for the service of man'.[7] Pinchot's tenet regarding humans' superior relationship to a servile environment is now so ingrained that it is rarely even noticed, let alone challenged. But, in fact, it is merely an assumption, one that we can change.

Indeed, Pinchot's contemporary John Muir provided eloquent observations to the contrary:

> When we try to pick out anything by itself, we find it hitched to everything else in the universe. One fancies a heart like our own must be beating in every crystal and cell ...[8]

The sun shines not on us but in us. The rivers flow not past, but through us ... The trees wave and the flowers bloom in our bodies as well as our souls ... [9]

The tension between Pinchot's view of conservation policy as a way to ensure humans' full and proper 'use' of the environment, and Muir's conservation ethic that respected the integrity of functioning, healthy ecosystems for the benefit of all, was played out most strikingly in the fight over Hetch Hetchy Valley. Pinchot supported damming this once-magnificent valley, part of Yosemite National Park, ostensibly to provide needed water to San Francisco.[10] Muir passionately decried the plan, speaking out against

> despoiling gain-seekers ... industriously, sham-piously crying, 'Conservation, conservation, pan-utilisation', that man and beast may be fed and the dear Nation made great ... Dam Hetch Hetchy! As well dam for water-tanks the people's cathedrals and churches, for no holier temple has ever been consecrated by the heart of man.[11]

Pinchot countered by assuring decision makers that 'the highest possible use which could be made of [Hetch Hetchy] would be to supply pure water to a great centre of population'.[12] Despite Muir's own appeals to lawmakers, and his identification of numerous alternative water sources, Pinchot's view eventually prevailed. Muir's Sierra Club visited Hetch Hetchy Valley for the last time in 1914; Muir died on Christmas Eve that same year.

Pinchot's utilitarian environmental metric remains the baseline today, reflected in trending policy methodologies as 'sustainable development' and 'greenhouse gas cap-and-trade'. It also is fully incorporated into our economic, accounting and corporate governance systems, which essentially ignore environmental costs and so magnify the impacts of this flawed worldview.[13] Yet, alternative metrics were in equally widespread use in the past. California again helps illustrate the evolution of our assumptions and attitudes about our place relative to the natural world, and our options for drawing new alternatives from past experiences.

Native inhabitants of present-day California numbered roughly 300,000 in the late 1700s, when the Spanish began immigrating to the area. California's native peoples understood their environment to have an intrinsic value of its own, one where '[n]ature was neither the enemy nor simply a means to an end or a commodity to be exploited for wealth or power'.[14] Indeed, many native California groups believed that plants and animals, as well as springs and trees, possessed thought and feeling.[15] While native Californians did make changes to their environment for their own livelihood, they did so with awareness that ill-advised changes would impact not only the environment but also themselves. For example, rather than diverting significant amounts of water, native peoples generally settled near water and adjusted their lifestyles to live in harmony with their environment.[16] Though they protected their water sources from other tribes, the concept of 'private' rights in the use of water was unheard of. Water was essential to life, and it could not be bartered or sold.[17]

The Spanish brought to California a new operative attitude: that nature was to be subdued. Historic Spanish law held that 'Man has the power to do as he sees fit with those things that belong to him according to the laws of God and man'.[18] This edict, however, was tempered in practice due to strong traditions of community survival and central authority towards water, arising from the practical reality of Spain's sparse rainfall and relative lack of significant rivers. This tradition evolved into the 'Plan of Pitic', a Spanish instrument of community water rights that emphasised water sharing 'for the common benefit'. The local *ayuntamiento*, or elected town council, assured allocation of water with 'equity and justice'.[19] Water wasting was outlawed and punished. Notably, the common right of town residents to an equitable share of water carried with it a responsibility to help build and maintain the water system – the *acequia madre* – a system still used in Santa Fe, New Mexico.[20]

Spanish law and tradition was far removed from native Californians' view of water as having rights and viability of its own. But Spanish law still recognised and respected the limits of the environment, and assured equitable human access to this essential element of life. Since the water was assumed to be for the common benefit, no one person had a superior right that could be exercised to another's detriment.

The 1849 Gold Rush upended all of these attempts to live in concert with the environment. Settlers pouring into California post-1849 were 'confident, impatient, entrepreneurial, defiant of life's limitations, and determined actively to possess and develop the enormous … expanse that had now opened before them'.[21] Gold mining required water, a lot of it, and the settlers fell back on a frontier tradition of 'first in time, first in right' to allocate it. Though no one could acquire the water itself, legal rights to 'beneficially use' it were rapidly locked up. As long as the water was used, the rights to use it continued, creating a built-in incentive to keep consuming. Unlike Spanish custom, priority of use – not equity or justice – became the deciding legal factor. Unlike native Californians' traditions, little considera-tion was given to the environmental impacts of the wildly wasteful and massively injurious water practices that quickly became commonplace. And unlike both traditions, the concept of private ownership in the use of water to the exclusion of others became firmly established.

California's dammed, dyked and diverted waterscape would be unrecognisable today to pre-1849 inhabitants, and still the damage grows. The state, incredibly, has now allocated formal rights to the use of over eight times more water from the vast Bay-Delta Estuary than actually exists.[22] Sentinel freshwater species such as the iconic salmon, which once swam so thick that people claimed they could cross rivers on their backs,[23] have all but disappeared. Pollution has steadily increased, resulting in more people forced to drink from contaminated wells or pay exorbitant sums for bottled water.[24] With snow packs dwindling under the glare of climate change, these seemingly intractable water dilemmas gather significant force.

Unfortunately, rather than exploring the spreading cracks in its water law foundation, the state's most recent response has been to pass a package of laws that take a few, tentative steps forward but ultimately call for more of the same – more dams, more water diversions, and more adherence to outdated and outmoded assumptions about our ability to endlessly manipulate a second-class environment to human benefit.[25] As a California Winnemem Wintu tribal leader remarked:

> The teachings of our Spiritual Leaders, and our inherent cultural beliefs, that the salmon are our relatives, are sacred, and necessary for the continuation of life -- makes us … [s]ad that … the people who had the responsibility to actually protect them – were in fact responsible … for their near total extinction. But, has that not been the case throughout water management in California? Nothing seems to be important to those that want to take, except how much more they can get.[26]

Flawed assumptions and false jurisprudence underlying modern-day systems of water laws have let us down the wrong path. History shows, however, that assumptions can change. As has been the case with science and ethics in recent years, the law can evolve to reflect our relationship with the natural environment and our responsibilities to support the rights of all of Earth's ecosystems and inhabitants to struggle, thrive and evolve together.

Science Evolves; Ethics Evolves. Law Glances up

Current attitudes and legal systems affecting the environment were shaped in large part by a massive transformation in Western science beginning at the start of the sixteenth century. Newton and others uncovered many of the natural world's operations with dizzying speed, radically inflating perceptions of humans' ability to predict and control the environment.[27] As the natural world was reduced to quantifiable and purportedly manipulatable chunks,[28] an inherent awareness of the integrated nature of humans and their environment was gradually lost. Science historian and philosopher Carolyn Merchant reflects that '[t]he mechanical view of nature now taught in most Western schools is accepted without question as our every day, common sense reality … None of this was common sense to our 17-century counterparts'.[29] Gifford Pinchot's 'conservation ethic' of 'controlled' use of the environment is a direct descendent of this mechanical view of nature, one that allows for its manipulation to human ends.

Science, however, has evolved in modern times to better address the subjective, integrated nature of relationships among humans and the natural world.[30] For example, quantum physicists have shown that, contrary to Newtonian scientists' almost exclusive focus on objectivism and mechanics,[31] subjectivity and integration in fact play significant roles in science.[32]

Like quantum mechanics, ecological science and systems theory emerged as scientific disciplines around the turn of the twentieth century and have rapidly grown in prominence in recent decades. Modern ecological science now demonstrates that that humans' own health and welfare is inextricably and equally bound up in that of the world's natural ecosystems. Along with systems theory, it also warns that the frantic pace of modern society, one increasingly distant from the rhythm of the Earth, has almost eliminated natural feedback mechanisms that otherwise would alert us to needed changes in behaviour.[33]

Just as a philosophy of human control over the natural world grew out of Newtonian science, a growing body of modern eco-philosophy is now evolving from twentieth century quantum mechanics, ecological science and systems theory. Reaffirming Muir's prescient observation about our connections with the universe, this body of philosophy and ethics calls for respect for ecosystem relationships and co-equal rights for all Earth's community members. Cultural historian and geologian Thomas Berry's core canon that '[t]he Universe is not a collection of objects, it is a communion of subjects'[34] has been reflected in the writings of Aldo Leopold,[35] Joanna Macy,[36] Arne Naess,[37] Gary Snyder[38] and many others.

These evolving bodies of science and ethics are helping to lay a new legal foundation, one grounded in the growing awareness of our mutual relationships. Just as our current, command-and-control body of law reflects the most widely-used science and ethos at the time it was created, so should an evolving body of 'Earth law' incorporate modern science and ethics to reflect the truth and heart of our place on Earth.

Recipe for Earth Law: Fold in Ecosystems; Clarify Relationships. Avoid Climate Microwave

Modern jurisprudence, or philosophy of law, rests on outmoded, injurious perceptions of humans' ability to predict and control the natural world. Its failure to grasp the full scope of the relationships that exist among humans and the environment means that it will fail to allow and constrain human behaviour as needed to promote healthy relationships.

Our ethical and practical survival now calls us to the table to develop a system of Earth-based law that reflects our growing scientific and ethical awareness about our place in the Earth community. Thomas Berry calls for an 'Earth Jurisprudence' based on cooperative relationships as the organising principle for law and governance.[39] Cormac Cullinan expands upon its implementation in *Wild Law*.[40] Both works call for laws that would explicitly recognise the interdependence among humans and the environment and respect both sides of that relationship.

Both works also address the importance of indigenous traditions and assumptions, such as the native California tribes' respect for water as an equal community member. Numerous examples exist of indigenous cultures around the world recognising their integral connection to the natural world, and the 'reciprocal nurturance' that takes place among Earth's community members.[41] The wisdom of such traditions, along with modern science and evolving ethics, can be brought to bear in the development of an Earth-based legal system that enhances our environment and, by extension, our own lives.

Communities around the United States already are starting to adopt laws reflecting these relationships. Driven to action primarily by outside attempts to injure local waterways and lands, U.S. municipalities are adopting ordinances that specifically recognise an 'enforceable right of natural communities and ecosystems to exist and flourish'.[42] On the larger world stage, Ecuador recently became the first nation to adopt a constitutional provision endowing nature with inalienable, enforceable rights.[43] Its new Constitution states that:

> Nature or Pachamama, where life is reproduced and exists, has the right to exist, persist, maintain and regenerate its vital cycles, structure, functions and its processes in evolution.[44]

The Constitution provides the natural world with a 'right to restoration' that is independent of humans' right to compensation.[45] It endows '[e]very person, people, community or nationality' the right and responsibility to 'demand the recognitions of rights for nature before the public bodies'. These rights and remedies reflect the recommendations of law professor Christopher Stone, who wrote that legal rights must be enforceable by or on behalf of the injured entity, and must include remedies that run to the benefit of the injured holder of the right – in this case, the environment.[46] Actions by the public on behalf of the natural world are essential to a rights-based, 'Earth law' system.

Our challenge is to take up and expand these nascent efforts and rework our legal system to reflect mounting scientific awareness of our integrated relationships with the natural world, and our corresponding ethical responsibility to nurture those relationships. California again lends itself to an example. Currently, California law addresses waterways' needs for healthy, clean flows only indirectly, through such methods as conditions in diversion and pollution permits. No waterways hold legal rights to the water they need to survive. As a result, California's ecosystem integrity increasingly relies on a last-gasp application of the Endangered Species Act when existing strategies ultimately fail. Climate change, inappropriate uses, and other threats are pushing California to the limit on water. Taking the environment's share only delays the day of reckoning with our own. We must break our dangerously well-trod path of use, overuse, environmental decline, then hasty and unplanned reaction, and instead chart a better course.

One way to better reflect the mutual rights of humans and the natural world in this example would be to allocate legal water rights to ecosystems, equivalent to those rights currently allocated to human uses. Formal, legal water rights would be allocated to enough water to allow the ecosystems to thrive. These rights would be enforced by independent legal guardians representing the ecosystems. This approach would allow ecosystems' water needs to be planned for and protected from the beginning, rather than after extensive damage has been done.[47]

Importantly, even this example falls short of the wisdom of native California traditions, which held the concept of private rights in the use of water as an anathema. Water rights allocations should be viewed only as a system of accounting among users, rather than a system of accumulation for individual benefit to the exclusion of all others. Allocating water rights to ecosystems is only an interim step toward the larger goal of developing and implementing a system of Earth law that acknowledges and nurtures the rights of the natural world, and respectfully allows the natural world to nurture us in turn.[48]

Letters from 2050

The wisdom gained from self-reflection prompts the inevitable question, 'How do we live?' Environmental prophet Aldo Leopold wrote that the 'virtue of a living democracy' is its ability not to avoid mistakes, but to learn and eventually benefit from them.[49] Creating a system of Earth law that recognises the equal, inherent rights of all Earth's community members to struggle, thrive and evolve together is a daunting task. Fortunately, our burgeoning efforts in this regard will begin to create their own feedback loop – one where science and ethics drives law, which drives culture, which drives further evolution in law, science and ethics – until the law and the culture meet, and we cannot envision a time where our laws relegated the natural world to second-class status.

This feedback loop will drive evolution of other disciplines as well. Our policy, economics, and financial and corporate governance systems similarly arose from a flawed foundation of 'control over nature'. The success of our efforts to remake our legal systems will drive, and be enhanced by, similar evolution in these other disciplines. Along this path, 'environmentalism' itself will evolve from a subset of the population acting to safeguard the planet, into a deeply-felt awareness in the hearts and minds of all individuals, an awareness that guides how we live our lives and make our daily choices.

Thomas Paine wrote of British control over the American territories that '[t]here was a time when it was proper, and there is a proper time for it to cease'.[50] Our current legal system grew out of an era in which society believed that humans could, and therefore should, control the environment. Modern science and ethics increasingly inform us differently, and we are beginning to appreciate the many benefits of respectful relationships with the natural world. We can similarly advance our legal system into a body of Earth law that embeds and implements such mutual, respectful rights. These efforts will lead us, at the end of the next forty years, toward a vibrant, joyous Earth Day every day.

Notes

1   American cartoonist Walt Kelly first used the quote 'We Have Met the Enemy and He Is Us' on the wildly popular 1970 Earth Day poster. In the forward to his earlier *The Pogo Papers*, Kelly expounded: 'There is no need to sally forth, for it remains true that those things which make us human are, curiously enough, always close at hand. Resolve then, that on this very ground, with small flags waving and tinny blast on tiny trumpets, we shall meet the enemy, and not only may he be ours, he may be us'. Walt Kelley, *The Pogo Papers*, Simon and Schuster, New York, 1953.

2 Secretariat of the Convention on Biologic Diversity, 'Status and Trends of Global Biodiversity' in *Global Biodiversity Outlook*, Montreal, 2001, available at: http://www.cbd.int/gbo1/chap-01–02.shtml.

3 ibid.

4 California Natural Resources Agency, *2009 Climate Change Adaptation Strategy*, Dec. 2009, p. 3, available at: http://www.energy.ca.gov/2009publications/CNRA-1000-2009-027/CNRA-1000-2009-027-F.PDF.

5 ibid. (citing 'tens of billions of dollars per year in direct costs' and exposure of '*trillions* of dollars of assets to 'collateral risk' resulting from climate change) (emphasis in original).

6 The Endangered Species Committee, or 'God Squad', was created by Congress in reaction to a U.S. Supreme Court decision that stopped the construction of the Tellico Dam because it jeopardised the endangered snail darter. *Tennessee Valley Authority v. Hill*, 437 U.S. 153 (1978); 16 U.S.C. § 1536(e). The 'God Squad' decides whether to grant exemptions for federal agency actions that would otherwise trigger species protection requirements under the Act. 16 U.S.C. §§ 1536(a)(2), (e), (h). See also Ted Gup, 'Down with the God Squad,' *TIME*, Nov. 5, 1990, available at: http://www.time.com/time/magazine/article/0,9171,971548,00.html ('Man cannot manage nature through a series of ad hoc rescue attempts, ignoring the underlying causes for the loss of biodiversity'). Endangered Species Act protections are being increasingly challenged by those most affected by the poor government planning that eviscerates species. For example, a proposed amendment to 2009 House appropriation bill H.R. 2847 by California Congressman Devin Nunes (Visalia) would have removed funding for court-mandated protections for endangered salmon. The amendment was defeated, but only in an extremely close June 2009 vote. See California Sportfishing Protection Alliance, *Hold your Congressman responsible: How the California delegation voted on the Nunes amendment to HR 2847*, June 19, 2009, available at: http://www.calsport.org.

7 Gifford Pinchot, *The Training of a Forester*, J.B. Lippincott Co., Philadelphia, 1914, p. 13.

8 John Muir, 'Mount Hoffman and Lake Tenaya' in *My First Summer in the Sierra*, Houghton Mifflin Co., Boston and New York, 1911, p. 211.

9 John Muir, 'Mountain Thoughts' in *John of the Mountains: The Unpublished Journals of John Muir*, ed. Linnie Marsh Wolfe, Univ. of Wisconsin Press, Madison, 1979, p. 92.

10 Efforts are underway to restore the Hetch Hetchy Valley; *see* Restore Hetch Hetchy at http://www.hetchhetchy.org/.

11 John Muir, *The Yosemite*, Century, New York, 1912, pp. 255–257, 260–262. Reprinted in Roderick Nash, *The American Environment: Readings in The History of Conservation*, Addison-Wesley Publishing Company, Reading, Mass., 1968.

12 Char Miller, *Gifford Pinchot and the Making of Modern Environmentalism*, Island Press, Washington D.C., 2001, pp. 139–40.

13 *See, e.g.,* E.F. Schumacher, *Small Is Beautiful,* Harper & Row Publishers, New York City, 1973; *see also* Marjorie Kelly, *The Divine Right of Capital*, Berrett-Koehler Publishers, San Francisco, 2001.

14 Norris Hundley, Jr., *The Great Thirst*, University of California Press, Berkeley, 2001, pp. 1–2.

15 ibid., p. 4.

16 ibid., p. 5.

17 ibid., p. 25.

18 ibid., p. 28, citing *Las siete partidas del sabio rey don Alfonso 1265*, pt. 3, title 28, law 1.

19 ibid., p. 41, referencing Plan of Pitic, art. 20 (English translation at John W. Dwinelle, *The Colonial History: City of San Francisco*, 4th ed., Towne and Bacon, San Francisco, 1867, addenda 7).

20 ibid., p. 43.

21 ibid., p. 67, citing Robert Kelley, *Battling the Inland Sea: American Political Culture, Public Policy and the Sacramento Valley, 1850–1986*, University of California Press, Berkeley, 1989, p. 14.

22 State Water Resources Control Board, *Water Rights within the Bay/Delta Watershed*, Sept. 26, 2008, pp. 3–4, available at: http://deltavision.ca.gov/BlueRibbonTaskForce/Oct2008/Respnose_from_SWRCB.pdf.

23 *See, e.g.*, Freeman House, *Totem Salmon: Life Lessons from Another Species*, Beacon Press, Boston, 1999, pp. 11, 123.

24 For example, 41% of drinking water wells sampled in Tulare County, California had illegal levels of nitrate, and bacteria were found in a third of the wells. California State Water Resources Control Board, *Domestic Well Project,* 2006, available at: http://www.swrcb.ca.gov/gama/voluntary.shtml#tularecfa. The poverty level of Tulare County is twice that of the state of California as a whole. U.S. Census Bureau, Tulare County (2007 data), available at: http://quickfacts.census.gov/qfd/states/06/06107.html.

25 *See, e.g.*, SB X7 2 (Cogdill), *Safe, Clean, and Reliable Drinking Water Supply Act of 2010* (signed into law Nov. 9, 2009), available at: http://www.leginfo.ca.gov/pub/09–10/bill/sen/sb_0001–0050/sbx7_2_bill_20091109_chaptered.pdf. *See also* Matt Weiser, 'Water package lacks clout to reverse Delta's decline', *Sacramento Bee*, Nov. 15, 2009, p. 1A, available at: http://www.sacbee.com/politics/story/2326156.html.

26 Gary Mulcahy, Winnemem Wintu Tribe, 'Judge Tosses Biological Opinion for Salmon and Steelhead in California', *Earthjustice Press Release*, April 16, 2008, available at: http://www.earthjustice.org/news/press/2008/judge-tosses-biological-opinion-for-salmon-and-steelhead-in-california.html.

27 Brian Goodwin, *Nature's Due: Healing Our Fragmented Culture*, Floris Books, Edinburgh, 2007, pp. 69–79.

28 For example, Newton wrote in 1722 that 'God created everything by number, weight and measure' (quoted in Roger V. Jean and Denis Barabé, *Symmetry in Plants*, World Scientific Publishing Co., Singapore, 1998, p. xxxvii); Descartes reduced the scientific understanding of the world to a 'problem of mechanics' (Alfred Weber, 'Descartes' in *History of Philosophy*, Charles Scribner's Sons, New York, 1908, referencing René Descartes, *Les Principes de la philosophie*, II., III., 1644); and Galileo called on scholars to '[m]easure what is measurable, and make measurable what is not so' (quoted in H. Weyl, 'Mathematics and the Laws of Nature' in *The Armchair Science Reader*, eds I. Gordon and S. Sorkin, Simon and Schuster, New York, 1959).

29 Carolyn Merchant, *The Death of Nature: Women, Ecology, and the Scientific Revolution*, HarperCollins Publishers, New York, 1980, p. 193.

30 Goodwin, *Nature's Due*, pp. 69–70.

31 It should be noted that Newton's Third Law ('for every action, there is an equal and opposite reaction') does acknowledge that there is no unidirectional force, and that forces are instead a series of interactions. Newton's *Principia*, Volume 1, reads: 'Whatever draws or presses another is as much drawn or pressed by that other. If you press a stone with your finger, the finger is also pressed by the stone. If a horse draws a stone tied to a rope, the horse (if I may so say) will be equally drawn back towards the stone ...' Implementation of the Third Law with regard to the natural world since Newton, however, has skewed this basic truth. Humans almost exclusively look at forces through the lens of human action, rather than also through the lens of the natural world, to the detriment of the environment.

32 For example, modern science has demonstrated that mere observation itself can influence physical processes taking place. Modern scientists have also shown that separate quantum particles (such as photons of light or atoms) can actually act as parts of the same entity, reflecting changes in the other individual particle instantly, as if communicating. For technical information related to quantum theories such as the Schrödinger's cat thought experiment, the double-slit experiment and quantum entanglement theory, *see* MIT's OpenCourseWare site at: http://ocw.mit.edu/OcwWeb/web/courses/courses/index.htm#Physics. More philosophical reflections and analysis can be found at The Dalai Lama, *The Universe in a Single Atom*, Morgan Road Books, New York, 2005, and Gary Zukav, *The Dancing Wu Li Masters*, William Morrow and Company, Inc., New York, 1979.

33 *See, e.g.,* Howard T. Odum, *Systems Ecology: An Introduction*, John Wiley & Sons, New York, NY, 1983. *See also* Joanna Macy, *Mutual Causality in Buddhism and General Systems Theory*, State University of New York Press, Albany, 1991. See also John D. Sterman, 'A Banquet of Consequences: Systems Thinking and Modeling for Climate Change Policy', MIT SDM Systems Thinking Conference 2010, Oct. 21, 2010, available at: http://sdm.mit.edu/systems_thinking_conference_2010/presentations/sterman.pdf.

34 Thomas Berry, *The Great Work*, Bell Tower, New York City, 1999.

35 Aldo Leopold, *A Sand County Almanac*, Oxford University Press, Oxford, 1949.

36 Joanna Macy, *World as Lover, World as Self*, Parallax Press, Berkeley, 1991.

37 Arne Naess, 'The Shallow and the Deep, Long-Range Ecology Movement', *Inquiry*, Vol. 16, 1973, pp. 95–100.

38 Gary Snyder, *The Practice of the Wild*, Shoemaker & Hoard, Emeryville, 1990. *See also* Jack Turner, *The Abstract Wild*, Univ. of Arizona Press, Tucson, 1996.

39 Berry, *The Great Work*.

40 Cormac Cullinan, *Wild Law*, Green Books, Totnes, 2003.

41 The Andean indigenous tradition of *uywa*, or 'reciprocal nurturance', acknowledges the give-and-take relationships that exist among equal partners in the human/natural world community. The ongoing suite of relational 'conversations' inherent in the practice of *uywa* 'dissolves the hierarchical relationship between humans and

nature so rooted in modern Western culture in which nature is at the service of man'. Grimaldo Rengifo Vasquez, 'Nurturance in the Andes' in *Rethinking Freire: Globalization and the Environmental Crisis*, eds C.A. Boewers and Frederique Apffel-Marglin, Lawrence Erlbaum Associates, Inc., Mahwah, N.J., 2005, p. 40. 'Nurturing in the sense of the word *uywa* ... is a conversation, affective and reciprocal between equivalents ... Conversation does not end in an action that is the responsibility of someone to be changed, but in mutual nurturing'. ibid., pp. 40–41. *See also* Frederique Apffel-Marglin and PRATEC, *The Spirit of Regeneration: Andean Culture Confronting Western Notions of Development*, Palgrave, New York City, 1998.

42 *See* Community Environmental Legal Defense Fund, www.celdf.org.

43 'Rights of Nature', *Constitución de la República del Ecuador*, Title II, Ch. 7, adopted Sept. 2008, available at: http://www.asambleanacional.gov.ec/documentos/constitu-cion_de_bolsillo.pdf.

44 ibid., Art. 71 (translated at CELDF, *Ecuador Approves New Constitution: Voters Approve Rights of Nature*, Sept. 28, 2008, available at: http://www.celdf.org/Default. aspx?tabid=548).

45 ibid., Art. 72.

46 Christopher D. Stone, *Should Trees Have Standing: Toward Legal Rights for Natural Objects* William Kaufmann, Inc., Los Angeles, 1974.

47 Ecosystems have the right not only to sufficient water, but also to clean water. Research shows that salmon die when exposed to combinations of pesticides that are harmless individually, exposing major flaws in our pollutant-by-pollutant regulatory system. *See, e.g.*, Cathy Laetz *et al*, 'The Synergistic Toxicity of Pesticide Mixtures: Implications for Risk Assessment and the Conservation of Endangered Pacific Salmon', *Environmental Health Perspectives,* Vol. 117, No. 3, March 2009, pp. 348–353, available at: http://www.eenews.net/public/25/9960/features/documents/2009/03/03/document_gw_01. pdf. Unfortunately, contaminants on an individual basis regularly exceed safe limits, increasing the danger to salmon – and humans – further. For example, toxic contamination is so ubiquitous in certain areas of the Central Valley that a USGS study found nervous system pesticides in all *rainfall* samples collected. Celia Zamora *et al*, 'Diazinon and Chlorpyrifos Loads in Precipitation and Urban and Agricultural Storm Runoff during January and February 2001 in the San Joaquin River Basin, California' in USGS, *Water – Resources Investigation Report 03–4091*, Sacramento, CA, 2003, available at http://pubs.usgs.gov/wri/wri034091/.

48 For further reflections on the interactions among water, humans and earth, *see* Theodor Schwenk, *Sensitive Chaos: The Creation of Flowing Forms in Water and Air* Sophia Books, Rudolph Steiner Press, East Sussex, 1965, rev'd translation 1996. 'Water and [the] spiritual activity of the human being belong together ... through water flows the wisdom of the universe'. ibid., p. 97.

49 Quoted in Julianne Lutz Newton, *Aldo Leopold's Odyssey*, Island Press/Shearwater Books: Washington D.C., 2006, p. 83.

50 Thomas Paine, *Common Sense* (1776).

Stories from the
Environmental Frontier

Mari Margil

It should come as no surprise to the readers of this book that our planet is dying, and our communities are dying along with it. By almost every measure, the environment today is in worse shape than when the major environmental laws were adopted in the United States over thirty years ago. Since then, countries around the world have replicated these laws. Yet, species are disappearing. The climate is on fire. Rainforests are being destroyed. Global fisheries are collapsing. Coral reefs are being wiped out.

Clearly something isn't working. And yet, traditional environmental laws continue to be instituted, as though if we just do enough of it, it will work.

There are those who think otherwise, who believe that the current structure of laws and regulations can't protect the environment, because in fact they were never designed to. And there's an organisation in the United States which – having worked within the existing structure of laws and regulations – decided to turn its back on that structure, recognising that it isn't working and something fundamentally different needed to replace it.

The Community Environmental Legal Defense Fund

The Community Environmental Legal Defense Fund was founded in 1995 in the State of Pennsylvania, to help communities protect the natural environment by stopping new incinerators, factory hog farms, and other projects. The Legal Defense Fund helped communities appeal to their state environmental agencies to stop them, but what we found was that the very agencies we looked to for help, were instead handing out permits to corporations to build those incinerators and factory farms. We helped hundreds of communities appeal these corporate permits – but even when we won, we lost. This is because the corporations would either re-write the law, or exhaust communities with permit application after permit application, until they could site their project. What our experience showed us was that our system of environmental laws and regulations *don't actually protect the environment*. At best, they merely slow the rate of its destruction.

After several years, we stopped doing that work. We weren't helping anyone protect anything. Our work has now fundamentally changed and over the past few years, we've had a chance to work with folks like Michael Vacca, a construction worker from western Pennsylvania who wanted to protect his community from coal mining; Jack O'Neill, a Vietnam War veteran and elected Selectboard member from Barnstead, New Hampshire, working to stop the privatisation of his community's water; and Alberto Acosta, former president of the Constitutional Assembly of Ecuador, who'd seen his country ravaged by multinational oil corporations. They're three people, facing three seemingly different problems, who found that they couldn't protect the places where they live because the environmental laws that they looked to for help seemed to have very little to do with actually protecting the environment.

Blaine Township, Pennsylvania

In 2006 our phone rang with a call from Michael Vacca. Michael lives in the tiny, rural Township of Blaine in western Pennsylvania deep in the heart of coal country. Over the past two decades, communities across the region have been devastated by something called longwall coal mining. Mining corporations drive their longwall machines underground, ripping out massive panels of coal over two miles long. When the coal is gone, the land above is unsupported and caves in. Houses, roads, and farmland fall into the mine. Rivers and streams run dry.

Michael wanted to stop the mining, but had seen other communities fight and lose their battles to stop it. As vice chairman of Blaine's Planning Commission, he wanted to see if he could use local zoning laws to block the mining. But instead he found himself stuck inside a box – the same box that thousands of communities across the U.S. have found themselves stuck in – in which the law doesn't give them the legal authority to say 'No.'

That first call with Michael lasted two hours, and at the end he invited us to hold a 'Democracy School' in Blaine. The Democracy Schools are three-day workshops at which we help communities examine how our structure of law works, and for whom.

All three elected Blaine Supervisors were at the Democracy School.

Darlene Dutton, one of the supervisors, asked us why it seemed that a mining corporation – in this case, Penn Ridge Coal headquartered nearly a thousand miles away in Tulsa, Oklahoma – was able to decide what happened in Blaine, rather than the people who actually live there. She then asked us why the Pennsylvania Department of Environmental Protection was actually giving coal corporations the legal authority to mine, when the devastating impacts from the mining couldn't have been more clear. These were not easy questions to answer. But they're being asked more and more, as people and communities across the country are finding that they don't have the legal authority to protect the environment.

In response to Darlene's questions, we talked about how our environmental laws work – how they're based on the idea that *nature is property*. Meaning our environmental regulatory laws merely *regulate* the rate at which nature is used. Knowing this, it's not so surprising then to learn that the major U.S. environmental laws – such as the Clean Air Act and the Clean Water Act – were passed under the authority of the Commerce Clause of the U.S. Constitution. Thus treating the environment merely as a natural resource necessary for *commerce*, rather than as ecosystems to be protected in their own right.

Some have compared how the law treats nature as to how we once treated slaves – as a thing to be used until it was no longer. This is because nature is considered *rightless*, and as such, the people of Blaine Township, in trying to protect nature, found that they could not defend the rights of the ecosystems in Blaine, because *there were no rights to defend*.

Scott Weiss, chairman of the Blaine Supervisors, then asked us about something called 'standing' – the legal requirement that you need to prove you have 'standing' in order to go to court to protect nature – meaning you've experienced some direct harm from logging, or the pollution of a river. This means that you have to prove that destruction of the environment somehow directly injures you. Damages are then awarded to you and not the ecosystem that's been destroyed. Much like a slave owner could receive monetary damages if someone beat his slave – the slave couldn't receive damages, but the slave owner could because damage had occurred to *his* property.

At the conclusion of the Democracy School, the Blaine Supervisors asked us to help them draft a set of local laws, or ordinances, which would ban longwall coal mining while declaring that ecosystems have rights within Blaine Township. Passed unanimously by the Supervisors in 2006, the ordinances did three things:

- First, they banned corporations from mining.
- Second, they recognised the rights of ecosystems.
- And third, they stripped corporations of their power to override the ordinances.

First, in support of the ban on mining, the ordinances declared 'that the Department of Environmental Protection's enabling of mining corporations has not been the exception in this state and nation, but a normal governmental practice'. Second, the ordinances established that ecosystems – including wetlands, rivers, and streams – possess 'inalienable and fundamental rights to exist and flourish within the Township of Blaine'. As well, the people of Blaine had the ability to defend the rights of ecosystems without having to prove standing, and damages were to be measured by harm caused to the ecosystem itself. Third, the ordinances stripped corporations of something called 'corporate constitutional rights'. Corporations, declared by the courts to be *persons* under the law, enjoy protections under the U.S. Constitution including First Amendment free speech rights, Fifth Amendment rights to due process, and Fourteenth Amendment rights to equal protection.

At the Blaine Democracy School, the supervisors asked us why corporate constitutional rights matter. Our answer was that they matter because corporations are able to wield these rights against you, against communities and laws that seek to protect the environment. These constitutional rights guarantee that corporations can lobby the U.S. Congress to let them build new coal-fired power plants. They use them to protect their ability to siphon off our water or longwall coal mine. They use them to stop us from doing anything in our activism that will actually change how the system of law operates.

In many ways, what the people of Blaine were doing was flipping the law on its head – so instead of the law protecting the rights of property and commerce, they were using the law to protect the rights of people, communities, and nature. We talked about what happens when people start to reject the system of law they're living under. What happened when the early Abolitionists began to organise – that they were declared treasonous and every effort was made to shut them down. What happened when the Suffragists fought for rights for women – that they were arrested and called radicals. How those who feel threatened by change will do everything they can to stop it.

In Blaine, that would be the mining corporations. And as expected, in the fall of 2008, two coal corporations sued Blaine Township to overturn their ordinances. They argued that the community didn't have the legal authority to ban mining, and that the ordinances violate their corporate constitutional rights. Instead of backing down or counting on the courts to save them, the people of Blaine decided instead to up the ante. They drafted a Home Rule Municipal Charter incorporating the rights of ecosystems and stripping corporations of constitutional rights. The charter constitutionalised the ordinances, and, if it had been adopted, it would have become the nation's first local sustainability constitution. Intense opposition fuelled by mining interests helped defeat the charter at the November 2009 general election. Blaine Township will now face the longwall machines.

Town of Barnstead, New Hampshire

Like Blaine, the town of Barnstead in central New Hampshire is rural and largely conservative. What they faced there wasn't mining of coal, but of water. Companies like Nestlé are targeting communities across the U.S. for their water. Just up the road from Barnstead, the USA Springs corporation had set its sights on the town of Nottingham. The company sought a permit from the state to withdraw over 400,000 gallons of water a day to bottle and sell overseas.

The people of Nottingham had fought for seven years to stop USA Springs from coming in and privatising their water. They appealed permits to the state Department of Environmental Services, circulated petitions, lobbied their state legislature, held protests, and filed lawsuits. They did everything 'right' through conventional, environmental organising, but somehow they still weren't winning.

Down the road at a Barnstead Democracy School, Jack O'Neill, a member of the town's Selectboard, asked us why the state environmental agency seemed to him, to be more interested in granting corporations permits to take their water, then helping people in the community protect it. Turns out, as we cover in the

Democracy Schools, there's a reason why that is. Over a hundred years ago the United States' first regulatory agency, the Interstate Commerce Commission, was created at the request of the railroad corporations – the Wal-Marts of their day. As the U.S. Attorney General, Richard Olney, told the president of Burlington Railroad back in 1893, the agency:

> is, or can be made, of great help to the railroads. It satisfied the popular clamor for government supervision … at the same time that the supervision is almost entirely nominal.

He went on to say that the agency acts as 'a sort of barrier between the railroad corporations and the people.'

As one Barnstead resident put it, it seemed as though nothing had changed in over a hundred years. To the folks in Barnstead it seemed that if they took the path of Nottingham, it was only a matter of time before a corporation came along and took their water. Because of that, Jack O'Neill and the other selectboard members asked us to draft an ordinance that would ban corporations from coming in and siphoning off their water, and which offered the best and highest protection for their aquifer. They also wanted the ordinance to strip corporations of their ability to override the community's law making.

We worked hand-in-hand with them to draft the ordinance. Like Blaine's, the Barnstead ordinance:

- Recognises that ecosystems have legally enforceable rights.
- Bans certain corporations from carrying out activities the community doesn't want.
- And lastly, strips corporations of constitutional protections.

Adopting the ordinance at a 2006 Town Meeting by a vote of 135 to 1, Barnstead became the *first community in the nation to ban corporations from privatising their water*. The Barnstead ordinance, in recognising ecosystem rights, stated:

> Natural communities and ecosystems possess inalienable and fundamental rights to exist and flourish within the Town of Barnstead. Ecosystems shall include, but not be limited to, wetlands, streams, rivers, aquifers, and other water systems.

Folks still struggling to protect their water in neighbouring Nottingham soon called us. They wanted us to draft them an ordinance modelled on Barnstead's. Gail Mills, who with her husband Chris became leaders in the campaign to pass the ordinance, explained their decision to turn their back on the environmental regulatory system that they'd fought in for so long. She said, 'We have to go out and make our own history, and not let others define it for us.'

In March 2008, the people of Nottingham made history. They voted to adopt the ordinance at their Town Meeting, banning corporations from privatising their water, recognising the inalienable rights of ecosystems, and stripping corporations of constitutional rights.

This work is spreading in New England as the threat from Nestlé and other corporations grows. Following in the footsteps of Barnstead and Nottingham, the towns of Shapleigh and Newfield recently became the first communities in the State of Maine to ban corporations from privatising their water and recognise the rights of ecosystems.

Ecuador

These stories from communities in Pennsylvania, New England, and elsewhere were shared with folks at the non-profit Pachamama Alliance – which has offices in San Francisco and Ecuador.

In 2007, Ecuador began the process of drafting a new constitution. For centuries, the people and landscapes of Ecuador have been exploited by outsiders. And in recent years, it was revealed that Texaco had dumped more than 18 billion gallons of toxic wastewater into the Ecuadorian rainforest.

The Pachamama Alliance invited us to Ecuador to meet with elected delegates to the Constitutional Assembly. We were not experts in Ecuadorian law, but there are similarities that cut across international lines. There, like in the U.S., the law treated nature as property. We told the delegates stories of Blaine and Barnstead, and how the people in those communities understood that without fundamentally changing how we treated nature in law they could not protect it. We shared with the delegates how we worked the communities to draft and adopt new laws recognising legally enforceable rights of ecosystems. We also had the opportunity to meet with the president of the Constitutional Assembly, Alberto Acosta.

We thought that we'd have an uphill battle trying to explain to this former minister of energy and mines why communities in the U.S. were adopting laws recognising ecosystem rights. But before we had a chance to say anything, he told us that to his mind, the law treats nature as a slave, with no rights of its own. We had found a meeting of the minds in one of the most unlikely, but most critical of places.

We were asked to draft language for the delegates, and over a series of months they shaped and expanded the language. And in September 2008, the people of Ecuador approved the new constitution, becoming the very first country in the world to recognise in its constitution rights of ecosystems. Article 1 of the chapter on Rights of Nature states:

> Nature or Pachamama, where life is reproduced and exists, has the right to exist, persist, maintain itself and regenerate its own vital cycles, structure, functions and its evolutionary processes.

Conclusion

In 1973, Professor Christopher Stone penned his famous article, 'Should Trees Have Standing?' He explained this idea of rights of nature, and why it's so hard for us to think about those without rights – the 'rightless' – as possibly having rights.

He described why every time a movement is launched to recognise rights for the rightless – like the abolitionists did and the suffragists did – the movements and the people involved are deemed treasonous and radical. He wrote:

the fact is, that each time there is a movement to confer rights onto some new 'entity' the proposal is bound to sound odd or frightening or laughable. This is partly because until the rightless thing receives its rights, we cannot see it as anything but a thing for the use of 'us' – us being, of course, those of us who hold rights.

The people in the communities we work with recognise that the structure of law was never intended to protect the environment, but instead to regulate its exploitation, and that they must write new structures of law – maybe writing their own constitutions – to replace it. These are not people who call themselves activists, or for that matter environmentalists. But they recognise that in order to change the existing structure of law a movement for nature's rights is necessary. It is time we heard their voices and joined their cause.

The Lorax asked, 'Who speaks for the trees?' The people of Ecuador, Blaine, Barnstead, Nottingham, and a dozen other communities have answered, 'We do.'

And now I ask all of you. Will you speak for the trees?

For if not you, then who?

And if not now, then when?

Section 3:
Ecological Conceptions of Property

In Western law, the natural world is defined as human property, which can be used and exploited to satisfy human preferences. Nature receives value through human use, and protection through human property rights. This situation presents a critical challenge for Earth Jurisprudence, which seeks to evolve our present understanding of property to recognise the inherent value of nature and our relationship with it. This section will consider a growing movement toward an ecological understanding of property and toward de-growth.

Owning the Earth

Nicole Graham

The Earth and jurisprudence are both systems. The Earth is a system of physical and interlinked relationships. Jurisprudence is a system of abstract laws. Jurisprudence is a human creation. As such, jurisprudence is a system that depends for its existence on the systems of Earth because the former is the creation of a species whose existence is of the latter. It is therefore important, indeed necessary, to situate the system of laws within the physical context of the Earth's systems, because although the law currently situates itself above or separate to the physical realm, in reality the converse is true. Humans are physical beings dependent on, and subject to, their only home and ultimate jurisdiction – Earth. In a discussion of the idea of Earth Jurisprudence it is necessary to consider the laws of land and water ownership and use in terms of their physical (or economic) viability. In other words, laws that regulate the ownership and use of the Earth and its resources would need to both facilitate and regulate human-earth relations that are consistent with the unilateral dependence of humans and their socio-economic systems on the Earth's systems.

This chapter considers the notions of human ownership and property in the Earth and its resources through an exploration of two questions. First, what is it, precisely, that is thought to be owned? Second, what does ownership mean? Section II of the chapter addresses the first question by exploring the object of ownership in modern property law. Dominated by the notion of 'rights', modern property law lacks any referent to the physical realm and thus, disturbingly, disconnects ownership from the physical conditions of the land. Section III challenges the concept of ownership as rights in modern property law. It discusses the potential cultural and environmental significance of embedding the concept of responsibility for that which is owned within the property law of an Earth Jurisprudence. Finally, Section IV of the chapter contrasts the notion of property as entitlement with property as responsibility in the context of the interaction of human systems with the earth's systems. It argues that positioning the knowledge of the land prior to the exercise of any rights and responsibilities of ownership would render property law viable and enduring. The achievement of an Earth Jurisprudence means the recognition and integration of the capacities and limits of the Earth's systems within the laws that regulate the ownership and use of the resources of those systems.

Owning the Earth as a Right

Property law would seem an obvious starting point for a discussion of the laws that regulate land use and ownership. Or would it? If you have studied a property law course you would have learned about different forms of title (e.g. Old System Title, Torrens Title, Qualified Title, Possessory Title and Native Title) and about how those forms of title are secured and lost against other claims to the title (priorities). You would have learned about whether and how those titles could be used to secure loans of money (mortgages). You would have learned about how holders of the title could sell pieces of their property to other people both temporarily (leases are the sale of the right of possession for a period of time) and permanently (easements, covenants and the sale of all or a portion of the property). You would have learned about the ways that people can share title to property with others (co-ownership). But you would not have learned about the ownership of land – because property is not about land.

In law schools across the Anglophone world, one of the first things that students are taught is that property law is not about real things such as land or water. Modern property law is about abstract 'rights' between 'persons'. So what you would learn in a course on property law has nothing to do with the ownership of land – it concerns only the ownership of abstract things, 'rights'. Property law regulates the relationships we have with each other, not the relationships we have with the Earth. What we 'own' is not, in a legal sense, land. What we 'own' is a 'right' against another 'person' – the land is irrelevant. Property law is not about the physical world – property law is about the metaphysical world of human creation. For this reason, property is referred to by lawyers and scholars as 'dephysicalised'.[1] The idea that property law is not about the real and physical world is a point repeated in the case law, in legislation and by eminent legal scholars. And it is a point that has been made for centuries.[2]

Does it matter that land is irrelevant to the law that governs its ownership? Or in legal terms, does it matter that property is 'dephysicalised'? If the ownership of land is not the same thing as the use of land, which is addressed by another area of law, environmental law, then why do these questions matter? It matters because ownership facilitates particular kinds of land and water uses and because these uses have physical, and potentially adverse, consequences for those lands and waters. The human ownership and use of various parcels of the Earth and its resources are directly related. Our jurisprudence or system of laws should therefore reflect this direct relationship through the alignment of the law that governs use and ownership – the alignment of property law and environmental law. But the law that regulates human relationships to the land in contemporary jurisprudence is divided into two. Both property law and environmental law abstract the relationship between people and place, between humans and the Earth. Both laws dephysicalise what is a physical relationship in a physical world.

Environmental law exists independently of the physical consequences of its jurisdiction and is based on the rules and regulations of the governments and councils of the day. Environmental law is subordinate to property law.[3] Property rights are important to the cultural identity, economy and law of modern Anglophone societies. In some jurisdictions, property rights are regarded as fundamental rights of citizenship, are equated with liberty,[4] and are protected in national constitutions. The task of governments, in environmental regulation, can thus be a difficult one that challenges the long-established cultural and legal priority of property law and its notion of ownership as 'rights'. Disputes between people over parcels of the Earth and its resources are often framed as one between the interests of society (freedom) and the needs of society (sustainable environment). These disputes pit proprietorship against environmentalism in a false and adversarial picture of a relationship between people and place that cannot be anything other than one of dependency. The fact is there would be no proprietorship, no property law, no concept of ownership, if there were no 'thing' to own – no Earth. The tension and conflict between property rights and environmental regulation forgets that we, and our system of laws, depend on the Earth's systems and cannot, truly, be thought to somehow triumph against it.

Given the legal priority of property rights, and given our cultural preoccupation with entitlement, the content of property rights requires some clarification. What kinds of rights, precisely, are property rights? Property rights, in modern law, provide the right-holder first and foremost with the right to exclude all others from the object of property. The right of exclusion is considered the foundation of the system of private property. The other important right held by the property right holder is the right to dispose of or alienate the property. The sale of part or all of one's property can be achieved only by the right to dispose of it as one chooses – to alienate the land from one's self. The right of disposition or alienation is also a central aspect of the system of private property. Another important right of a property right-holder is so important that it almost goes without saying. The holder of a property right may very well believe that their right entitles them to a diverse range of activities on and uses of one's 'property' notwithstanding the views of others. These activities and uses are regarded as the purpose of ownership – the benefits or profits of ownership. Without these, the very idea of ownership seems pointless.

In Australian and American case law, cultural and judicial perceptions of property are expressed and debated in terms of the right to use one's property in a way that ensures the benefits and profits of ownership. Courts hear and uphold arguments that without the benefits and profits, the property right is worthless. The conflation of the land with its commercial value and potential is all but unchallenged both in our culture and at law. In Australia, the High Court case of *Newcrest Mining*,[5] for example, saw the argument of a mining company that although their property right had not been compulsorily acquired or taken by the government of the day, it may as well have been because changes to land use law that prohibited mining activities in the land in question, effectively 'sterilised' the property right itself. The sterilisation of the land by the mining activity was not at issue. The claim was for compensation for the loss of the property right and the Court accepted this claim. In the United States Supreme Court case of *Lucas*,[6] the same argument was made out by an individual against the local council that had prohibited further development on an area of land subject to erosion, accretion and catastrophic flooding. David Lucas claimed that although the council had not taken his land, they may as well have because without the right to develop the land he owned, the land was worthless to him. He argued for compensation and the Court accepted this claim. The value of property in both these cases is a right to a commercial benefit or profit. The value of the land corresponds directly to that benefit and profit, and thus without the right to use the land in a way that achieves that profit, the property holder believes they hold nothing. The Courts' acceptance of this argument in both jurisdictions are two examples of the law's facilitation and protection of a dephysicalised idea of property that renders land irrelevant to the laws of ownership.

The focus of modern property law on rights and entitlement is an obstacle to the development of an Earth Jurisprudence. The notion of rights does not adequately account for and respond to the interaction between the systems of the earth and the systems of human society, of which law is one. The privileging of rights above other kinds of relationships between people and place renders invisible and irrelevant the 'things' that make life possible. Laws of ownership that fail to enquire, understand and accept the capacities and limits of the earth's systems fail to achieve their ultimate purpose – to regulate viable land and water use practices on an enduring basis. How might property be reconceived to facilitate a more enduring and physically responsive regulatory framework for people-place relations? If we thought about ownership in broader and deeper terms of responsibility, would property law achieve its purpose? Is that all that is required of an Earth Jurisprudence – a change of vocabulary? The following section of the chapter explores the concept of ownership in terms of responsibility and participation in the use and management of the earth's resources.

Owning the Earth as a Responsibility

Rather than reinventing the wheel, an effort to reform the rights-based law of ownership could learn a great deal from the long-established and successful systems of law or jurisprudence of indigenous peoples.[7] Indigenous Australian jurisprudence, for example, offers numerous and diverse laws of people – place relations that are based on notions of responsibility for the land, often described as 'caring for country'. In recognition of the dependence of human systems on the earth's systems, Indigenous Australian law is structured around the 'laws of reciprocity and obligation'.[8] The concept of ownership in this legal system is not one of right and entitlement as in the Anglo-Australian legal system, but custodial. The relationship between people and place is not proprietary in the sense of the land being subordinate or irrelevant to the owner, rather the land is regarded as the source of life and law. 'Country is central to the identity of an Aboriginal person, providing physical, cultural and spiritual nourishment'.[9] The emphasis in Indigenous land laws on responsibility to 'look after our home country and protect it'[10] is indicative not only of a different construction of the idea of property and ownership to the dominant paradigm of rights-based property, but also indicates a different worldview.[11]

The worldview that underpins Indigenous jurisprudence comprises a world in which there is a continuum and integrity of the earth's systems that includes the human species. Within this worldview, there is not the separation of people and place that characterises the worldview of Anglo-Australian jurisprudence.[12] In English, the definition of the word 'environment' is 'the aggregate of surrounding things' and this reflects, to an extent, a worldview in which people are positioned at its centre and everything else around them. In this view, culture is separate from nature so it is unsurprising that law, being cultural, does not regard itself as derivative of nature. The separation between the physical and metaphysical, between place and people, is almost antithetical to Indigenous jurisprudence. Indigenous Australian legal scholar Irene Watson wrote:

> The non-indigenous relationship to land is to take more than is needed, depleting *ruwi* [land] and depleting self. Their way with the land is separate and alien, unable to understand how it is we communicate with the natural world. We are talking to relations and our family, for we are one.[13]

Watson's account of a familial bond with the land is shared by many other Indigenous Australian lawyers and elders. Eualeyai Elder, Paul Behrendt, employed the metaphor of the parent–child relationship to explain the connection to, and specifically, the responsibility to country that attaches to Indigenous Australian law:

> Ownership [of land] for the white people is something in a piece of paper. We have a different system. You can no more sell our land than sell the sky ... Our affinity with the land is like the bonding between a parent and a child. You have responsibilities and obligations to look after and care for a child. You can speak for a child. But you don't own a child.[14]

Watson and Behrendt articulate a system of law that interacts with the earth's systems in a way that rationally and consciously responds to the dependence of the former on the latter. More accurately, they articulate not the interaction of two separate systems, cultural and natural, but the integrity of a single system. Perhaps this alignment of culture and nature, of the system of human laws and the systems of the earth, holds the key to the recognition of an appropriate formulation of an Earth Jurisprudence for non-Indigenous communities. But as Deborah Bird Rose observed two decades ago,

(In) spite of … many eloquent statements by American Indians, Aboriginal Australians and others, we have very little idea of what a non-human-centred cosmos looks like and how it can be thought to work.[15]

If we accept Rose's point, the important question to ask is what prevents the possibility of a non-human-centred cosmos taking shape in Anglo-centric law and culture? Could it be that the concept of a 'right' is itself part of an inward focus that places a priority on the identification and protection of the needs and interests of the self? Watson's work suggests this is the case. In her critique of native title law, Watson remarks that 'granting title to land has never been our question.'[16] The notion of entitlement she argues 'is the domain of those who want it named and determined for their short time and space on earth.'[17] Watson links property-as-right and the concept of entitlement to greed,[18] which she says is different from and antithetical to 'caring for country'. If Watson is correct, then the development of an Earth Jurisprudence will need to critically reflect on the anatomy of the concept of ownership in modern property law. It is not sufficient to know that modern property law is dephysicalised and rights-based, nor is it sufficient to know that what those rights consist of. The worldview from which property rights follow must be part of the audit. How is the concept of 'right' at the origin of modern property law manifest in its day-to-day operation?

In the practice and experience of modern property law, decisions are made about land and water use by property rights-holders predominantly on the basis of the most commercially profitable use available to them. Where lands, waters and the resources therein are shared across boundaries or are indeed public lands, decisions that are made about their use are complicated by the competing desires and needs of multiple property rights-holders, government agencies and community members. In natural resource planning and management literature, these people and groups of people are referred to 'stakeholders'. Where decisions about land and water use arouse the interests and needs of multiple stakeholders, they are referred to in the literature as 'wicked problems'. Wicked problems are those where there are

> complex interconnected systems linked by social processes, with little certainty as to where problems begin and end, leading to difficulty in knowing where and how constructive interventions should be made and where the problem boundaries lie.[19]

The difficulty of environmental decision-making, it seems, is that the decisions are made with a number of rights in conflict. It is the existence of the concept of the 'right' itself that is at the heart of the difficulty. Were the relationship characterised by responsibility, would the same complexity, competition and conflict arise? Were the decisions on the use of land and water based on knowledge of the capacity and limit of the land to support that use, would it be simpler? Environmental decision-making literature sheds light on how the concept of rights influences and indeed dominates modern property law in its daily operation.

In Montana, U.S., natural resource management scholars Paul Lachapelle and Stephen McCool found that ownership can be conceptualised and approached in terms of participation and responsibility.[20] Focusing on forest management, they argue that the interest of both title-holders and non-title holders in decisions about forests was experienced differently because of the conflation of property rights with control or power over the use of land. Where the entire community, not just the dominant right-holding stakeholder, is engaged in the decision-making process 'a sense of ownership ... is created, leading to greater chances for political support and implementation'.[21] Their argument, however, does not subvert the concept of ownership as a series of rights in the earth's systems and resources; rather it extends those rights to people beyond those on the formal legal title. As a democratisation of a process, this is an interesting and valuable point. However, their discussion of responsibility for environmental decision-making does not replace the priority of the 'right' – it complements it. For them, responsibility is not for the land but for the decision, and hence ownership is not of the land but of the decisions about and over the land which remain, ultimately, separate from the human community.

Could ownership of land be regarded in the same way we think about owner-ship of human behaviour? In the discipline of psychology, ownership refers to the taking of responsibility for one's actions and their consequences. How would this meaning of ownership work in application to the land? In particular, how would one own or take responsibility for the actions and consequences of mining land when the purpose of the ownership (the conditions of the mining lease) is contingent on mining activities? Surely the content of property rights, specifically the rights of exclusion and alienation, prohibit the appointment and acceptance of ownership of the consequences of the relevant land use activities? It is at this point that knowledge of consequences becomes important to a discussion of redefining ownership to include responsibility.

In modern non-Indigenous society, the source of knowledge of land and water features and processes is science. Lachapelle and McCool refer to a 'technocentric' approach to natural resource planning in which science and scientific method dominate the decision-making process. They critique this approach for its failure to recognise, validate and perhaps accommodate a range of human interests in the land notwithstanding scientific fact. They argue that 'science alone does not address the desirability of the conditions, since these are normative decisions based on value judgments'.[22] In other words, information about the earth and its systems, while important, does not and cannot override the human interest in decisions about the resources of those systems. The argument that knowledge of place may describe the capacities and limits of the land, but that human needs and interest remain paramount, if only for pragmatic reasons, is concerning. The displacement of the knowledge of place is a key feature of 'rights' in modern property law. It is to the question of the relationship between knowledge and ownership that the chapter next turns.

Ownership and Knowledge

Modern property rights exist independently of the knowledge of the capacities and limits of the land over which those rights are exercised. Indeed, some property rights, for example: the right to graze livestock, the right to irrigate, the right to mine and the right to develop coastal and estuarine landscapes, may be exercised despite clear and long-standing evidence that the capacities and limits of the lands over which they have been exercised have been exceeded. Clearly, there are physical limits to the status quo. The hope for effective environmental protection and the establishment of viable and enduring laws that regulate the relationship between people and place, between humans and their home, the Earth, must take cognisance of the artificial and unhelpful separation of questions of ownership and use and endeavour to once more align the two into an integrated land law. This would achieve a 'rephysicalising' of property law and restore the centrality of the physical world to jurisprudence.

One of the best-known failures of the current global economy has been the method used to allocate the costs of the negative or adverse consequences of the production and consumption of the earth's resources. The 'externalisation' of these costs is one of the most significant issues facing governments around the world. A popular response has been the promotion of the idea of environmental markets – in carbon, for example. By integrating a knowledge not only of the capacities (or benefits) of land and water to produce certain resources, but also of the limits (or burdens) of such production, the total cost of ownership is thought to be calculable. The disposition or alienation of property may gradually be understood as being more than the right to any profitability of the land, to include the responsibility or 'cost' of restoring any damage and/or the cost of protecting the land and its resources from any harm. [23]

The development of property and ownership laws in an Earth Jurisprudence would contain and indeed be based upon the knowledge of the earth's systems. Such knowledge can provide information about the capacities and limits not in general terms but in specific terms of place-based contexts. Certainly, the overarching biophysical system must be understood as an entirety and the interaction between its five components[24] is elementary to sound environmental decision-making. Feedback processes, both positive and negative, are also important to understand, particularly the impact of human actions on these spheres, in terms of energy and waste in particular. This would help the law to integrate its authority with the laws of the physical world.

An important part of knowledge of the earth's systems is not limited to space and place. It also involves an understanding of time – the physical change within and between the earth's systems both in terms of repeating patterns and rhythms and in terms of permanent change. The changes of the Earth and its systems underline that Earth is not a state, but a living thing. Each part of the Earth's system affects other parts in complex processes of feedback. All species must, of necessity, constantly adapt to changes in their ecological conditions such as water supply and changing temperature, which in turn are changing or adapting in response to other factors in the feedback process. Failed adaptations lead to extinction as well as to speciation. Creating laws land use that take into account the dimension of time, as well as space, is vital to those laws being viable and enduring. For example, in the allocation of property rights in water, knowledge of the El Niño–Southern Oscillation variations of the wet La Niña and the dry El Niño, in addition to knowledge of seasonal rainfall patterns, would help make those rights and the laws that regulate them functional. Without incorporating that knowledge into water law, over-allocation has led to desertification (anthropogenic drought), financial hardship, family and community crisis and economic damage to society. Whether change is temporary or permanent, property laws that are dephysicalised lack the capacity and flexibility to accommodate change in the physical conditions of those rights and thus become dysfunctional and meaningless.[25]

The viability of knowledge-based land laws is evident in the long-established and successful Indigenous Australian legal system. The system is not inherently superior nor was it rapid in development. The Indigenous Australian legal system linked the knowledge of place to law through sheer experience of specific geographical conditions, over a very long period of time and across a vast continent of diverse and changing climatic conditions. The point is not to essentialise and racialise law but to identify and respect the intellectual integrity and practical success of laws that have been and remain locally viable and authoritative. The intellectual and practical opportunities of learning about Indigenous land laws are opportunities to learn about many complex systems, patterns and relationships that connect people and place. Ownership cannot exist without responsibility, and responsibility cannot exist without knowledge. How can we assume responsibility for things we don't know and understand? How can we claim entitlement to things we assumed no responsibility for maintaining? Modern property law conceptualises and articulates limits to its application in terms of jurisdiction and authority. Yet this authority and jurisdiction derives not from the specific physical conditions of local places, but from itself, in a circuitous and irrational fashion. As modern property law increasingly exceeds the physical conditions of its own existence – what local authority can it be said to have?

Conclusion

The principle objective of modern property law is not the regulation of human-earth relations but the regulation of human-human relations and the distribution of the earth's resources between people as tradeable commodities. This anthropocentric and dephysicalised approach to defining and regulating the ownership of the land is a potent obstacle to the development of Earth Jurisprudence. It renders invisible and irrelevant the actual physical capacities and limits of the Earth to the model of ownership of land and consequently facilitates maladapted land use practices. Anglo-Australian property law is devoid of a vocabulary of responsibility to and for land. By contrast, Indigenous-Australian land laws take the concept of custodianship as the foundation of their systems. The geophysical success of such a regime is a helpful starting point for reflection and reform of the dominant dephysicalised property of Anglo-Australian law. In particular, the centrality of the knowledge of place to Indigenous land law is a helpful lesson in reshaping a more rational and functional law of ownership. So when we think about what it is that we claim to own, we speak of a place, not an income, and a place with geographically specific conditions and geologically specific histories that are not finished. Land exists in place and in time. Our knowledge of those dimensions is important to our knowledge of the consequences of our actions on and to the land long after we are gone. Such knowledge might constitute a law of ownership befitting a body of Earth Jurisprudence.

Notes

1 See N. Graham, 'Dephysicalisation and Entitlement: legal and cultural discourses of place as property' in *Environmental Discourses in International and Public Law* eds B. Jessup and K. Rubenstein, Cambridge University Press, Cambridge, 2010.

2 See N. Graham, 'Restoring the 'Real' to Real Property Law: a Return to Blackstone?' in *William Blackstone: Life, Thought, Influence* ed W. Prest, Hart Publishing, Oxford, 2009, pp. 151–167.

3 See S. Coyle and K. Morrow, *The Philosophical Foundations of Environmental Law: Property, Rights and Nature* Hart Publishing, Oxford, 2004.

4 See V. Been, 'Lucas v. The Green Machine: Using the Takings Clause to Promote More Efficient Regulation?' in *Property Stories* eds G. Korngold and A.P. Morriss, Foundation Press, New York, 2004.

5 *Newcrest Mining (WA) Limited v Commonwealth* (1997) 147 ALR 42.

6 *Lucas v South Carolina Coastal Council* 505 U.S. 1003 (1992).

7 See page 214 of this book.

8 I. Watson, 'Buried Alive', *Law and Critique*, Vol 13, 2002, p. 267.

9 L. Behrendt and L. Kelly, *Resolving Indigenous Disputes*, Federation Press, 2008, p. 1.

10 L. Behrendt and L. Kelly, *Resolving Indigenous Disputes*, Federation Press, 2008, p. 1.

11 M Graham, 'Some Thoughts about the Philosophical Underpinnings of Aboriginal Worldviews', *Australian Humanities Review*, Vol 45, 2008, p. 186.

12 See N. Graham, *Lawscape: Property, Environment, Law*, Routledge, 2010, Chapters 1 and 2.

13 I. Watson, 'Buried Alive', *Law and Critique*, Vol 13, 2002, p. 256.

14 P. Behrendt, cited in L. Behrendt and L. Kelly, *Resolving Indigenous Disputes*, Federation Press, 2008, 89.

15 D. Rose, 'Exploring an Aboriginal Land Ethic' *Meanjin*, Vol. 2, 1988, p. 379.

16 I. Watson, 'Buried Alive', *Law and Critique*, Vol 13, 2002, p. 260.

17 Ibid.

18 See I. Watson, 'Indigenous Peoples' Law Ways: Survival Against the Colonial State', *Australian Feminist Law Journal,* Vol 8, 1997, pp. 39–58.

19 R. Harding, C, Hendriks, and M. Faruqi, *Environmental Decision-Making,* Federation Press, 2009, 21.

20 P. Lachapelle and S. McCool, 'Exploring the Concept of 'Ownership' in Natural Resource Planning', *Society and Natural Resources,* Vol 18, 2005.

21 Ibid, p. 280.

22 Ibid, p. 280.

23 See N. Graham, 'The Mythology of Environmental Markets' in *Property Rights and Sustainability* eds D. Grinlinton and P. Taylor, Martinus Nijhoff, 2010.

24 The atmosphere, hydrosphere, cryosphere, lithosphere and biosphere.

25 For example, '[f]armer Malcolm Holm accidentally cut off his left hand in a grain machine the day after being notified that his pre-purchased water allocation was to be reduced by thirty-two per cent due to the low level of the Hume Weir.' *Beyond Reasonable Drought,* The Five Mile Press, Scoresby, p. 162.

Private Rights in Nature: Two Paradigms

Eric T. Freyfogle

One way to talk about private property rights in land is to do so abstractly, by pondering the rights an owner has or ought to have in a hypothetical land parcel, unattached to any real place. Another approach, a better one, is to begin in a living landscape, taking note of its terrain, waters, and problems and then posing basic questions about how private property ought to operate in that landscape. This second approach gives nature a voice in the lawmaking process. And it leads, or can lead, to understandings of property more respectful of the land community, to forms of private ownership better suited for people wanting to live without degrading their natural homes.

My inquiry, then, starts with the soils, waters, plants, animals, people, and cultures that surround my home in the central part of the United States, near the heart of what is known, commonly and aptly, as the Corn Belt.

The Setting

The lands of east-central Illinois are among the most heavily used in the world. The soil is rich and deep, the rainfall plentiful, and the land close to flat – a combination nearly perfect for growing corn and soybeans. English-speaking pioneers reached the area early in the nineteenth century. Arriving with them were surveyors, who divided the region into square-mile sections with boundaries drawn, insofar as possible, on cardinal directions. Sections were arranged into townships, and townships into counties, with occasional range lines to supply points of beginning. Nature played little role in this geometric fragmentation except for the big rivers that demanded attention and the correction lines required to accommodate the Earth's curvature. Now, roads track most of these original section lines, forming a rectangular grid system that frames both the landscape and the minds of its residents.[1]

Farm sizes in Illinois began at a quarter-section (160 acres) early in the nineteenth century. In the generations since they have swelled considerably. With a bit of seasonal assistance and with a few tasks done by contractors a single farmer today can tend 1000 or even 2000 acres. With this expansion has come a radical biological simplification. Pastures, hayfields, orchards, gardens, woodlots – almost all are gone, along with the associated livestock. Bulldozers have removed fence-rows; abandoned farmhouses have disappeared.

Champaign County lies in the centre of this high-yielding region.[2] Covering 1000 square miles it is home to 190,000 people. A full 83 per cent of the county is devoted to row crops, with 99 per cent of these acres in corn and soybeans. The tallgrass prairie that once dominated the region is at best a faint memory; statewide, the prairie has declined by 99.99 per cent, from 23 million acres to 2300. Given that loss and the fact that forests occupy only 1% of the county, the change in land cover has been almost total. This transformation was slowed in the nineteenth century because of the region's poor natural drainage. Full use of the county had to await the development of steam shovels to help farmers install subsurface drainage tiles. The steam shovels also deepened and straightened miles of meandering streams, creating ditches that lowered water tables. In combination, these ditches and tile drains enable farmers to begin planting in April or early May. By speeding up water flows, however, the drainage exacerbates flooding and droughts, increases stream bank erosion, and diminishes the capacity of streams to break down pollutants.

Nearly all of Champaign County is in the hands of private owners. Indeed, setting aside roadways and the remnants of a now-abandoned air force base on the county's north edge (mostly off-limits due to contamination), the public owns less than 1 per cent of county land. Fields of corn and soybeans begin at the edge of outer housing subdivisions. The view looking outward from this edge is particularly stark once harvest ends in October or November. Buildings and trees are few and scattered; the line of sight is nearly unbroken. Even grasses along the section lines are typically mowed by farmers or road crews. As habitat for wildlife the typical land section is largely worthless.

Despite this lack of trees and privacy Illinois residents have taken to building homes in farm landscapes. The typical home builder purchases several acres along a paved road and constructs a house near the centre. The remainder of the residential tract is left in lawn grasses (some exceptions, thankfully), which require hours of weekly mowing during the growing season. Mowing typically goes right to the edge of adjacent crops, creating a crisp look of human control and with predictable effects on wildlife. As for the one per cent of Champaign County covered by forest, it survives along the county's several rivers. River corridors remain tree-lined (except the mowed ditches), yet the trees create merely a leafy fringe along river edges. Homes are increasingly situated among these trees, sometimes almost to the riverbank. Land-use controls are few.

Owning the Land: The Industrial View

Driving and shaping this pattern of land use has been the institution of private property, American style. Property is a legal institution, sanctioned by law and supported by the police powers of the state. It is also, and importantly, a cultural institution; a constellation of cultural values and understandings; an emblem and embodiment of freedom and core American values.

As commonly understood, private property gives owners vast powers to use the things they own. That power is particularly potent when it comes to altering nature. To own land in the dominant view is to have the power to exploit it fully, even to consume and destroy it. The public's only legitimate interest is in limiting harmful spill-over effects; that is, in controlling actions by a landowner that cause material harm to neighbours.[3]

This view of private property is particularly strong outside urban areas. With neighbours few and distant, rural landowners see little reason why they cannot exploit their lands to the fullest. This means, in Illinois farm country, a farming cycle that begins each spring by killing essentially every macroscopic plant and animal on the land, planting one genetically modified plant species, and then keeping out all other plants and animals. Ownership includes the right to drain land as one sees fit; to remove trees along stream banks; and to fill-in low spots in fields. Producers rarely coordinate their land uses, which means that land management occurs parcel by parcel.

Concerns about environmental degradation have brought criticism of these land-use patterns (although not in local newspapers or television, which avoid challenging rural audiences).[4] Critics point to the water pollution and to radically altered water flows, which exacerbate floods and droughts, erode stream banks, and materially harm aquatic life forms. They point to the near total loss of native wildlife habitat and to resulting losses and declines in wild species (and the rise of a few species to pest status). A few critics also point to the soil, perhaps the most vital of all natural resources, which is slowly eroding and declining in structure and quality. Despite this criticism little gets done: the lobby for industrial agriculture successfully resists all change. Farm-subsidy programs sometimes attach strings to payments, giving farmers incentives to avoid breaking sod and to leave remaining wetlands intact. Beyond that, however, there are only intermittent conservation-reserve programs, which pay farmers to leave land untilled or to plant trees.

An important side effect of payment programs is that they affirm and strengthen prevailing views of private property.[5] When government pays land-owners to leave land along waterways untilled or to leave patches for wildlife, the implicit message is that landowners could legally do otherwise. Farmers could, in the exercise of their property rights, plow every square meter and cut down every tree. Payments thus endorse a line of thinking that has become widely accepted: conservation by farmers benefits the public, which means the public in fairness ought to pay for it. From this perspective it only makes sense for farmers to demand payment when they alter tillage practices that generate climate-changing gases. The prevailing reasoning goes like this: farmers by right can do as they please, with no accountability for harm. It is only appropriate, then, that they get paid when they curtail emissions. Indeed, when they reduce tillage they perform the service of 'storing' carbon in the ground. A much different story emerges when the reasoning starts from a different point. If the beginning point is farmland left alone, with nature taking its course, then the producer's role changes considerably. A shift to better tillage practices means, at best, that producers are allowing the soil to regain some of the carbon the producers themselves have released – carbon storage that nature could do on its own, better, and for free.

The cultural force of this vision of ownership stems from more than just economic realities. It is linked also to the widespread assumption that private property is somehow a basic individual right.[6] The idea here is that private property arose originally – or at least it exists today – outside the law, without government involvement. Government's function has been to protect these private rights. This tale of private property's origins in fact makes no sense given how property necessarily operates. Property is inherently a social institution based on law. Without law landowners would enjoy no protection against interference – the core of what ownership means. Before property arose all people could use nature freely. Once a community embraces private property, some individuals – the landowners – gain the power to exclude outsiders and to demand payment if not homage from the people they allow to use their lands. The embrace of private property, that is, entails a massive restriction on the liberty of non-owners.

To see this interplay of property, state power, and dominance is to see why the familiar institution of private property is in fact morally problematic.[7] What justifies the use of state power to curtail the liberties of people otherwise free to wander at will? Why do residents of Champaign County, Illinois, when they reach the edge of town, have no place to continue walking except along the narrow ribbons of highway (which never include sidewalks and, in the case of interstates, actually prohibit pedestrian traffic)? Why must a tenant farmer who produces a crop relinquish, at year's end, half the crop to another person who did none of the work?

So familiar is private property that few people take time to think about how it works. Few people realise that private property is the dominant institution of landscape governance, particularly in the United States. Few also know much about its history.[8] In Illinois we have forgotten that rural landscapes were open to public use for many decades after the state entered the union in 1818. People could freely hunt, forage, gather firewood, and, most significantly, graze livestock on all land, public or private, that was not protected by a strong fence. Property law in early Illinois also differed significantly in that it tended to protect sensitive land uses and to prohibit more intensive industrial activities. Rights to divert and pollute water were limited. Owners had little power to alter natural drainage to the detriment of neighbours. Only over the course of a century of gradual legal change, propelled by the forces of industrialisation, did owners gain rights to use their lands more intensively, in the process giving up some of their rights to complain about harms caused by neighbours.

It is vital to remember the twists and turns in the legal elements of land ownership because they highlight an important truth. Private property is very much a human social creation. What can be owned and what it means to own are products of human choice, exercised through law-making processes. Different cultures at different times have had widely differing ideas about even the main elements of ownership. Another truth here is that the only legitimate justification for private property – given how it empowers some people to dominate other people – is that a well-designed system of private property can generate widespread benefits for nearly everyone. That is, private property can be morally justified in terms of the good overall consequences that it generates. To earn this moral support, though, a scheme of private property must actually promote the common good. The law must define ownership so that landowner actions generate net benefits overall. And this justification based on overall benefits needs to occur on an element-by-element basis, looking at all the ways landowners can use what they own. Should land ownership include the right to farm in ways that gradually degrade soil and push wild species toward extinction? Landowners should not have such powers unless the public gains.

Owning the Land: An Ecological View

The industrial view of property did not really come together until rather late in the nineteenth century. It was then that industrial values and the collective urge to dominate nature reached their peaks.[9] Within a few decades urban reformers were pushing for greater social control over land uses to abate urban ills. Reform movements gained ground, leading to widespread land-use controls in nearly all cities by the 1960s and 1970s. This movement made little progress, however, when it confronted land-use ills in rural areas. The 1990s saw the rise of a powerful environmental backlash movement, which included as a key element the zealous presentation of an extreme view of landowner powers.[10] In this view, landowners could develop any and all lands they might own, without regard for ecological effects. If government restricted their options, government should pay for the lost income.

The conservation movement in the United States resisted this backlash more or less successfully. It did so, though, by stressing the values of environmental regulations. It did not, unfortunately, take time to put forth an alternative vision of private ownership. It did not, that is, counter the vision of ownership put forward by the self-titled 'property rights movement' with an ecologically grounded alternative that similarly endorsed private property but insisted that landowners act responsibly.[11] The ensuing battle, as a result, seemed to pit property rights on one side and environmental benefits on the other, rather than, as it might have, an outdated industrial version of private ownership versus an updated ecological version.

An alternative, green vision of owning nature could have started by recognising the moral value of nonhuman life forms, perhaps by vesting legal entitlements in species, rare communities, free-flowing rivers, and other parts of nature. Alternatively and more practically, it could take shape by drawing upon terminology and ideas already embedded in Anglo-American property law and legal history. It could, that is, accept the main legal components of private ownership but then revise and reform them to meet current needs – just as prior generations of lawmakers have done.

What might a new, ecological vision of ownership entail?[12] What rights should landowners have to use and alter nature so that society gains the benefits of private property while also enjoying the benefits of ecologically healthy lands and waters? And how should these private entitlements be governed so that land-use decisions consider larger spatial scales, future generations, and other life forms?

The circumstances of Champaign County usefully highlight the most significant needs for change. Too much county land is tilled every year, leading to erosion, rapid water runoff, loss of wildlife habitat, and other ecological costs. Farm chemical usage is excessive and drainage is handled poorly. The county needs to reconstruct wetlands to perform their important ecological functions. Wildlife habitat also needs restoration, particularly along river corridors. Crop diversity needs to return along with brushy areas, trees, and other natural features. New home construction should be clustered to reduce the economic and ecological costs of dispersion.

This incomplete list of problems and needs, based simply on the rural lands (and ignoring, for instance, water-use issues), provides keys to the reform of private property. For starters we can return to the central element of ownership in common law jurisprudence, the rule that landowners must use what they own so as to cause no harm. Generations of lawmakers have adapted this do-no-harm principle to the circumstances and values of their age, using it to shape private rights to foster the common good as they saw it. They have done so by giving specific meaning to the vague idea of harm, defining it so as to prohibit activities they disliked and to allow activities they favoured.

In an ecologically informed system of private property, the do-no-harm principle could limit the powers of landowners to use land in ways that amounted to degradation. Harm could include development on lands ill-suited for it, such as wetlands, barrier islands, floodplains, and unstable slopes. It could cover specific practices, such as farming in ways that unduly degrade soil or harm wildlife habitat. It could include removing trees along a riparian corridor and even excess mowing. Perhaps most of all, the term can include harm to the land itself and thus to future users of the land; it need not be limited to harms that disturb neighbours. The fact that an activity was not deemed harmful a generation or century ago is no reason why we cannot view it as harmful today.

In defining harm, ecologically minded lawmakers will encounter a particular challenge that their predecessors rarely faced. Harmful activities typically have caused problems in themselves and based on where they occur. In the case of Champaign County, in contrast, much of the harm arises from activities that are reasonable when undertaken on appropriate scales but harmful when too many people engage in them. Too much land, as noted, is tilled, and too much is thoroughly drained. The removal of vegetation has gone much too far. Fertiliser and pesticide applications are too extensive. Crop diversity is too low. These are carrying capacity harms, and need to be understood and described as such. The solution is to find ways to curtail the activities overall. Various approaches are possible, some more complex than others. They key point is simply to understand these special types of harm and find ways to talk about them. Is it possible to develop some type of fair-share approach that requires pro-rata cutbacks by all landowners? Is it possible instead to craft a trading rights scheme that uses market forces to identify low-cost solutions? These and other options present themselves once the basic problems are understood.

Such shifts in the definition of land-use harm would likely refine private rights in ways that pay greater attention to nature; that in effect tailor the rights of owner-ship to the natural features of a land parcel. Thus, rights would vary not among landowners per se, but among land parcels. Land uses could be allowed only on land parcels that are ecologically suited for them in terms of soils, drainage, and vegetation. This basic idea – property rights tailored to the land itself – could provide a strong counterpoint to the prevailing tendency to define private rights abstractly, as if land parcels were identical.

Two further elements of a new ecological vision of ownership should begin from the little-mentioned realities that, under longstanding legal precedents, both wild animals and water are owned by the people collectively. They are components of nature that states hold in trust and must manage for the benefit of all rather than just a few. These legal precedents have considerable potential. Water is central to natural systems and many land uses disrupt natural water cycles. Similarly, wild-life need places to live and many private actions significantly harm wild species. The public thus has legitimate interests at stake when private landowners act in ways that harm water and wildlife. To revive the ideas of public ownership and of state trustee duties is to add needed complexity to a discussion that has centred for too long simply on the wants of individual owners.

One of the chief problems in Champaign County arises from the narrow focus that dominates drainage management, particularly by the numerous drainage districts. Ditches are viewed as single-purpose water conveyances, not as vital parts of ecological systems that could provide home for wildlife and links in fertility cycles. They could be managed, but are not, to help mitigate pollution loads that inevitably escape farm fields. Change here needs to come in the management of such districts, in the ways the districts use their drainage easements, and in the links between the districts and individual landowners.

The problems posed by these districts raise broader issues about needs for better methods of collective governance at the landscape scale, particularly better methods for landowners to work cooperatively to address shared problems. In truth, many environmental problems require remedial action at scales well above the individual land parcel. Local owners need to play roles in this work, even as they are strongly pressured from above to achieve measurable standards of progress. Coordination is needed to design and protect habitat for wild species. It is needed to manage large waterways and activities in large catchment basins and along shorelines. The approach used in recent generations has been to vest responsibilities in governments with landowners then usually resisting. Government is the actor, landowners are the regulated parties. A better vision is needed, a vision that enlists the energies and knowledge of landowners in confronting shared problems. Landowners could, for instance, formulate plans to improve wildlife habitat in farm landscapes, leading them to see and question the effects of many of their practices (e.g., excessive mowing). Many landowners will still deny problems. But surely it will help to give them more chances to participate, not just as speakers at hearings but as principals who prepare plans.

A looming challenge in the case of many of these issues has to do with the trade-off between defining private rights with specificity and defining them instead more vaguely. Specific definitions tell landowners what they can and cannot do; they help landowners plan, and reduce disputes. But many problems cannot be understood in advance and nature itself changes over time. Good management often requires adaptability and experimentation. For many reasons, then, future lawmakers may prefer to use vague standards when describing property rights, including vague statements of duties. Vague standards can work when lands are incorporated into governance arrangements that translate the vague limits into more specific requirements. A property right to make reasonable use of water, for instance, is a vague standard. It can work well if a suitable institution exists to translate the vague reasonableness limit into specific water uses, taking account of competing uses and ecological effects. Designing such institutions needs to draw attention from lawmakers and conservationists.[13]

New Ideas, New Rhetoric

A central inadequacy of the conservation movement in the United States has been its failure to produce new ways of thinking and talking about private property. Critics have attacked conservation as hostile to private property and willing to sacrifice core individual rights in the name of protecting nature. The conservation side has offered no response, in effect conceding the merit of the attack. A better response is possible.

Private property need not stand as an impediment to the promotion of healthy landscapes. Private property is not the problem. The problem is the way the institution is talked about and defined by law. New rhetoric can arise. And as for the laws, they are as flexible and subject to change as all other laws.

New rhetoric should draw upon the history of property law, reminding audiences of its long evolution to the present and thus its considerable flexibility. It particularly needs to remind audiences that property is a morally problematic institution that is best understood – and indeed is only morally legitimate when shaped – as a tool that promotes the common good. Then there is the critical need to decide when landowners deserve payment and when they do not, and ensure that poorly designed payment programs do not honour and strengthen misguided views of landowner rights. At bottom, the conservation movement simply must address private property directly, studying it carefully and then formulating a well-considered vision. It needs to stand tall in favour of private rights, updated by ecological realities and clearly tied to a vision of responsible land use.

Notes

1   A thoughtful meditation on the effects of this rectangular grid system, particularly on perceptions of nature and attitudes toward land use, is offered in C. Meine, *Correction Lines: Essays on Land, Leopold, and Conservation*, Island Press, Washington, D.C., 2004, pp. 187–209.

2   Statistical information on Champaign County is drawn from the 2004 Champaign County Statistical Abstract, available at http://www.ccrpc.org/planning/stats/statisticalinfo.php (last visited November 17, 2009).

3   In influential presentation of this view by an academic is R. Epstein, *Takings: Private Property and the Power of Eminent Domain*, Harvard University Press, Cambridge, MA, 1985.

4   An example of sharp criticism is A. Kimball, ed., *The Fatal Harvest Reader: the Tragedy of Industrial Agriculture*, Island Press, Washington, D.C., 2002.

5   The effects of payments, and the need to decide when payments are and are not appropriate, are considered in E. Freyfogle, *On Private Property: Finding Common Ground on the Ownership of Land*, Beacon Press, Boston, 2007, pp. 105–30.

6   The principal approaches to private landownership in the United States are reviewed in E. Freyfogle, *Bounded People, Boundless Lands: Envisioning a New Land Ethic*, Island Press, Washington, D.C., 1998, pp. 91–113.

7   The challenge of justifying property morally is addressed in L. Becker, *Property Rights: Philosophic Foundations*, London, Routledge & Kegan Paul, 1977.

8   The discussion here is drawn chiefly from E. Freyfogle, *The Land We Share: Private Property and the Common Good*, Island Press, Washington, D.C., 2003, pp. 22–99, and Freyfogle, *On Private Property*, pp. 29–60.

9   Freyfogle, *The Land We Share*, pp. 65–99.

10 J. Echeverria and R. Booth, eds., *Let the People Judge: Wise Use and the Private Property Rights Movement*, Island Press, Washington, D.C., 1995.

11 I criticise the U.S. conservation movement for failing to address this issue and others in E. Freyfogle, *Why Conservation Is Failing and How It Can Regain Ground*, Yale University Press, New Haven, CT, 2006.

12 My suggestions on this issue are set forth more fully in Freyfogle, *The Land We Share*, pp. 203–81, and Freyfogle, *On Private Property*, pp. 84–156.

13 I discuss the possible benefits of blending the categories of public land and private land by developing intermediate categories in E. Freyfogle, *Agrarianism and the Good Society: Land, Culture, Conflict and Hope*, University Press of Kentucky, Lexington, KY, 2007, pp. 83–106.

How We Control the
Environment and Others

Paul Babie

The 'commodification' and 'propertisation' of carbon appear to be the latest fashion in 'solving' human-caused (anthropogenic) climate change. Implicit in the weak political agreement reached at the end of 2009 at the Copenhagen Climate Conference are proposed pieces of 'cap-and-trade' legislation that allow greenhouse gas emitters to purchase private property interests in the right to emit carbon dioxide. Aside from the fact that such schemes target reductions in only one of the greenhouse gasses for which humans are responsible (leaving other, much more potent greenhouse gasses – methane and nitrous oxide – entirely unregulated), the main difficulty with these enactments is that they place faith in the concept of private property, the concept largely responsible for the problem in the first place.

The climate change crisis implicates two harmful aspects of the concept of private property. First, from the perspective of Earth Jurisprudence, which seeks to move Western understandings of law from an anthropocentric to an ecocentric focus,[1] it commodifies nature. As this chapter will show, part of the very essence of private property is the ownership of goods and resources. This is shorthand for saying that private property allows humans the control of the animate and inanimate environment – animals, trees, rivers, airspace, and so forth. Private property, then, presents a significant challenge for Earth Jurisprudence.

The second dimension of private property implicated by climate change is the way that the commodification of nature allows humans to use the environment as a tool against other humans. Other authors in this volume address the commodification of nature, and this is crucial in redefining our relationship with the natural world. As part of that redefinition, however, humans must continue to play a significant part. The concern of this chapter, therefore, is the use of goods and resources, both natural and manufactured, as a tool of environmental and so human subjugation. More than that, it offers an alternative way of conceiving of our relationship to the natural world and others, through private property, from an unlikely source – religion.

The chapter contains four parts. The first part describes private property as a legal-social relationship making possible the choices which lie behind anthropogenic climate change. The second part argues that private property makes possible a physical-spatial relationship, which I call the 'climate change relationship'. The third part argues that the climate change relationship reveals an important two-fold weakness in theorising about private property: first, that it is the 'idea' of private property, and not the concept (on which theorists tend to focus), that is the real culprit behind anthropogenic climate change; second, this idea, when coupled with the 'sovereignty' that private property is, allows humans to produce consequences or 'externalities' that affect not only the global environment, but also people the world over. This means that private property, seemingly a private law concept limited to its jurisdictional boundaries, in fact has a global, as opposed to a national territorial, reach. The final part argues that given the implications of this sovereignty, alternative ideas of private property are necessary; religion offers a long and rich source of these alternatives. This part explores what the three monotheistic religious traditions – Judaism, Christianity and Islam – say about the way we ought to conceive of private property in relation to the environment and so others. More to the point, rather than conflicting *ideas* (plural) their teachings about private property tend to converge to offer a uniform *idea* (singular).

Private Property: A Legal-Social Relationship

The dominant contemporary concept of private property, found in all modern legal systems, emerges from liberalism. Liberal theory concerns itself with the establishment and maintenance of a political and legal order which, among other things, secures individual freedom in choosing a 'life project' – the values and ends of a preferred way of life. And according to liberalism, in order for this project – in other words, life – to have meaning, some control over the use of goods and resources is necessary; and private property is the liberal vehicle which ensures this control to individuals, allowing them to fulfil their chosen life.[2] We can unpack the liberal conception a bit more.

Liberal private property consists of a 'bundle' of legal relations, or rights, between people in relation to the control and use of goods and resources.[3] At a minimum, these rights include use, exclusivity, and disposition. Thus, I can use my car (or, with few exceptions, any other tangible or intangible good, resource, or item of social wealth), for example, to the exclusion of all others, and may dispose of it in any way. Of course, the car involves an intermixture of a number of other goods and resources, themselves manufactured from raw material. But consider, too, the raw materials or, rather, the earth, nature, the environment. Nature can be used just as much as I use my car: I can cultivate soil and use nitrogen-based fertiliser to grow crops, or cut down trees in order to cultivate, or clear the land (destroying habitat for native flora and fauna) in order to build the factory that manufactures the car in the first place. In other words, private property allows for the commodification of nature in order, instrumentally, to pursue certain goals. And the holder may exercise these rights in any way they see fit, to suit personal preferences and desires. Or, we might put this in a way that comports more with the language of liberal theory – rights are the shorthand way of saying that individuals enjoy choice, the ability to set agendas[4] about the control and use of goods and resources – the environment, nature, the earth, in accordance with and to give meaning to a chosen life project.

We began, though, by saying that these rights are in fact legal *relations*; the rights exist as a product of relationship between individuals. Wesley Hohfeld first brought this to our attention in the notion of 'jural opposites'; simply, if there is a right (choice) to do something, there is a corresponding duty (a lack of choice) to refrain from interfering with the interest protected by the right.[5] Rights would clearly be meaningless if this were not so. The liberal individual holds choice, the ability to set an agenda about the earth (a good or resource), then, while all others (the community, society) are burdened with the lack of choice about that good or resource. Perhaps C. Edwin Baker best summarises the idea of right and relationship this way: '[private] property [i]s a claim that other people ought to accede to the will of the owner, which can be a person, a group, or some other entity. A specific property right amounts to the *decisionmaking authority* of the holder of that right'.[6] And in this web of 'asymmetrical'[7] legal relationships we find the liberal concept of private property.

Thus, choice, agenda setting, or 'decisionmaking authority' is not merely about the control and use of goods and resources, but also, and more importantly, about controlling the destiny of the natural world and the lives of others. Private property and non-property rights overlap, and the choices made by those with the former have the potential to create negative outcomes for those without choice in relation to the good or resource in question. Relationship, then, also describes the outcomes, the consequences, the 'externalities' of choice. Every legal system acknowledges this, 'tak[ing] for granted that owners have *obligations* as well as rights and that one purpose of property law is to regulate property use so as to protect the security of neighbouring owners and society as a whole'.[8]

Thus, while it legitimately exerts its power to protect the decisionmaking authority of liberal private property vested in the individual, more importantly, through regulation, the state mediates the socially contingent boundary between private property and the non-property rights of others. Choice, in other words, is not entirely unlimited and unfettered. Rather, because it operates within a network of social relationships, every system of private property mediates between those who hold choice and those who do not through moral imperatives, duties and obligations. Thus, while '[private property] ... initially appears to abhor obligation ... on reflection we can see that it requires it. Indeed, it is the tension between [unfettered private property rights] and obligation that is the *essence* of [private] property'.[9]

Herein resides a final important, and paradoxical, dimension of choice. While we know that choice allows individuals to decide what to do with goods and resources, we also know that context limits that choice through state power in the form of inherent regulation. But we began this discussion within the broader

framework of liberal theory, and that is important, for the freedom that liberalism secures to the individual to choose a life project means that in the course of
doing that, the individual also chooses one's own context – the laws, relationships,
communities, and so forth that constitute the political and legal order. In other
words, in the province of politics and adjudication people also choose their contexts
(through electing representatives, who enact laws and appoint judges who interpret
those according to ideological agendas[10]), which in turn defines the scope of one's
rights – decisionmaking authority – and the institutions that confer, protect and
enforce it. Individuals as much choose the regulation of property (through political
and judicial processes) as how to control goods and resources (through private
management and market transactions).[11]

Climate Change: A Physical-Spatial Relationship

The externalities of private property produce, then, many other types of relationship, not legal-social, but physical-spatial. Joseph William Singer provides an apt
summary:

> [private] property owners and the public are linked to each other through indi-
> vidual actions [choices] and laws affecting the use of [private] property (which can ...
> be both beneficial and detrimental). From this perspective, we could conceive of
> [private] property as a type of ecosystem, with every private action and legislative
> mandate potentially affecting the interests of other organisms.[12]

Anthropogenic climate change is a stark example of such a relationship. While
the science of anthropogenic climate change is complex, it is clear enough that
humans, through their choices, produce the greenhouse gasses that enhance the
natural greenhouse effect that heats the earth's surface. Among other effects, this
enhancement results in drought and desertification, increased extreme weather
events, and the melting of polar sea ice (especially in the north) and so rising sea
levels.

And it is private property which allows individuals – both human and corporate – to choose, to set agendas, about the use of goods and resources in such a
way that emits these greenhouse gasses. Agendas cover the gamut of our chosen
life projects: where we live, what we do there, and how we travel from place to
place. Corporate choices are equally important, for they structure the range of
choice available to individuals in setting their own agendas, thus conferring on
corporations the power to broaden or restrict the meaning of private property in
the hands of individuals. Green energy (solar or wind power), for instance, remains
unavailable to the individual consumer if no corporate energy provider is willing
to produce it.

These choices do not end at the borders, physical or legal, of a good or resource;
as we have seen, they are not made in a vacuum, but take place within a web of relationships, not legal and social, but physical and spatial. Who is affected? Everyone,

with the poor and disadvantaged of the developing world disproportionately bearing the brunt of the human consequences of climate change[13] – decreasing security, health problems, food shortages and increased stress on available water supplies. Consider security: this will be decreased both within countries affected directly by climate change, and in those indirectly affected through the movement of large numbers of people displaced by the direct effects of climate change. In the case of rising sea levels, for instance, sixty per cent of the human population lives within 100 km of the ocean, with the majority in small- and medium-sized settlements on land no more than 5 m above sea level. Even the modest sea level rises predicted for these places will result in a massive displacement of 'climate' or 'environmental refugees'. The human externalities of anthropogenic climate change reveal a physical-spatial relationship that we can call the 'climate change relationship'. The next section offers three reflections about the weaknesses of the concept of private property revealed by this relationship.

Weakness: 'Idea' and 'Sovereignty'

If private property is self-seeking choice, then it matters what individuals – not the theoretical liberal individual, but real, socially-situated, flesh and blood people – *think* that they have when faced with making a decision about where they live, how they get there, what they will wear, and so forth. I call this the *idea* of private property: it consists of images, stories, and legends about what private property *means*. Who can forget 'possession is 9/10ths of the law', 'finders, keepers – losers, weepers'. And that is precisely the point – we *can't* forget these idealised portrayals of private property, because:

> [f]rom the earliest moments of childhood, we feel the urge to assert ourselves through the language of possession against the real or imagined predations of others. 'Property' as an assertion of self and control of one's environment provides human beings with a place of deep psychological refuge. With its concreteness and its unfailing assurances, property promises to protect us from change and from our fear that we will leave no evidence of our passage through this world.[14]

All of which pushes us inexorably to one conclusion: the layperson understands private property as an individual and absolute entitlement to a thing, a part of the natural world or something produced from it – Blackstone's 'sole and despotic dominion'[15] – which cannot be challenged by any other person, and which the state will protect. According to the idea of private property, an individual or corporation holds absolute rights (choice) to the use of a thing – land, water, trees, car, house, factory, patent, and so forth – with which no one, not even our chosen government, can interfere. This remains deeply and intuitively embedded in the human psyche.[16] One associates with it words like 'mine', 'yours', 'castle' – 'within the borders of one's land [or in relation to a thing, whatever it is], [one] is supreme and can do whatever [one] wishes'[17] – or 'desert' – 'one who applies ... labor to the

earth deserves to reap [what was] sowed'.[18] In short, we might summarise the idea
of private property as Felix Cohen did:

> [T]hat is property to which the following label can be attached:
>
> To the world: Keep off X unless you have my permission, which I may grant or
> withhold.
>
> Signed: Private citizen
>
> Endorsed: The state[19]

So long as choice persists – and as long as liberalism underpins contemporary
political, economic and social life, it will – then it matters *how* a person understands what that choice *means*. So long as a person, when directly faced with a
clear and specific choice – car or not, green house or not, coal-powered electricity
or not – thinks first of themselves, as an individual, free to choose to suit themselves without any regard for others, negative consequences, externalities, inevitably follow. And so long as that is the case, the *idea* is the culprit behind the role
played by private property in allowing the activities that produce anthropogenic
climate change. Regulation might control, and even prevent, *some* choices, but it
cannot prevent all choice (unless, of course, property, or liberalism itself, is removed
entirely).

And all of this is endorsed by the state, which leads to the second part of the
weakness revealed by the climate change relationship in contemporary theorising
about private property. Choice is also a form of power, as Morris Cohen, in a
seminal article, noted early in the 20th century when he described private property
as 'sovereignty'. By this, Cohen meant private property is *really* a state delegation
of *power* permitting the individual to do as one pleases with a particular good or
resource; more significantly, though, 'we must not overlook the actual fact that
dominion over things is also *imperium* over our fellow human beings'.[20] This
insight cannot be gainsaid. Choice is exercisable not only in relation to the good
or resource, but also, significantly, in relation to others. There is, in other words, a
state-enforced asymmetry here – a choice made by one individual about a good or
resource (the natural world) has the potential to affect a great number of others.[21]

None of which is a concern, of course, when the state which endorses these
grants of sovereignty also has the power, through its own sovereignty, to protect
others by using its power to regulate. But here's the problem. The liberal concept
of private property, as with all Western jurisprudence, developed in a post-West-
phalian world, or at least one in which arbitrary national boundaries were treated
as more important than human-caused phenomena that may transcend those
boundaries. In fact, there was probably very little recognition that individuals could
even produce trans-boundary consequences and, as such, so it was thought, the
state could enforce both the holding of choice through private property and ensure
the limitation of negative externalities, because all of that would occur within
national legal boundaries.[22] Indeed, as William Twining writes

most of the leading Western jurists of the twentieth century have focused very largely on municipal state law, have had strong conceptions of sovereignty, and have assumed that legal systems and societies can be treated as discrete, largely self-contained units. They have either articulated or assumed that jurisprudence and the discipline of law is or should be concerned only with two kinds of law: the domestic municipal law of nation states and of public international law ...[23]

The history of private property and theorising about it exhibits no break in this pattern. As we have seen, though, climate change puts paid to the truth that whatever the holders of private property may do to others, it is contained by national jurisdictional boundaries.

Seen in this light, the asymmetry of private property takes on a new meaning (although it was always there, we just never realised it until something like climate change revealed it). The sovereignty (power) to affect the control and use of goods and resources and so the lives of others within the jurisdiction which conferred the choice is, in fact, 'supreme'[24] in the fullest sense of that word. What is conferred by one state on one individual has the potential to allow for untold consequences for others who reside outside of the legal jurisdictional boundaries of the state that conferred that choice. As we have seen, the externalities of climate change bear disproportionately, *asymmetrically*, on those of the developing world and not on those of the developed world where most of the individual decisions about goods and resources that produce greenhouse gasses occur.

And more troubling still, this sovereignty cannot be limited by the very people who are subject to it – those who live beyond the legal jurisdictional borders of the state that delegated it. The countervailing power that one might have to choose one's own context through political processes is meaningless. The citizens of Sudan or of Tuvalu, whose problems are in part the consequence of anthropogenic climate change, are powerless to choose the political-legal context that affects them. Rather, those in developed nations who hold the sovereignty conferred by private property choose that context for them.

What we are faced with in the climate change relationship, then, is an absolutist idea of private property coupled with a 'supreme' sovereignty (power) over the natural world and others. The outcome? The externalities of climate change – decreased human security and food supplies and increased health risks and water stress. It may be impossible to alter the nature of the sovereignty delegated by states to their citizens. But could it be possible to alter the idea that animates the exercise of that power? That is the argument advanced in the next part. And to do that, I draw on the three monotheistic religions as one possible source for alternative ideas.

Religion: Alternative Ideas

Religion tends to be left out of the debate and dialogue about the solutions to contemporary problems. That is a shame, for these traditions offer sophisticated and nuanced understandings about how humans interact with the natural world and with each other, and how they ought to act towards both. They offer existing models of community (in the broadest sense, encompassing both the environment and humanity), developed over many years, which draw upon principles related to the holding of private property. In short, they offer refreshingly novel alternative 'ideas' to the liberal absolutist one.[25]

This part considers the models presented by Judaism, Christianity, and Islam. One of gratifying aspects of this exercise is not simply the uncovering of new ideas about private property found in religion, but also that those ideas tend to be very similar. Rather than a plurality of ideas found in different religions, the teachings of those traditions tend to converge on one idea (one that runs entirely counter to the absolutist idea): *obligation*, both towards the environment and towards others.[26]

Judaism

While Judaic law and its underlying religious values devote significant attention to private property as a means of allocating resources, by far the dominant focus is on relationship and obligation. Much of the Jewish law relating to the obligations one has to others through the use of goods and resources stems from Leviticus 25:14: 'do not, any of you, oppress your brother or your sister'. This is captured by three admonitions found in the Tanakh, or the Hebrew Scriptures: *tzedakah*, gleanings, and jubilee (these latter two are specific dimensions of *tzedekah* aimed at the obligations one has towards the land).[27]

Tzedekah stems from the command to share property with those who have none – over and over, people are commanded to look out for the widow, the orphan, the stranger and the poor. This obligation extends to everyone in the community, not as a matter of charity, but as a matter of justice. *Tzedakah* means righteousness or justice, and it means that the duty to provide support from what one has – largely gleaned from the land – for those in need is just that, a duty, a matter of justice, and not one of choice. It is an obligation.[28]

Gleanings and Jubilee might be called specific dimensions of *tzedakah*. Throughout the Tanakh and the Talmud one finds agricultural practices that take up the obligation of justice to the poor and needy. Gleanings, such as leaving harvested sheaves in the field, or planting to the very edge of the field and leaving that area unharvested, are aimed at providing for those in need. Jubilee is the requirement to leave the land fallow at certain times – every seven years, which is called Sabbath – and then in the fiftieth year following seven Sabbath cycles of years. These cycles provide not only for the land itself, but also for those in need, for during the Sabbath years, what grows is owned by all, while during the Jubilee year there is a renouncing of debt and redistribution of land.[29]

Christianity

In seminal work developed over the last twenty years, Orthodox theologian John Zizioulas draws upon the Christian Trinitarian understanding of God to argue that the human person can only truly exist, ontologically, within relationship and community. While the liberal *individual* is concerned with oneself, the human *person* is a member of community and so is concerned with others.[30]

This ontology of the person has profound implications for understanding the numerous admonitions found in the New Testament concerning private property. Consider just one, the parable of the rich fool found in Luke 12:16–21:

> Then [Jesus] spoke a parable to them, saying: 'The ground of a certain rich man yielded plentifully. And he thought within himself, saying, "What shall I do, since I have no room to store my crops?" So he said, "I will do this: I will pull down my barns and build greater, and there I will store all my crops and my goods. And I will say to my soul, 'Soul, you have many goods laid up for many years; take your ease; eat, drink, and be merry.' " But God said to him, "Fool! This night your soul will be required of you; then whose will those things be which you have provided?" So is he who lays up treasure for himself, and is not rich toward God.'

Taken within the context of the broader importance of relationship and community to the existence of the human person, such admonitions tell us something about the choices that one makes about private property. Choice is still important in the Christian worldview, yet those choices harmful to the community are in fact detrimental to one's own personhood. From a Christian perspective, when making a choice about private property, one ought always to have in mind relationships with others, especially the poor and the oppressed.

Yet, in order to care for other humans, one must also practise sound stewardship of goods and resources, the natural world. The parable of the rich fool captures this notion. And while stewardship may be a choice, to avoid or reject it not only harms the environment and thus the community, but also oneself. Thus, stewardship of the natural world and concern for others becomes obligatory if one is to be truly human.

Islam

There is little in Islamic theology or jurisprudence that outlines the concept, as distinct from the law, of property, although it is clear that to a great extent, Islam mirrors the liberal conception.[31] But as with Judaism and Christianity,[32] while the concept is implicit, obligation within a broader social context is explicit in the Islamic legal focus.[33] Indeed, in many books on Islamic law, those chapters that deal with charity precede those on the acquisition and disposal of property.[34]

The source of Islamic obligation concerning private property flows from two sources. First, all of what is belongs to and is given by God and so is held by humanity on trust by way of property. Second, this trust operates within a broader social context, one that embraces all of humanity.[35]

From these sources emerge two specific aspects of private property obligation. On the one hand, there exist obligations in the exercise of property rights themselves, such as the obligation of honesty, fair dealing and equity in doing business,[36] the prohibition of *riba*, or unlawful gain,[37] the prohibition on *gharar*, or uncertainty (undertaking a venture blindly without sufficient knowledge, or undertaking a risky transaction),[38] and the prohibitions against waste and prodigality.[39]

Far more important than obligations imposed upon the exercise of rights, though, is the overarching concept of charity, the second pillar of Islam and thus an ethical principle that applies to all that is done.[40] Broadly conceived, charity in Islam ranges from great deeds such as the emancipation of slaves, feeding the poor, and doing good to humanity in general, to the smaller acts of benevolence, such as the use of kind words.[41] And in the context of private property it takes the form of almsgiving, both obligatory and voluntary.

Zakat, the obligatory approach, is a form of charity mandated as a calculated amount on assets of a value above the legal threshold,[42] usually a minimum of 2.5% of one's surplus wealth and earnings, after satisfying regular needs and expenses, at the end of the year.[43] *Sadaqah*, the voluntary form, is over and above *zakat*; although nominally voluntary, there is a moral obligation to give *sadaqah*.[44] These obligations are based on the notion that individuals are part of the community. Islamic social responsibility requires that private property be used not only for one's personal advantage and benefit – preference-satisfaction – but also for the advantage and benefit of the community. The rights of the needy thus qualify the right of private property.[45] Thus, *zakat* and *sadaqah* 'act not only as a levelling influence but, also a means of developing higher sentiments of [humanity], the sentiments of love and sympathy towards [others] …'[46] and so 'the divine imperative to provide for the needy, the thrust towards community, is strengthened'.[47]

Religion is an Alternative Idea
This part began by arguing that the world's religious traditions, and the monotheistic religions specifically, offer alternative ideas (plural) of private property. Yet, what emerges from this brief review of those traditions is not a range of options, but a uniform *idea* (singular): obligation rather than *imperium*. This univocal idea can be broken into four components. First, it places great emphasis on personhood and community. Christianity suggests that the human person only exists within community and relationship with others, and Judaism and Islam make very clear that ignoring the community diminishes the individual.

Second, it requires that the individual ought to take a specific stance towards the community. Christianity provides admonitions of stewardship of goods and resources for the poor and the oppressed that allow the individual to prioritise and so foster community. Judaism, while recognising that choice exists, largely rejects it as an organising idea in private property, opting instead for mandatory obligations to share one's wealth with others, especially the poor and the oppressed. Islam provides for both mandatory and voluntary charity, both of which foster a robust model of community. This idea of private property, then, diminishes the importance of choice and prioritises relationship and community; choice is important, but obligation towards the community is more so.

Third, and perhaps even more significant, is the fact that, seen this way, it is impossible to act as a steward towards others if one is using the environment as a tool to harm them. This is the truth revealed by the climate change relationship. In teaching stewardship towards others, then, the religious idea of private property also requires stewardship of and care for the natural world. This of course is nothing new – it is well known that religion teaches stewardship of the environment;[48] what is novel here is the reason for that stewardship – obligation towards other humans, not just across the street or in the next city or town, but the world over.

Finally, while religion certainly causes division and conflict, at least as concerns private property and the environment, it also has within it the seeds of cooperation, the potential to unite, to allow humans to act as stewards of the natural world and so to foster rather than destroy community. Together, the monotheistic traditions do this through teaching the importance of relationship and community, promoting those ideals through stewardship and charity – a concern for others rather than oneself – be it voluntary or obligatory.

Conclusion

If we are serious about addressing the challenges posed by anthropogenic climate change, then we must confront the truth that the *idea* of private property, when coupled with the *imperium* which the state confers on individuals though that concept, is a rather significant culprit behind the crisis. From the individual perspective, it seems to provide an unfettered and uncontrollable power over the natural world and the earth's resources, making the environment a tool with which to control every other person on the planet.

This requires a reassessment of concepts like private property, not to abolish them, but to find new ways to invoke them. This may involve seeking alternatives to the absolutist liberal idea of private property that currently holds sway. And while some would see it as an unwelcome interloper, religion provides a source for such alternatives. Indeed, the religious *idea* of private property is one grounded in obligation, towards the natural world and towards others.

Yet, even if one cannot ultimately accept religion as offering a viable alternative, the very process of searching there for new ideas to inform the exercise of choice over the natural world and others will allow us to '...see how we can use the idea of climate change – the matrix of ecological functions, power relationships, cultural discourses and material flows that climate change reveals – to rethink how we take forward our political, social, economic and personal projects over the decades to come'.[49] If that is all religion can do for us, it has still done much.

Notes

1   T. Berry, *Evening Thoughts: Reflections on Earth as a Sacred Community,* Sierra Club Books, San Francisco, 2006, pp. 19–20; C. Cullinan, *Wild Law: A Manifesto for Earth Justice,* Green Books, Devon, 2006, pp. 160–70.

2   G.W.F. Hegel, *Philosophy of Right,* Clarendon Press ed (T. Knox (trans), Oxford, 1942, first suggested this.

3   See J.W. Singer, *Introduction to Property,* Aspen Publishers, New York, 2005, p. 2.

4   L. Katz, 'Exclusion and Exclusivity in Property Law', *University of Toronto Law Journal,* vol. 58, 2008, p. 275.

5   W.N. Hohfeld, 'Some Fundamental Legal Conceptions as Applied in Judicial Reasoning', *Yale Law Journal,* vol. 23, 1913, p. 16; *Yale Law Journal,* vol. 26, 1917, p. 710; W.N. Hohfeld, *Fundamental Legal Conceptions as Applied in Judicial Reasoning,* Yale University Press, New Haven, 1919; W.N. Hohfeld, *Fundamental Legal Conceptions as Applied in Judicial Reasoning II,* Yale University Press ed (Walter Wheeler Cook (ed)), New Haven, 1923.

6   C.E. Baker, 'Property and its Relation to Constitutionally Protected Liberty', *University of Pennsylvania Law Review,* vol. 134, 1986, pp. 741, 742–3 (emphasis added).

7   D. Lametti, 'The Concept of Property: Relations Through Objects of Social Wealth', *University of Toronto Law Journal,* vol. 53, 2003, p. 325.

8   J.W. Singer, 'How Property Norms Construct the Externalities of Ownership', *Harvard Law School Public Law Research Paper No. 08–06,* 2008, <http://ssrn.com/abstract=1093341> at 2 January 2009, 3 (emphasis in the original).

9   J.W. Singer, *Entitlement: The Paradoxes of Property,* Yale University Press, New Haven, 2000, p. 204 (emphasis added).

10   D. Kennedy, *A Critique of Adjudication {fin de siècle},* Harvard University Press, Cambridge, 1998; K. Rittich, 'Who's Afraid of the *Critique of Adjudication?*: Tracing the Discourse of Law in Development', *Cardozo Law Review,* vol. 22, 2000, p. 929.

11   I am most grateful to Joseph William Singer for bringing this crucial point to my attention.

12   J.W. Singer, 'The Ownership Society and Takings of Property: Castles, Investments, and Just Obligations', *Harvard Environmental Law Review,* vol. 30, 2006, pp. 309, 334, n 82.

13   United Nations Intergovernmental Panel on Climate Change, *Climate Change 2007 – Impacts, Adaptation and Vulnerability: Working Group II Contribution to the Fourth Assessment Report,* (2007) <http://www.ipcc.ch/ipccreports/ar4-wg2.htm> at 14 January 2010, 7.

14   L.S. Underkuffler, *The Idea of Property: Its Meaning and Power,* Oxford University Press, Oxford, 2003, p. 1.

15   Private property is '… that sole and despotic dominion which one man claims and exercises over the external things of the world, in total exclusion of the right of any other individual in the universe'. W. Blackstone, *Commentaries on the Laws of England: Volume II, Of the Rights of Things (1766),* University of Chicago Press ed, Chicago, 1979, p. 2. See also D. Schorr, 'How Blackstone Became a Blackstonian' *Theoretical*

Inquiries in Law, 2009, <http://ssrn.com/abstract=1066621> at 14 January 2010, 1–3; R.P. Burns, 'Blackstone's Theory of the 'Absolute' Rights of Property', *University of Cincinnati Law Review,* vol. 54, 1985, p. 67; C.M. Rose, 'Canons of Property Talk, or, Blackstone's Anxiety', *Yale Law Journal,* vol. 108, 1998, pp. 601, 603.

16 J. Williams, 'The Rhetoric of Property', *Iowa Law Review,* vol. 83, 1998, pp. 277, 280–2; B.R. Berger, 'What Owners Want and Governments Do: Evidence from the Oregon Experiment', *Fordham Law Review,* vol. 78, 2009 p. 1281; J.R. Nash and S.M. Stern, 'Property Frames', *Washington University Law Review,* 2010 forthcoming <http://ssrn.com/abstract=1463782> at 14 January 2010; M.J. Radin, *Reinterpreting Property,* University of Chicago Press, Chicago, 1993, p. 123.

17 Singer, *Ownership Society,* p. 317.

18 Ibid., p. 322.

19 F.S. Cohen, 'Dialogue on Private Property' *Rutgers Law Review,* vol. 9, 1954, pp. 357, 374, 378–9.

20 M.R. Cohen, 'Property and Sovereignty', *Cornell Law Quarterly,* vol. XIII, 1927, pp. 8, 13.

21 Lametti, *Concept of Property,* p. 7.

22 W. Twining, 'Law, Justice and Rights: Some Implications of a Global Perspective' Draft 1/07, <http://www.ucl.ac.uk/laws/academics/profiles/index.shtml?twining> at 15 January 2010, 4.

23 Ibid., pp. 7–8.

24 Katz, *Exclusion and Exclusivity,* p. 275.

25 See J.W. Singer, *The Edges of the Field: Lessons of the Obligations of Ownership,* Beacon Press, Boston, 2000, pp. 41–2.

26 Indeed, while this chapter focuses only on the three monotheistic traditions, others reveal that most world religions teach the same things when it comes to obligation towards the environment: see eg A.J. Maidin, 'Religious and Ethical Values in Promoting Environmental Protection in the Land Use Planning System: Lessons for Asian Countries', *Journal of Islamic Law Review,* vol. 2, 2006, pp. 53, generally, and especially 70–6; R.S. Gottlieb (ed), *The Oxford Handbook of Religion and Ecology,* Oxford University Press, Oxford, 2006.

27 Singer, *Edges of the Field,* pp. 42–56.

28 Ibid.

29 Ibid.

30 J.D. Zizioulas, *Being as Communion: Studies in Personhood and the Church,* St Vladimir's Seminary Press, Crestwood, New York, 1985, pp. 27–65; J.D. Zizioulas, *Communion & Otherness: Further Studies in Personhood and the Church,* T&T Clark, London, 2006. And see C. Yannaras, *The Freedom of Morality,* St Vladimir's Seminary Press, Crestwood, New York, 1984, pp. 13–27; D.H. Knight (ed), *The Theology of John Zizioulas: Personhood and the Church,* Ashgate, Aldershot, England, 2007.

31 J. Hussain, *Islam: Its Law and Society,* Federation Press, Leichhardt, NSW, 2nd ed, 2004, pp. 169–70; H.P. Glenn, *Legal Traditions of the World,* Oxford University Press, Oxford, 3rd ed, 2007, p. 183 and see specifically the references cited at nn. 65–7; W.B.

Hallaq, *An Introduction to Islamic Law*, Cambridge University Press, Cambridge, 2009, p. 16; A.A. Fyzee, *Outlines of Muhammadan Law*, Oxford University Press, New Delhi, 4th ed, 1974, pp. 217–332 and 355–467; M.M. Ali, *The Religion of Islam: A Comprehensive Discussion of the Sources, Principles, and Practices of Islam*, The Ahmadiyya Anjuman Isha'at Islam (Lahore) USA, Dublin, Ohio, 1990, pp. 509–27.

32 Islam itself sees the concept of charity as being a basic principle of every religion: Ali, *Religion of Islam*, pp. 344–5.

33 Hussain, *Islam*, p. 170.

34 See Ali, *Religion of Islam*, Part III, chapters II and VII.

35 R. Greaves, *Aspects of Islam*, Darton, Longman and Todd, London, 2005, pp. 34–8 and 75–95.

36 Hussain, *Islam*, pp. 170–1.

37 Ibid., pp. 171–3.

38 Ibid., p. 173.

39 Glenn, *Legal Traditions*, p. 183, citing; Ali, *Religion of Islam*, pp. 510–2.

40 Ali, *Religion of Islam*, pp. 573–5.

41 Ibid., p. 341 and pp. 573–5.

42 Hussain, *Islam*, p. 170, citing N. Baydoun, *Notes on Islamic Financial System* (1997).

43 Ali, *Religion of Islam*, pp. 347–50; Hussain, *Islam*, pp. 176–7.

44 See Ali, *Religion of Islam*, p. 342.

45 Ali, *Religion of Islam*, pp. 343–4 and 350–3.

46 Ibid., p. 346.

47 Greaves, *Aspects of Islam,* p. 37.

48 See Paul Babie, 'Private Property, the Environment and Christianity' *Pacifica: Australasian Theological Studies*, vol. 15, 2002, p. 307; Gottlieb, *Religion and Ecology*; Maidin, *Religious and Ethical Values*, generally, and especially pp. 70–6.

49 Mike Hulme, *Why We Disagree About Climate Change: Understanding Controversy, Inaction and Opportunity*, Cambridge University Press, Cambridge, 2009, p. 362.

Earth Jurisprudence and the Ecological Case for Degrowth

Samuel Alexander

> If we do not change direction, we are likely to end up where we are going.
> – Chinese Proverb

Earth Jurisprudence seeks to redefine the relationship between human beings and the environment by abandoning the anthropocentrism of much contemporary legal theory in favour of an ecocentric perspective. This shift in consciousness is seen as necessary because existing legal systems, by generally treating nature as a 'resource' to be exploited for human gratification, have failed to maintain the health and integrity of the ecosystems upon which the entire community of life depends.[1] What is urgently needed, and what Earth Jurisprudence promises, is a deep ecology of law, one normatively grounded on the idea that human law ought to reflect and respect the bio-physical laws of nature.

This chapter explores the Earth Jurisprudence of economic growth, an area that has received almost no attention in this new body of literature[2] but one in which the themes and insights of Earth Jurisprudence are highly relevant. I will begin by describing the 'growth model of progress' which is implicit to most modern political ideologies, whether on the left or the right, and which we will see shares certain assumptions with law-and-economics scholarship. Put simply, the growth model assumes that the overall well-being of a society is approximately proportional to the size of its economy, in terms of Gross Domestic Product (GDP) per capita, since more money means that more social and individual 'preferences' can be satisfied via market transactions. From this perspective the answer to almost every problem – including environmental problems – is *more economic growth*. The notion of a macroeconomic 'optimal scale' is all but unthinkable. A bigger economy is simply better.

I wish to unpack the ecological consequences that flow from the growth model and argue that Earth Jurisprudence, by situating itself outside the growth model, has a very important role to play in the critique of growth and in the provision of an alternative conceptualisation of 'nature' in law, especially property law where human-centred, economic analysis dominates. I base my analysis on the notion of 'degrowth,' an emerging term that designates a radical critique of economic growth. My basic argument is that when an economy has grown so large that it exceeds the regenerative and absorptive capacities of Earth's ecosystems, then lawmakers ought to initiate a 'degrowth' process of planned economic contraction. Only by doing so, I will argue, can over-consuming societies begin repaying their ecological debts to the Wild.

The substantive part of this chapter begins, as I have said, by outlining the growth model of progress. This dominant model propagates the pernicious myth of an infinitely expanding economy on a finite planet, a myth Earth Jurisprudence serves to undermine. Laying the foundations for an Earth Jurisprudence of economic growth, I then offer a brief history of 'growth scepticism,' the view that eventually an economy will reach an 'optimal scale' beyond which it should not grow. Since the scientific consensus indicates that the 'optimal scale' has already been surpassed, I conclude that law-makers in over-consuming societies ought to initiate a 'degrowth' process of planned economic contraction. I discuss this possibility in the context of property theory, speculating about what might become of the property systems of advanced capitalism should economic growth lose its privileged position as the touchstone of policy and institutional success.

Property, Wealth Maximisation, and the Growth Model of Progress

Daniel Bell once noted that 'Economic growth is the secular religion of advancing industrial nations,'[3] suggesting that those nations have developed an unconditional faith in the God of Growth and are now prepared to sacrifice everything in its name. More recently, Clive Hamilton has observed:

> Nothing more preoccupies the modern political process than economic growth. As never before, it is the touchstone of policy success ... [Political] parties may differ on social policy, but there is an unchallengeable consensus that the overriding objective of government must be growth of the economy.[4]

The defining assumption of our age, in other words, is that despite Western affluence much more economic growth is still needed to improve our lives. It is thought that more growth – and only more growth – can eliminate poverty, reduce inequality and unemployment, and properly fund schools, hospitals, the arts, scientific research, environmental protection programs, etc. Our great social problem, according to this popular narrative, is that even the richest nations do not have enough money to live well. And thus economic growth is heralded across much of the political spectrum as the goal towards which we must direct our collective energy.

According to this 'growth model of progress,' the overriding objective of the property system is to facilitate economic growth as efficiently as possible,[5] a goal which is widely assumed to be best achieved when property rights are traded in 'free markets'. In the context of legal theory, 'free markets' have received their most systematic (and extreme) defence in the scholarship of law-and-economics, epitomised by the work of Richard Posner.[6] Making a representative statement on the economic analysis of property rights and market structures, Posner writes:

> If every valuable (meaning scarce as well as desired) resource were owned by someone (universality), ownership connoted the unqualified power to exclude everybody else from using the resource (exclusivity) as well as to use it oneself, and ownership rights were freely transferable or as lawyers say alienable (transferable), value would be maximised.[7]

Leaving to one side the dubious assumptions about ownership, notice that 'value maximisation' (or 'wealth maximisation,' to use Posner's better known phrase) is held out as the proper function of the property system. Posner explains that wealth maximisation is achieved when goods and other resources are in the hands of those who value them most, and someone values a good more only if he or she is both willing and able to pay more in money to have it.[8] The most efficient path to wealth maximisation is through 'free markets,' Posner argues, in which private property is 'universal,' ownership is 'unqualified,' and ownership rights are 'freely transferable'.[9]

It should be noted at once that 'wealth maximisation,' as the phrase is used in law and economics, and 'maximising GDP' are not equivalent. Posner, in fact, has expressly distinguished them in his work, noting that 'wealth cannot be equated to GDP or any other *actual* pecuniary measure of value'.[10] Nevertheless, by looking to 'willingness to pay' (measured by actual pecuniary exchange) as the indicator of how various things are 'valued' in the marketplace, Posner ends up judging the 'the Good' and 'the Right' in terms of dollars,[11] which is what the 'growth model' does by assuming economic growth is the best measure of 'progress'. Accordingly, although it would be a confusion to conflate 'wealth maximisation' and the 'growth model of progress,' it is fair to say that the latter is based upon many of the assumptions of the former, meaning that there is some overlap in criticism.[12]

Since human beings are narrowly conceived of as self-interested, rational, independent consumers whose individual 'preferences' can be better satisfied with ever more money,[13] the economic analysis of property aims to grow the economic pie as big as possible. From this view, where *efficiency* is the governing ethical norm, more economic pie is always better; and efficiency is defined as 'exploiting resources in such a way that 'value' – human satisfaction as measured by aggregate willingness to pay for goods and services – is maximised'.[14] The anthropocentrism here is stark and unabashed.

This neoclassical perspective accepts that there is an 'optimal scale' at the *micro*economic level – which is to say, it accepts there will eventually come a point where growth in an individual business's production will cost more than it is worth (and therefore be judged 'uneconomic' growth). However, there is no place in neoclassicism for an 'optimal scale' at the *macro*economic level, no 'optimal scale' of the economy as a whole. This is because technological and allocative efficiency are thought to allow for an infinitely expanding economy, despite the obvious fact that we have but a single, finite planet to 'exploit'. Technological efficiency, it is assumed, will continually allow human beings to consume a finite set of resources more efficiently or, better yet, to consume a set of resources hitherto inaccessible. Allocative efficiency, it is assumed, will ensure that market mechanisms continually move resources into the hands of those who will 'exploit' them best. Upon these assumptions, the growth model purports to show that an economy can and should continue growing indefinitely. Economic growth, one is led to believe, is always and everywhere the most direct path to increased human well-being.

In reality, however, economic growth has led to ecological crisis, the details of which need not be reviewed here.[15] Let it simply be noted that the scientific community has concluded that the expanding global economy is now consuming natural capital and diminishing the capacity of Earth to support life in the future. Plainly, then, we must rethink mainstream attitudes to economic growth, for we are losing the planet.

A Brief History of Growth Scepticism

The sceptical attitude towards growth that I am advocating has its roots at least as far back as John Stuart Mill, who in 1848 asked: 'Towards what ultimate point is society tending by its industrial progress?'[16] This powerful question is significant for three reasons. First, because it acknowledges, implicitly, that industrial progress (or economic growth) is only of instrumental value and not of any intrinsic value; second, because it raises the possibility that there might come a time when further economic growth no longer advances any worthwhile purposes; and third, and perhaps most importantly, because it prompts us to consider not only how much economic growth is *enough*, but also the prior question of what we want growth *for*.

Mill believed that if there came a time when economic growth stopped contributing to well-being (or began undermining those things upon which well-being depends), the most suitable form of government would be what he called 'the stationary state'. By this he meant a condition of zero growth in population and physical capital stock, but with continued improvement in technology and in what he called 'the art of living'.[17] Mill was warning posterity – us, it would seem – to respect the limits to growth before the consequences of going beyond those limits compelled respect, in one way or another.

Later, in 1967, the economist Ezra Mishan published *The Costs of Economic Growth*,[18] in which he condemned the growth model for assuming that economic growth could increase infinitely on a finite planet, and for assuming that increased consumption (or a higher 'standard of living') always meant a better 'quality of life'. Mishan highlighted the fact that economic growth had *costs*, and he argued that those costs were beginning to outweigh the benefits. A similar critique, though with a closer environmentalist focus, was advanced in 1972 by the Club of Rome, in their treatise, *Limits to Growth*.[19] In 1976, another critique, this time with a sociological focus, was advanced by Fred Hirsch, in his text, *Social Limits to Growth*.[20]

Since the early 70s, ecological economist Herman Daly has built upon and significantly developed a number of these ideas.[21] Of particular relevance to the present discussion is the work he pioneered with J.B. Cobb developing a post-growth measure of progress, called the Index of Sustainable Economic Welfare (ISEW).[22] As a model of progress, this index and others like it take into consideration important social and environmental factors that economic growth alone does not reflect. For example, the ISEW and similar indexes begin with total private consumption expenditure and then make deductions for such things as resource depletion, pollution, income inequality, loss of leisure, etc., and make additions for such things as public infrastructure, volunteering, and domestic work.[23] Merging environmentalist and economic perspectives, Daly argues that sustainable development in the developed world necessarily entails a radical shift away from 'growth economies' towards what he calls a 'steady-state economy'.[24] By this he means (updating and refining Mill) an economy that continues to develop in response to new technologies and changing cultural and market forces, but without growing in a bio-physical sense.

More recently, a forceful critique of growth has come from Tim Jackson,[25] who argues that the process of getting richer is now, paradoxically, causing the very problems that we seem to think getting richer will solve. A fundamental attack on the neoliberal growth project, Jackson urges the advanced capitalist societies to begin exploring the possibility of 'prosperity without growth'.

Some of the most rigorous and provocative expressions of growth scepticism have emerged over the last decade out of the Degrowth Movement in Europe. Regrettably, this rich literature has barely made an appearance in Anglo-Saxon academic or public dates, although perhaps the tide may eventually turn as a result of the First International Degrowth Conference in Paris 2008, some of the proceedings from which are freely available online and predominantly in English.[26]

Though not a unified doctrine by any means, an emerging consensus within the Degrowth Movement has resulted in a Declaration of Degrowth.[27] This document (to paraphrase) calls for a paradigm shift from the general and unlimited pursuit of economic growth to a concept of 'right-sizing' both global and national economies. At the global level, right-sizing means reducing the global ecological footprint (including carbon footprint) to a sustainable level. In countries where per capita footprint is greater than the sustainable global level, this right-sizing implies a reduction to this level through the process of voluntary economic contraction (i.e. degrowth).[28] In countries where severe poverty still remains, right-sizing implies increasing consumption to a level adequate for a decent life. This will involve increasing economic activity in some cases, but redistribution of income and wealth both within and between countries is a more essential part of the process.[29]

Responding to those 'free-market environmentalists' or 'technological optimists' in the mainstream who claim that there is no conflict between growth and sustainability (i.e. 'development'),[30] the Degrowth Movement points out that although techno-efficiency solutions have been widely applied, flows of material and energy (or *throughput*) are still increasing.[31] This increase in throughput despite the increases in efficiency is largely due to 'rebound effects,' as Francis Schneider neatly explains by way of example: 'a car that consumes less gasoline per km leads to financial savings that may be spent on longer car distances ... A so-called secondary rebound exists when a house is better insulated and that reduced expenditure on heating is reinvested in buying a second car or travelling by plane'.[32] In short, efficiency 'creates revenue that can be spent on the same (primary rebound) or other commodities (secondary rebounds)'.[33] Efficiency, for this reason among others, is a fatally flawed solution to the problems of growth and over-consumption.

The tradition of growth scepticism outlined above, from Mill to 'degrowth,' may strike some as an unspeakable heresy and perhaps be dismissed in advance of honest consideration. But what seems clear is that if we are genuinely committed to the idea of a sustainable society – of each generation of life on Earth meeting its needs without jeopardising the prospects of future generations to meet their needs – then the pursuit of more economic growth by those nations which are already consuming an unsustainable share is ethically objectionable. Time has come, that is, for the over-consuming nations to give up the pursuit of growth, for it is looking ever-more like an unambiguous act of violence.

There have been and are other growth sceptics, arising out of various disciplines and focusing on different issues, but the focal point of this chapter concerns various issues that arise when *property rights* are considered within a post-growth framework – the framework that I claim best represents the position implicit to Earth Jurisprudence. With an admittedly broad brush, I will now bring this chapter to a head by offering a few thoughts on how the property systems of advanced capitalism might change – and how our thinking about those systems might change – and with what consequences, if economic growth were to lose its privileged position as the touchstone of policy and institutional success.

Property beyond Growth

We have seen that the 'growth model of progress' assumes that the overriding objective of a property system is to facilitate economic growth as efficiently as possible, a goal which is thought to be best achieved when 'unqualified' property rights are traded in 'free markets'. It is my contention that by neglecting the problems of over-consumption and the costs of economic growth, this neoliberal perspective on property is naively entrenched in an outdated context and dangerously informed by an outdated set of values. The ecological crisis, in particular, as well as unprecedented Western affluence, places us in a context so radically different to that of earlier generations that the justifiability of inherited property rights, shaped for a different age, must not be taken for granted. Indeed, they should be subjected to a fundamental reassessment and, if need be, a fundamental revision. Justification, after all, as historicists from Hegel to Foucault have shown, is context-dependent. And things have changed.

Looking beyond the 'growth model,' a post-growth property system would be one in which property rules, market structures, and tax policies were designed explicitly to achieve welfare-enhancing objectives more specific than the efficient growth of GDP. These more specific objectives would include protecting the environment or eliminating poverty, and in pursuing these objectives the efficient growth of GDP (or lack thereof) would be treated as a by-product of secondary importance. In essence, *sufficiency* would be privileged over *efficiency*, and in over-consuming regions of the world this would need to involve planned economic contraction, or 'degrowth'.

Reforming Western property systems in the spirit of degrowth could be initiated in a variety of ways. The following proposals, rather than aiming to grow the economy, would aim to promote sufficiency and sustainability, resist over-consumption, foster community and democracy, protect the vulnerable, and distribute wealth more broadly.

Proposed Reforms: (1) Unconditionally guarantee a minimal, state-funded, Basic Income for all; (2) Establish a system of progressive income tax that culminates in a democratically determined 'maximum wage;' (3) Facilitate the emergence of more communitarian and ecologically sensitive corporate structures, such as Worker Cooperatives or Employee Stock Ownership Plans; (4) Regulate, with explicit reference to the common good, powerful and socially influential 'private' institutions, such as banks, mass media, and mega-corporations; (5) Impose stricter and more precautionary environmental rules on land use and development, such as prohibiting 'intensive' land use and other forms of unsustainable production/ manufacture; (6) Devise land use rules that are more socially beneficial, such as those which would curtail 'urban sprawl' or protect common spaces and assets; (7) Increase taxes (especially inheritance taxes, luxury taxes, and wealth taxes) as (a) a corrective/redistributive measure; (b) to channel the forces of money toward socially and environmentally beneficial places; and (c) to better fund public services (health, education, public transport, the arts, foreign aid, etc.); (8) Establish more 'green' taxes to internalise externalities and price commodities at their 'true cost'. All this is merely suggestive of a vague (and no doubt controversial) alternative framework. My aim here is obviously not to provide a blueprint.

Although such reforms may well slow economic growth – even to the point of 'degrowth' – and thereby not maximise a nation's material 'standard of living,' my underlying thesis is that the reforms would at the same time increase 'quality of life'. Such a thesis, of course, is incommensurable with the 'growth model of progress,' since that model effectively conflates 'standard of living' and 'quality of life'. But as the ecological (and social)[34] critiques of growth make clear, it is wrong to conflate those very distinct concepts.

To conclude, I acknowledge that any transition to a post-growth economy would require extensive structural reform of the property system over time. And I accept that this will give rise to the culturally powerful charge of 'inefficiency,' always levelled at those who propose 'interventionist' or 'regulatory' structural reform of the property system.[35] But I believe that this objection, due to its very nature, loses much of its force when considered in light of the ecological critique of growth. In short, the 'inefficiency' objection is that post-growth reform of the property system would not maximise economic growth, since it would involve interfering in the supposedly 'efficient' operation of so-called 'free markets'. Based on the environmental (and social) science – even based on the economics! – the post-growth response is that not maximising economic growth is partly or beside the point. In a sense, then, I would respond to the objection by embracing it, not as an indictment but as a defence.[36]

In summary, I have argued that when an economy has grown so large that it exceeds the regenerative and absorptive capacities of Earth's ecosystems, then lawmakers ought to initiate a 'degrowth' process of planned economic contraction. I have tried to show that this argument has particular relevance to the jurisprudence of property, and I hope that my preliminary contribution may provoke some interest in how growth scepticism might impact, not only on property law, but on legal issues more broadly. It is my contention that legal scholars will find in growth scepticism the seeds of much fruitful and important research.

Notes

1   In 2005 the United Nations published a report, compiled and reviewed by over 1300 leading scientists, which revealed that 60 per cent of global ecosystem services are 'being degraded or used unsustainably,' resulting in 'substantial and largely irreversible loss in the diversity of life on Earth'. Millennium Ecosystem Assessment, *Ecosystems and Human Wellbeing: Synthesis* (2005) 1 <http://www.millenniumassessment.org/documents/document.356.aspx.pdf> at 20 June 2009.

2   The only study I have found is J. Guth, 'Cumulative Impacts: Death-Knell for Cost-Benefit Analysis in Environmental Analysis', *Barry Law Review*, vol. 11, 2008, p. 23.

3   D. Bell, *The Cultural Contradictions of Capitalism*, Hienemann, London, 1976, p. 237.

4   C. Hamilton, *Growth Fetish*, Allen & Unwin, Crows Nest, 2003, pp. 1–2.

5   For a recent expression of growth fetishism, see B. Freidman, *The Moral Consequences of Economic Growth*, Random House, Toronto, 2005.

6   See generally, R. Posner, *Economic Analysis of Law*, Little Brown, Boston, 1986.

7   Ibid 32.

8   See R. Posner, 'Utilitarianism, Economics, and Legal Theory' *J Legal Studies*, vol. 8, no. 1, 1979, p. 103; and 'Wealth Maximization Revisited', *Notre Dame L J*, vol. 2, 1985, p. 85.

9   See Posner, above n 7.

10   See Posner, 'Utilitarianism, Economics, and Legal Theory,' above n 8, p. 120 (emphasis in original)

11 R. Posner, *Economics of Justice* (Cambridge, Mass.: Harvard University Press, 1981).

12 Despite criticising several key assumptions of law-and-economics, I wish to emphasise that the thesis of this chapter is directed more specifically at the 'growth model'.

13 For one formulation of this 'economic' conception of humankind, see R. Posner, 'The Economic Approach to Law', *Tex. L R*, vol. 53, 1975, p. 761, where he states: 'The basis of an economic approach to law is the assumption that the people involved with the legal system act as rational maximisers of their satisfactions'.

14 Posner, *Economic Analysis of Law*, 10.

15 For state-of-the-art environmental research, see WorldWatch Institute at www.worldwatch.org. See also, above n 1.

16 J.S. Mill, *Principles of Political Economy*, Augustus M. Kelly, Clifton, 1973 [1870].

17 Ibid.

18 E. Mishan, *The Costs of Economic Growth*, Weidenfeld & Nicolson, London, 1993 (revised ed.).

19 D. Meadows et al, *Limits to Growth: The 30 Year Update*, Chelsea Green Publishing, White River Junction, 2004.

20 F. Hirsh, *Social Limits to Growth*, Harvard University Press, Cambridge, Mass., 1976.

21 H. Daly, *Steady-State Economics*, Island Press, Washington, D.C., 1991 and *Beyond Growth: The Economics of Sustainable Development*, Beacon Press, Boston, 1996.

22 H. Daly and J. Cobb, *For the Common Good: Redirecting the Economy Toward Community, the Environment, and a Sustainable Future*, Beacon Press, Boston, 1989.

23 Other post-growth measures of progress include the Genuine Progress Indicator and The Human Development Index. These tend to suggest that many advanced capitalist societies have stopped progressing, in terms of genuine human well-being, despite continuing to get richer. For a review of some of the research, see A. Offer, *The Challenge of Affluence: Self-Control and Wellbeing in the United States and Britain since 1950*, Oxford University Press, New York, 2006. See also, J. Talberth, C. Cobb, and N. Slattery, 'The Genuine Progress Indicator' (2006) <www.rprogress.org> at 20 May 2009.

24 Daly, *Steady-State Economics*.

25 Tim Jackson, *Prosperity without Growth*, 2009 <http://www.sd-commission.org.uk/publications/downloads/prosperity_without_growth_report.pdf> at 20 May 2009.

26 See www.degrowth.net.

27 Ibid.

28 S. Latouche, 'Degrowth Economics: Why Less Should Be Much More' *Le Monde Diplomatique*, 2004, <http://mondediplo.com/2006/01/13degrowth> at 20 May 2009.

29 New Economic Foundation, 'Growth Isn't Working: The Unbalanced Distribution of Benefits and Costs from Economic Growth', 2006, <http://www.neweconomics.org/> at 20 May 2009 (showing that First World growth is an ineffective response to global poverty.)

30 Unpacking this view, see W. Sachs (ed.) *Development Dictionary*, Zed, London, 1992.

31 See F. Schneider, 'Macroscopic Rebound Effects as Argument for Economic Degrowth' in *Proceedings of the First International Degrowth Conference*, 2008, <www. degrowth.net> at 20 May 2009.

32 Ibid 29.

33 Ibid.

34 See S. Alexander, 'Law, Economics, and the Social Critique of Growth: An Inquiry into the Correlation between Material Wealth and Well-being' (unpublished manuscript on file with the author).

35 For example, see R. Epstein, 'Why Restrain Alienation?', *Colum. L Rev*, vol. 85, 1985, p. 970.

36 I have explored some of these ideas further in S. Alexander, 'Property beyond Growth: Toward a Politics of Voluntary Simplicity' in David Grinlinton and Prue Taylor, ed., *Property Rights and Sustainability*, Nijhoff/Brill, Netherlands, 2010.

Part Four

International Law and Governance

Towards a Garden of Eden

Polly Higgins

There are certain principles of universal validity and application that apply to civilisation as a whole. They are the principles that underpin the prohibition of certain behaviour, for example apartheid and genocide. Such abuses arose out of value systems based on a lack of regard for fellow humanity and are now universally outlawed. The rendering of such action as illegal is premised on the advancement of a higher morality that operates without caveat of qualification, a morality based on the sacredness of human life. In a world aspiring to sacredness of life, it is still necessary to identify the crimes to prevent those who fail to live by similar values. But what of the wellbeing of *all* life – not just that of humanity, but of all who inhabit a territory over which one has certain responsibilities?

It was the humanitarian crisis of the Second World War that prompted the creation of the United Nations organisation, whose stated aims are to facilitate cooperation in international law, international security, economic development, social progress, human rights, and the achieving of world peace. The Charter of the United Nations (UN Charter) declared in 1945:

> We the peoples of the United Nations, determined to save succeeding generations from the scourge of war … to promote social progress and better standards of life in greater freedom.[1]

In advancement of peace, the term genocide was soon given international legal recognition to describe the enormous deliberate destruction of human life, such as the holocaust of World War 2. Trials were held in Nuremberg to prosecute perpetrators. However, it took over 50 years for the creation of the International Criminal Court (ICC) to provide a permanent international enforcement tribunal, as set down by the provisions in the Rome Statute and ratified in 2002.[2] Jurisdiction is limited to prosecution of individuals of the four *'most serious crimes of concern to the international community as a whole'*.[3] Referred to collectively as crimes against peace, they are: genocide, crimes against humanity, war crimes, and the crime of aggression.[4] Now another type of international crime against peace has arisen; that crime is ecocide.

The Crime of Ecocide

The neologism ecocide is already in use to a limited extent, denoting large-scale destruction, in whole or in part, of ecosystems within a given territory.[5] Ecocide is in essence the very antithesis of life. It can be the outcome of external factors, of a *force majeure* or an 'act of God' such as flooding or an earthquake. It can also be the result of human intervention. Economic activity, particularly when connected to natural resources, can be a driver of conflict. By its very nature, ecocide leads to resource depletion, and where there is escalation of resource depletion, war comes chasing close behind. The capacity of ecocide to be trans-boundary and multi-jurisdictional necessitates legislation of international scope. Where such destruction arises out of the actions of mankind, ecocide can be regarded as a crime against peace; a crime against peace for all those who reside therein. In the event that ecocide is left to flourish, the twenty-first century will become a century of 'resource' wars.[6]

For the purpose of international law, I propose the following definition for ecocide:

> the extensive destruction, damage to or loss of ecosystem(s) of a given territory, whether by human agency or by other causes, to such an extent that peaceful enjoyment by the inhabitants of that territory has been severely diminished.

There are two categories of ecocide: non-ascertainable and ascertainable ecocide. Non-ascertainable ecocide describes the consequence, or potential consequence, where there is destruction, damage or loss to the territory per se, but without specific identification of cause as being that which has been created by specific human activity.

Ascertainable ecocide describes the consequence, or potential consequence, where there is destruction, damage or loss to the territory, *and* liability of the legal person(s) can be determined. The destruction of large areas of the environment and ecosystems can be caused directly or indirectly by various activities, such as nuclear testing, exploitation of resources, extractive practices, dumping of harmful chemicals, use of defoliants, emission of pollutants or war. Examples of ascertainable ecocide territories of sizeable note include the deforestation of the Amazonian rainforest[7], the proposed expansion of the Athabasca Oil Sands in north-eastern Alberta, Canada[8] and polluted waters, which account for the death of more people than all forms of violence including war.[9]

In any given example of ecocide, the extent of 'destruction, damage or loss' suffered requires analysis. Whereas 'destruction' and 'loss' are easy to ascertain by way of data, what constitutes 'damage' for the purpose of establishing the crime of ecocide is more complex. Size, duration and significance of impact of damage to a territory in most instances shall be of relevance to determine whether the crime is made out. The Rome Statute sets out an extended definition of damage to the environment, specifically as a consequence of war crimes, which provides useful assistance. Article 8(2)(b)(iv) criminalises:

widespread long-term and severe damage to the natural environment which would be clearly excessive in relation to the concrete and direct overall military advantage anticipated.

The wording used in this section was adopted from the 1977 United Nations Convention on the Prohibition of Military or any other Hostile Use of Environmental Modification Techniques (ENMOD). ENMOD specifies the terms 'widespread', 'long-lasting' and 'severe' as

(a)   'widespread': encompassing an area on the scale of several hundred square kilometers;

(b)   'long-lasting': lasting for a period of months, or approximately a season;

(c)   'severe': involving serious or significant disruption or harm to human life, natural and economic resources or other assets.

These expanded definitions, which are already embedded in international laws of war, offer an existing basis upon which the international crime of ecocide can be seated at the table of the ICC. The word 'ecocide' bestows the missing name and fuller comprehension of the crime of unlawful damage to a given environment. As a crime that is not restricted to the confines of war alone, the categorisation of ecocide as a crime against peace is appropriate. Thus, for the purpose of defining ecocide 'damage', determination as to whether the extent of damage to the environment is 'widespread, long-term and severe' can be applied to ecocide in times of peace as well as in times of war.[10]

Eco-Colonisation

The land grabs for resource exploitation of today by international corporations are a repeat of the past colonial conquering of 'virgin land' for commercial exploitation. Colonialism may be relegated by many as a subject matter of mere historical interest, but in truth colonisation is very much alive. Whilst the focus has shifted from human slavery to plundering of ecological resources, the mechanics have not. As in the past, resource-rich territory is distributed among the corporate interlopers, their control is registered, secured in legal title, and administered with the sole and self-advancing purpose of profitable gain. This is the reality of colonisation in the twenty-first century; it is no longer confined to the enslavement of people but enslavement of the planet. In the process, extensive damage is caused without recourse to or remedy for the well-being of either the territory or its inhabitants. The deal now, as then, is secured by long-term contract, and thus the wrongful conduct to the inhabitants by a corporation is sanctioned and legitimised by the state.

The global reach of international corporations is such that they surpass many states in economic or territorial stature. Eco-colonisation can and is happening in territories sometimes the size of nations. A pending UN report puts the conservative cost of global ecocide by the world's top firms at $2.2 trillion for 2008, a figure bigger than the national economies of all but seven countries in the world.[11] Such is the extent of ecosystem destruction on a global scale that similar principles and legal recognition on a par with genocide is now necessary for outlawing ecocide before further resource depletion triggers more war. It is sobering to reflect that not all colonisation was effected solely by nations: the discovery and occupation of *terrae nullius*[12] or the establishment of title of inhabited territories by other means (often by force) was also initiated by charter companies such as the British East India Company. This particular company had its own army to ensure control of its resources was effectively policed. Today lobbying (and sometimes closer to ground activities) ensures effective policing to ensure power remains vested with the colonising corporation. Thus, in reality, colonisation of old differs little from the colonisation of today – the pursuit of profiteering at the expense of exploitation of another's resources wreaks tragic consequences to people and planet.

Territories and their boundaries change over time, as do those who have governance of those territories. However it is the value ascribed by those who hold the governing responsibilities of a given territory to those who inhabit those territories, which governs the conduct and actions towards the inhabitants. The inhabitants of a territory can and do fluctuate. Nevertheless, it is the habitat for all those who reside there at any given point in time. It is an environment that can be regarded as a home not solely for human occupants but also can include animals, essential minerals, water and fertile land.

Responsibility is a Mantle Worn By All

It is important to exercise enforcement and deterrence of ecocide on individuals as well as states. Unlike the primary UN judicial organ, the International Court of Justice, whose main functions are to settle inter-state legal disputes and provide advisory opinions, the ICC is a permanent tribunal to prosecute individuals for crimes against peace. Thus, by incorporating ecocide as a fifth crime against peace under the Rome Statute, extensive damage to the environment is actionable against persons. It was the Nuremberg Principles that established individual responsibility under international law. The International Tribunal at Nuremberg held:

> Crimes against international law are committed by men, not by abstract entities, and only by punishing individuals who commit such crimes can the provisions of international law be enforced.

Similarly, Article 4 of the Convention on the Prevention and Punishment of the Crime of Genocide provides that genocide is punishable as a crime irrespective of whether those committing it are '*constitutionally responsible rulers, public officials or private individuals*'. This is an important tenet that has been retained by the ICC: responsibility is a mantle worn by all.

It is proposed that ecocide be a crime of strict liability, one without the requirement of *mens rea*.[13] The reasons are four-fold. Firstly, ecocide is a crime

of consequence. It is often not the conduct itself that is in question but the consequences of the conduct. For instance, a company in the business of creating and
generating energy may be at risk of committing ecocide depending upon where it
procures its energy. Use of extractive practices would render the operators liable,
whereas procurement from renewable sources would not. Secondly, the gravity
and consequence of extensive damage and destruction to the environment justifies conviction without proof of any criminality of mind. Historically, courts had
assumed that since a corporation could not have a criminal state of mind in isola-
tion from it's directors[14], it could only be guilty of an offence which did not include
any mental element. Strict liability would therefore ensure application of international governance of corporate created ecocide. Thirdly, without absolute liability
for ecocide, the legislation would be rendered largely ineffective.

The fourth reason is the rationale that strict liability places the focus on the
exercise of preventing the harm, not on the blame of the accused. In the case of
ecocide, as with all crimes against peace, the focus is ultimately on war prevention. Further depletion of resources will rapidly dissolve into violent conflict over
allocation of resources. By creating a pre-emptive binding obligation, the crime
of ecocide is focused on prevention from the outset. It creates a quasi-crime, a
regulatory offence, rather than an ordinary criminal offence. The concept of fault
for regulatory offences is based upon the reasonable care standard, which does
not imply moral blameworthiness in the same manner as criminal fault. Thus,
conviction for breach of a regulatory offence suggests nothing more than that the
defendant has failed to meet a prescribed standard of care, albeit a standard of
care of great exactitude. Rather than starting from a premise of punishing past
wrongful conduct, regulatory measures are generally designed to prevent future
harm through the enforcement of minimum standards of conduct and care. In
doing so, regulatory legislation involves the shift in emphasis from the protection
of individual interests to the protection of public and societal interests.

True recognition of the cost of environmental damage begins when the full
long-term implications are given proper weight; only when prosecution sentencing
has recourse to remedies that impose duties to restore and remediate will we begin
to balance the scales of justice. Fines do not ultimately provide satisfactory recourse
to those affected, nor does it prevent further illegal activity. The failure to govern
illegal logging of the Amazon amply demonstrates this point: fines are merely
factored in by the offending company as an externality, to be paid if and when
caught. Thus responsibility continues to be sidestepped time and again. However,
where ecocide becomes subject to criminal prosecution, the corresponding
breach of duty of care presents an alternative remedy. As an international crime
ecocide will be an imprisonable offence, moreover it will impose other remedies.
Restorative justice is a far more powerful legal tool than mere pecuniary justice.
Imposing extensive restoration provisions ensures duty of care is not evaded by
those who have derogated their responsibilities. In this manner, the acquisition of

land becomes governed by the counterbalancing responsibility to the Earth's right to life, a duty owed by those who seek profit by ecocide.

The adoption of ecocide as a fifth peace crime to be governed by the ICC would compel state parties to the Rome Statute and individuals therein to abide by their international legal responsibility to prevent ecocide being wreaked under their tenure. In doing so, the prevention of ecocide would attract the legal status of *erga omnes* (Latin: 'towards all') meaning an obligation flowing to all. Accordingly, *erga omnes* obligations are owed to the international community as a whole. When a principle, in this case the sacredness of all life, achieves the status of *erga omnes* the rest of the international community is under a mandatory duty to respect it in all circumstances in their relations with each other. An *erga omnes* obligation exists to prevent the breach of a primary crime. Ecocide would therefore be included as an example of an *erga omnes* norm, alongside piracy, genocide, and crimes against humanity such as slavery and racial discrimination.

Under the Rome Statute's complementarity principle, the court is designed to complement existing national judicial systems: it can exercise its jurisdiction only when national courts are unwilling or unable to investigate or prosecute such crimes. Primary responsibility to investigate and punish crimes is therefore left to individual states. Thus many signatories to the Rome Statute (but not all) have implemented national legislation to provide for the investigation and prosecution of crimes that fall under the jurisdiction of the ICC. Hence, by implementing ecocide as a crime at international level, the pressure is immediately created for the crime to be speedily implemented at national level.

There is an additional reason for seeking international recognition of ecocide: until we have correctly identified the problem, we are unable to provide the corresponding solutions. International law evolves in response to the changing world, and is by no means a perfect beast, growing and changing direction as it expands. But it is an arena that must develop, by necessity. Such is the extent of global ecosystem destruction that principles and legal recognition on a par with genocide are now urgently required. Voluntary corporate governance, market trading and offset mechanisms have proven to be manifestly unsuccessful in halting the ecocide.[15] Creation of the crime of ecocide creates a pre-emptive obligation to act responsibly before damage or destruction of a given territory takes place. In doing so the burden shifts dramatically, sending a powerful global message to the world, of a premise that applies to us all, not just to those involved in business or during war, to take responsibility for the well being of all life.

Conclusion
We stand at a unique point in time and history. Now, in full cognisance of the extensive damage already wreaked upon the planet, it is pleaded here that it is necessary to place ecocide on the same legal footing as the four international crimes of peace. Now we have the knowledge of a new colonisation – instead of people it is of the Earth, and she has no

recourse to justice. It is a silent, ignored and faceless crime, one currently without recourse to legal remedy. Amendment of the Statute of Rome to include the crime of ecocide will provide both voice and a route to prevention, restoration and remedy. More importantly, it sets the legal requirement for states and corporations to take individual and collective responsibility. Now more than ever it is so necessary to provide protection and assistance to territories at the receiving end of ecocide, whether through eco-colonisation or by the 'hand of god'. Furthermore, such a law would restore our collective duties and obligations to the planet to facilitate the advancement of 'better standards of life in larger freedom'.

Knowledge, once gained, places a legal duty of care on those who carry the burden of superior responsibility. We all carry that burden and we all are complicit in a system that does not work. Destroy our Earth and we destroy our homeland. General principles of equity inform our understanding of what we are morally obligated to do. Legally and morally there is opening before us the opportunity in which wise decisions can now be made. The failure or omission to do otherwise would be an abrogation of our responsibilities. The legal recognition of ecocide in the international arena is but one bridge to take us towards the Garden of Eden, a garden where one day laws will no longer be required because we all are responsible for our Earth. On that day the rulebook will no longer be necessary. On that day, by consensual agreement of the greater Earth community, governance shall simply be by the laws of the Earth. We are now finally beginning to identify what those laws, duties and responsibilities are and how we can change the rules of the game. They are some of the steps which we can take to evolve our new laws. History will not judge us well if we fail to embrace our responsibilities and ignore that which we now know.

Notes

1   Preamble, UN Charter, 1945.
2   The Rome Statute of the International Criminal Court entered into force on 1st July 2002. As of March 2010, 110 member states have ratified the Rome Statute, and a further 39 have signed but not ratified. A number of states including China, Russia and USA have not yet joined.
3   Article 5, Rome Statute.
4   The statute defines each of these crimes except for aggression which is currently under consideration, due to be determined in 2011.
5   The Longman Dictionary of Contemporary English defines ecocide as 'the gradual destruction of a large area of land, including all of the plants, animals etc living there, because of the effects of human activities such as cutting down trees, using pesticides etc [= ecological genocide]'.
6   See *Lessons Unlearned*, Global Witness Report, 2010: an analysis of UN inability to prevent resource wars.
7   Currently resulting in destruction, damage and loss of a territory the size of France.

8   If proposed expansion proceeds, tar sand extraction (known as 'dirty oil' due to it's
     excessively damaging outcomes) will result in the loss of vast tracts of boreal forest and
     muskeg (peat bogs) of a territory the size of England.

9   Globally, two million tons of sewage, industrial and agricultural waste is discharged
     into the world's waterways and at least 1.8 million children under five years-old die
     every year from water related disease, or one every 20 seconds. See: *Sick Water? A
     Rapid Response Assessment*. UNEP, UN-HABITAT report, 2010

10   It is of note that the Field Manual 27–10, The Law of Land Warfare, 1956, which
     governs the US Army's understanding of what constitutes a war crime under interna-
     tional law, includes at paragraph 498: '*Such offences in connection with war comprise:
     a. Crimes against peace; b. Crimes against humanity; c. War Crimes*'. Thus, although the
     USA State refuses to ratify the Rome Statute, US members of the military can and are
     prosecuted for crimes against peace.

11   '*The Economics of Ecosystems & Biodiversity*' (TEEB), UNEP. 2010 Final results to be
     presented at CBD COP-10, October 2010.

12   Land deemed owned by no-one and therefore susceptible to acquisition.

13   Crimes of strict liability are crimes where the *mens rea*, the 'mental element', is irrel-
     evant. Only the *actus reus,* 'the act of doing', need be established.

14   In certain circumstances of crime, where there may be a mental element, the company
     will have liability through the 'directing mind' of for example its directors or the
     CEO.

15   History repeats itself. The very same mechanisms were proposed by an industry faced
     with the prospect of prohibition of slavery 200 years ago. The difference then was the
     justifiable rejection of the proposals and slavery was abolished. Today, it is the very
     implementation of market-led proposals that has facilitated the continuance of such
     disastrous consequences.

The Earth Charter, Covenants, and Earth Jurisprudence

J. Ronald Engel & Brendan Mackey

Earth is a fragile home suspended in the great void. The thin atmosphere (most of its gases are found in the first 10 kilometres) is all that stands between us and dark, lifeless eternity. Despite what Hollywood tells us, Earth is the only planet known to support life and it is now scientifically established that we are pushing our planet's ecological limits and degrading and changing Earth's environment in ways that threaten to disrupt the very life support systems that have maintained this planet in a condition fit for life, and as humans have known it, for the last 180,000 years since our species emerged from ancestral life forms.[1]

Threats to Earth's global ecological integrity represent a new challenge for humanity, being a problem we have not had to face before in our history. In our species' relatively brief time on this planet we have been spared the kinds of natural disturbances that have in the geological past caused so-called mass extinction events.[2] And, until the last 200 years or so, the environmental harm caused by humans was limited in its geographical footprint. We caused only small scars and there were always other lands to move on to and exploit. However, humanity's current ecological footprint and our capacity to cause environmental damage is now so extensive, comprehensive and powerful, that we are putting at risk the future safety and well-being of humankind, and a large number of the other species with whom we share Earth as home.

In so doing, we risk breaking a covenant that can be rightfully considered sacred. This is the implicit covenant each generation has with those that follow; the promise of a planet, full of nature alive and a healthy environment that sustains the human endeavour and the flourishing of life. A covenant to secure, in the words of the Earth Charter, 'Earth's bounty and beauty for present and future generations', and that requires, among other things, a commitment to protect and restore the integrity of Earth's ecological systems, along with biological diversity and the natural processes that sustain life.[3] We have never needed to think about this covenant in such an explicit and intentional way, nor in such all-encompassing and ultimate terms. We are confronted with this inescapable question: what do we believe is the fundamental covenant underpinning our human civilisation and upon which the true security of future generations depends?

Covenants are open, unconditional commitments to be faithful to others regarding our most fundamental values and behaviours; they are the most profound and powerful social bonds we know. Covenants are our voluntary commitments to those relationships with other persons, nature, and the creativity of the universe recognised as embodying the goodness, rightness and truth of our being. They express our moral obligations to maintain and fulfil these relationships in the midst of the inevitable uncertainties of historic and physical forces. These commitments also apply to the community of those who are faithful to the covenant.

Our lives are enmeshed in a network of covenants, sometimes competing, sometimes implicit; marriage, friendship, religious belief, canons of professional practice, ethnic and gender identity, institutions of political governance. Even our understanding of ourselves as a distinct species can be considered as a covenant of our *common humanity*. Much of history is a clash and contest of covenants, and takes the course it does by virtue of the covenants that men and women, communities, and nations choose to honour. In the broadest ethical terms, if there is a plot line weaving its way over the course of human history, it is the struggle to make the covenants of civilisation more explicit and accessible to criticism, more inclusive in membership, more respectful and caring of individual rights and responsibilities, more holistic in their grasp of the multiple moral concerns that must be met for humans and nature to thrive; in sum, more adequately expressive of the goodness, rightness, and truth of being.

Sustainability is the latest manifestation of the quest for life-giving covenants that has dominated all human societies. The American philosopher John Dewey posed this quest in the form of a question which may be paraphrased thus: 'How can we make the good life (and all those qualities of physical and cultural existence we value) more secure and widely shared, in the face of ever-present change and uncertainty?'. He believed this question could not be adequately answered until the gap between science and ethics was bridged. As he wrote in his 1929 manifesto of pragmatic naturalism, *The Quest for Certainty*: 'The problem of restoring integration and cooperation between man's beliefs about the world in which he lives and his beliefs about the values and purposes that should direct his conduct is the deepest problem of modern life'.[4]

In his seminal 1979 book, *The Imperative of Responsibility*, the German-American philosopher Hans Jonas took Dewey's analysis one step further by showing that, in light of our contemporary knowledge of the world, the key assumptions made by traditional ethics are no longer valid, namely that the *conditio humana* is unchangeable, that human behaviour can be calculated on the basis of rational analysis and that human responsibility is limited to societal relationships.[5] Instead, as he argued, in light of the impact of human civilisation on the biosphere, we are now faced with a profound vulnerability and finitude of life, including humanity. No previous ethics had to consider the global condition of human life and the far-off well-being of the future. He found the paradigm for how we need to act if we are to respond to what Dewey called the 'problems of men', the defining covenantal choices of contemporary human history, in the care of parents for their young, in the promises we make, and the responsibilities we take for future generations.

The Defining Issues of Our Age

The unprecedented challenge that sustainability brings is framed by globalisation and the increasing constraints of living on a finite planet whose environmental life support systems are rapidly unravelling. The ancient human quest for certainty and life-sustaining covenants is redefined by the challenge of sustainability thus: how can we configure the human endeavour so that a predicted 9–12 billion people can live within the planetary boundaries needed to maintain Earth's environment and our planet's special qualities that we value, including the rich diversity of life and biosphere with which we have co-evolved?

The aggregate environmental impact of humans on Earth's environment can be measured by using various approaches. One of the most widely used is the *IPAT* formula ($I = P \times A \times T$), where I is impact, P is population, A is affluence, and T is technology.[6] The *IPAT* formulation suggests that there are various ways in which humans can configure how they live sustainably on this planet; more or fewer people, more or less affluence, dirtier or greener technology, and combinations thereof. But however we configure ourselves, there is a specific set of scientifically determined planetary boundary conditions we have to live within if Earth is to remain in a condition fit for life. Recent research has identified nine interlinked planetary boundary conditions that define the safe operating space for humanity with respect to the Earth system, and has empirically established that we have already overstepped three of these: climate change, rate of biodiversity loss, and nitrogen cycle.[7]

The climate change problem is currently the most compelling manifestation of the sustainability challenge we face. But the global biodiversity extinction crisis is no less urgent than the climate change problem and we confront a similar level of crisis in the global supply of freshwater and global food security. Unfortunately, the sustainability challenge is further complicated by the fact that these are not unrelated phenomena. Rather, they are deeply connected – climate, biodiversity, water, food – and what we do to one affects the others in complex ways that we are only beginning to scientifically comprehend or practically experience.

As if these issues of environmental sustainability were not complex and demanding enough when considered alone, it is clear that they cannot be resolved apart from progress in addressing the great social and political issues of the age. This was vividly demonstrated by the failure of the nations of the world at the 2009 Copenhagen conference on climate to reach the full agreement needed to address climate change.[8] Communities everywhere are claiming the modern ethical ideal of democracy, with its principles of equality, freedom, community, and universal human rights, as their birthright. Because sustainability is a matter of achieving social, economic and political justice within and between nations as much as achieving ecological preservation and restoration, comprehensive and integrative approaches are needed.

Every generation must choose the covenants that will define its overarching way of life and its answer to the question of how to construct our lives together such that all life flourishes. What covenant can we now choose, must we now choose, which will 'Secure Earth's bounty and beauty for present and future generations' in face of the overwhelming environmental and social uncertainties we are now experiencing in every part of our planetary home?

The Faustian Pact

The modern worldview is perhaps characterised by the assumption that we can do without covenants; all we need are contracts that promote efficiency. There are certainly many truly emancipatory accomplishments of the modern age of which we can be justly proud: our struggles for human rights and the rule of law, our enshrinement of critical reason, our dedication to the use of technology in the service of relieving human suffering, and our spectacular demonstrations of artistic creativity. But alongside these historic advances there is a dark side that suggests that we have already made a choice regarding one covenant, albeit implicitly and without recognising what it commits us to. Some hints as to its form and consequences can be gleaned by considering the well-known literary demonic covenant of unlimited human ambition and satisfaction commonly known as the *Faustian pact*.

The *Historia von Johann Fausten*, published in Frankfurt in 1587, was an effort to hold up a mirror to where the world was heading at the beginning of the modern era.[9] Faust makes his compact with Mephistopheles as follows:

> But fear not that I shall break this compact. What I promise is precisely what all my energies are striving for ... The Great Spirit has spurned me; Nature shuts against me ... I have long loathed every sort of knowledge. Let us quench our glowing passions in the depth of sensuality ... man's proper element is restless activity.[10]

Faust thereby denies any moral responsibility for and connection with future generations, the limits and needs of the natural world and the rest of humanity, or the application of reason in the exercise of judgment in the conduct of human affairs. Instead, he simply commits himself to the pursuit of maximising his personal pleasure, power, and knowledge within the remaining twenty-four years of life that Mephistopheles grants him. The Faustian covenant is a denial of every relationship with other persons, nature, and that which we have reason to believe embodies the goodness, rightness and truth of our being, as well as every moral obligation required to maintain and fulfil these relationships in the midst of the inevitable uncertainties of history and the physical environment.

There is a familiar ring to the Faustian covenant in some of the premises and consequences of neo-classical economic theory, especially (1) the central place given to maximising personal preferences[11] and (2) the treatment of negative environmental impacts as 'externalities' (although these can in theory be quantitatively accounted for and *internalised* into market mechanisms).[12] The latter facilitates viewing Earth and the biosphere as merely a quarry (a source of raw materials and energy for production of manufactured capital) and a dump for the wastes of our systems of production and consumption. As John Maynard Keynes wrote in the midst of the Great Depression of the 1930s:

> For at least another hundred years we must pretend to ourselves and to everyone that fair is foul and foul is fair; for foul is useful and fair is not. Avarice and usury ... must be our gods for a little longer still. For only they can lead us out of the tunnel of economic necessity into daylight.[13]

Are we exaggerating to suggest that the current path of unsustainable development is based on a contemporary version of Faust's compact with Mephistopheles? Like Faust, we who live in the modern globalised economy tend to *discount* the future, imagine we can live without regard for the natural world around us, are dismissive of scientific knowledge when we find it inconvenient, place great value on maximising short-term personal preferences, and ignore if not reject the intrinsic value of a *nature alive*. Yet this Faustian covenant is proving to be resilient in spite of all the evidence that it is an ultimately suicidal commitment. It is sobering to reflect on how rapid, substantial and coordinated the international community's response to the global financial crisis has been compared with the lethargy and procrastination around climate change negotiations.

The Earth Charter

A contemporary Faustian covenant cries out for a change of heart, mind and will; which, if it is to be lasting, must take the form of a new covenant that is truly promising for the greater community of life and future generations. There have been many attempts to state the terms of such a covenant and make it the foundation for a new regime of international governance and Earth Jurisprudence since the founding of the United Nations. The 1948 *Universal Declaration of Human Rights*, 1972 *Stockholm Declaration*, and 2000 *Millennium Development Goals* are among the best known and most influential in the international community.[14] Less well known are the 1983 *World Charter for Nature*, adopted by a majority vote of the members of the United Nations General Assembly, 1991 *Caring for the Earth*, the second world conservation strategy adopted by the World Conservation Union, 1992 *Declaration of the Parliament of World Religions*, 1996 *Earth Covenant*, circulated by Global Education Associates and signed by over two million people worldwide, 1997 *Declaration on the Responsibilities of the Present Generations Towards Future Generations*, 2000 *A Manifesto for Earth*, and 2002 *A Manifesto for Life*.[15]

The *Earth Charter* warrants special attention as a candidate covenant to meet the challenges of our age. The Earth Charter is the outcome of a decade-long, worldwide, cross-cultural dialogue on common goals and shared values. While it began as a United Nations initiative, it was born from a global civil society initiative. The Earth Charter was finalised and launched as a people's charter in 2000 by the Earth Charter Commission, an independent international entity. The drafting of the Earth Charter involved the most inclusive and participatory process ever associated with the creation of an international declaration, which is the primary source of its legitimacy as a guiding ethical framework. [16] It has been formally recognised by over 4800 organisations, including many governments and international organisations such as the IUCN and UNESCO. Interwoven into its text are principles drawn from both soft and hard international law, including the Universal Declaration of Human Rights, from numerous nongovernmental declarations and people's treaties issued over the past thirty years such as those noted above, contemporary science, the wisdom of the world's great religions and philosophical traditions, the declarations and reports of the seven UN summit conferences held during the 1990s, the global ethics movement, and best practices for building sustainable communities.[17]

The *Earth Charter* needs to be read in its entirety to appreciate how fully it sets forth the meaning of a covenant to future generations through its four foundational principles of (1) *Respect and Care for the Community of Life*, (2) *Ecological Integrity*, (3) *Social and Economic Justice* and (4) *Democracy, Non Violence and Peace*, which are in turn supported by an extensive series of subsidiary principles. Here we simply highlight its primary covenantal affirmations by matching following selective quotations with the key aspects of covenants noted above.[18] The *Earth Charter* is:

(1) Our voluntary, unconditional commitment
    We stand at a critical moment in Earth's history, a time when humanity must choose its future … We urgently need a shared vision of basic values to provide an ethical foundation for the emerging world community (Preamble)

(2) To those relationships with other persons, nature, and the creativity of the universe recognised as embodying the goodness, rightness and truth of our being
    Principle 1a. Recognise that all beings are interdependent and every form of life has value regardless of its worth to human beings.

(3) The moral obligations required to maintain and fulfil these relationships in the midst of the inevitable uncertainties of history and the physical environment
    Principle 12. Uphold the right of all, without discrimination, to a natural and social environment supportive of human dignity, bodily health, and spiritual well-being, with special attention to the rights of indigenous peoples and minorities.

Principle 16f. Recognise that peace is the wholeness created by right relationships with oneself, other persons, other cultures, other life, Earth, and the larger whole of which all are a part.

(4) And the community of those who are faithful to the covenant.
   Towards this end, it is imperative that we, the peoples of Earth, declare our responsibility to one another, to the greater community of life, and to future generations. (Preamble)

Keeping Promises Across the Generations

A covenant between the generations must be carried across the generations. The comprehensive and holistic promise the Earth Charter makes to future generations is a historically new commitment, but its roots go deep into the past. As Rudd Lubbers, a member of the Earth Charter Council and former Prime Minister of the Netherlands, stated in 1998, 'fulfilling history is about the special talent of listening to our ancestors, possessing awe for nature, and caring for future generations'.[19] There are many covenantal promises we could cite that have prepared the way for the Earth Charter. One is the covenant with nature that the prophet Hosea proclaimed on behalf of the god of creation.[20] Another is the *Thanksgiving Address* announced by the prophet of the Haudenosaunee tribal federation (commonly known as the Iroquois) of North America and carried over many generations by a covenant referred to as a *silver chain*.[21]

However, if our covenant to each other, the greater community of life, and future generations, of a healthy Earth where life flourishes is to be kept, we must find ways of putting into place, in every sector and endeavour of contemporary society, actions that embody the Earth Charter's vision, values and principles. One such sector is constitutional and international treaty law. As a covenant, the Earth Charter must become a matter of principled and effective Earth Jurisprudence.[22]

One person who has taken exemplary leadership for this task is Dr. Parvez Hassan of Pakistan, whose life can serve as an inspiration for all of us who want to be serious about our promises to future generations. Hassan received his legal education at Yale and Harvard universities, where he became interested in international jurisprudence and wrote his S.J.D. dissertation on the right to not be unlawfully detained in the International Covenant on Civil and Political Rights (1966) and the earlier Universal Declaration of Human Rights. Not long after establishing his law practice in Lahore he came to appreciate the importance of ecology when he saw the fast degradation of the environment and its impact particularly on dwindling supplies of clean water available to the impoverished rural communities of his native Pakistan. The recognition that the well-being of future generations depend upon our care of the environment today launched him on what was to quickly become a distinguished legal career in constitutional and international environmental law and his present position as advocate before the Supreme Court of Pakistan.[23]

In the late 1980s, as chair of the Commission on Environmental Law of the World Conservation Union (IUCN) Hassan led the effort to draft an international hard law treaty, titled *Draft International Covenant on Environment and Development*, which is both a summary and further extension of existing international environmental law, and is based on a set of strong universal ethical principles.[24] If adopted by the United Nations, it would have international treaty status comparable to the international covenants on human rights. Hassan served on the drafting committee of the *Earth Charter* and enabled the cross-fertilisation of ethics and law in the drafting process. The *Earth Charter* indirectly references the draft International Covenant in its concluding section, *The Way Forward*:

> In order to build a sustainable global community, the nations of the world must renew their commitment to the United Nations, fulfill their obligations under existing international agreements, and support the implementation of Earth Charter principles with an international legally binding instrument on environment and development.

In March 2007, President Pervez Musharraf suspended the Chief Justice of Pakistan, who had intervened repeatedly to prevent development schemes of the government and private developers which were environmentally unsound; in November, he suspended the Constitution, imposed a state of emergency, closed all TV stations except one supportive of the government, and dismissed almost the entire superior judiciary of the Supreme Court and the High Courts – about 70 judges – to ensure that there was no possibility of judicial reversal of his action. The Chief Justice, joined by the Court, declared these actions illegal, precipitating the now famous outpouring of street protests by members of the black-suited legal profession.[25] Parvez Hassan took part in these protests and was arrested, jailed, and mercifully released. In the midst of the conflict, supporters of the Earth Charter initiative issued a public statement calling for the restoration of constitutional law in Pakistan on the basis of the ethical principles of the *Earth Charter*.[26]

Covenants are made by individuals and kept by individuals as they act in their personal, professional and social worlds. The dedicated work of Parvez Hassan is an inspiring example of how one individual is living out the *Earth Charter*'s promise to future generations and in the process transforming our systems of national and global governance.

Conclusion

The future architecture of global governance remains an important matter of ongoing debate and dialogue. Will the world community continue to negotiate new legal instruments building upon the UN Charter and associated treaty processes or will entirely new forms of global governance emerge? Whatever direction and mode Earth governance takes, a critical component of Earth Jurisprudence will be the articulation of a foundational covenant like the Earth Charter that makes explicit the shared vision, values and principles needed to guide our individual and collective behaviours toward a more just, sustainable and peaceful world. By analogy, the Earth Charter's principles can be thought of as set of 'protocols' that the international community can use as a moral compass to help direct the evolution and implementation of global governance. Building upon the legitimacy conferred by the drafting process and the growing recognition that it is acquiring the status of a global soft law document,[27] the Earth Charter can function as a covenant that is morally (not legally) binding on those who endorse and adopt it. Thus, the Earth Charter can serve as a universal guide to public policy for more sustainable ways of living, provide a source of paralegal principles for jurisprudence, and form the basis for the subsequent development of hard law.

Notes

1   Two key scientific reports supporting this claim are (a) the *Millennium Ecosystem Assessment 2005* reports available at www.millenniumassessment.org/en/index.aspx and (b) the *2007 Intergovernmental Panel on Climate Change 4th Assessment Report* are available at http://www.ipcc.ch.

2   R.J. Twitchett, 'The Palaeoclimatology, Palaeoecology and Palaeoenvironmental Analysis of Mass Extinction Events', *Palaeogeography, Palaeoclimatology, Palaeoecology*, vol. 232, 2006, pp. 190–213.

3   The text of the *Earth Charter* in 49 languages may be read at www.earthcharterinaction.org.

4   J. Dewey, *The Quest for Certainty,* Putnam Publishers, New York, 1929, p. 255.

5   H. Jonas, *The Imperative of Responsibility: In Search of an Ethics for the Technological Age*, translated by Hans Jonas with the collaboration of David Herr, University of Chicago Press, Chicago, 1984 (first publication in Germany 1979).

6   The origins of IPAT are documented in (a) P.R. Ehrlich and J.P. Holdren, 'Impact of population growth', *Science*, vol. 171, no. 3977, 1971, pp. 1212–1217; for more recent formulations see (b) P.E. Waggoner and J.H. Ausubel, 'A framework for sustainability science: A renovated IPAT identity', *PNAS*, vol. 99, no. 12, 2002, pp. 7860–7865.

7   P.E. Waggoner and J.H. Ausubel, "A framework for sustainability science: A renovated IPAT identity", *PNAS*, vol. 99, no. 12, 2002, pp. 7860–7865

8   UNFCCC Press Briefing on the outcome of Copenhagen and the way forward in 2010; comments by UNFCCC Executive Secretary Yvo de Boer; published online 20 January 2010 at http://unfccc.int/files/press/news_room/statements/application/pdf/unfccc_speaking_notes_20100120.pdf.

9   D. Wootton, ed., *Christopher Marlowe, Doctor Faustus with the English Faust Book,* Hackett Publishing Company, Indianapolis, 2005.

10 J.W. von Goethe, *Faust*, BiblioLife, 2009.

11 C. Arnsperger and Y. Varoufakis, 'What Is Neoclassical Economics: The three axioms responsible for its theoretical oeuvre, practical irrelevance and, thus, discursive power', *Real World Economics Review*, issue no. 38, 2006, pp. 2–12.

12   P. Sonderholm and T. Sundqvist, 'Pricing environmental externalities in the power sector: ethical limits and implications for social choice', *Ecological Economics*, vol. 46, 2003, pp. 333–350.

13   J.M. Keynes, *Essays in Persuasion*, W.W. Norton, New York, 1933, 1963, pp. 363–73.

14   (a) J. Morsink, *The Universal Declaration of Human Rights: Origins, Drafting, and Intent*, University of Pennsylvania Press, Philadelphia, 1999; (b) B. Weston, R. Falk, H. Charlesworth, and A. Strauss, eds, *Supplement of Basic Documents to International Law and World Order*, 4th edition, Thompson West, St. Paul, Minnesota, 2006; (c) United Nations Resolution 55/2, *United Nations Millennium Declaration*, available at http://www.un.org/millennium/declaration/ares552e.pdf.

15   (a) W. Burhenne and W. Irwin, *The World Charter for Nature: A Background Paper*, Erich Schmidt Verlag, Berlin, 1983; (b) World Conservation Union, the United Nations Environment Programme, and the World Wide Fund for Nature, *Caring for the Earth: A strategy for sustainable living*, IUCN, Gland, Switzerland, 1991; (c) H. Küng and K Kuschel, eds, *A Global Ethic: the Declaration of the Parliament of the World's Religions*, Continuum, New York, 1993, available at http://www.light-party.com/Visionary/EarthCovenant.html; (e) *Declaration on the Responsibilities of the Present Generations Towards Future Generations*, adopted by the General Conference of the United Nations Educational, Scientific and Cultural Organization, meeting in Paris from 21 October to 12 November 1997 at its 29th session, Available at http://home.um.edu.mt/fgp/Declaration.html; (f) T. Mosquin and S. Rowe, 'A Manifesto for Earth', *Biodiversity*, Vol. 5, No 2, 2000, pp. 3–9; (g) J. Mayr et al, 'A manifesto for life: in favor of an ethic of sustainability', *Capitalism, Nature, Socialism*, vol. 13, no 4, 2002, pp. 121–125.

16   'What is the Earth Charter?', *Earth Charter Initiative Handbook*, 2008, Earth Charter International Secretariat, available at www.earthcharter.org.

17   See *Earth Charter Briefing Book*, Earth Charter Initiative, San Jose, Costa Rica, 2000.

18   See J.R. Engel, 'A Covenant Model of Global Ethics', *Worldviews* vol. 8, no. 1, 2004, pp. 29–46.

19   R. Lubbers, 'The global Sovereignty of the People', *Newsletter*, Boston Research Center for the 21st Century, 1998.

20   'Then I will make a covenant ... with the wild beasts, the birds of the air, and the things that creep on the earth, and I will break bow and sword and weapon of war and sweep them off the earth, so that all living creatures may lie down without fear'. (*Hosea* 2:18).

21   'I ask these fine human beings gathered here of many colors, many shapes, some tall, some short ... That we, in our diversity, will be one and we will bring our minds as one mind ... And we will say to the spirit of the rivers and the streams and the lakes and the oceans who quench our thirst ... we thank you for this day and this life and life ... on behalf or our children, we are grateful'. J. Ransom, ed, *Words that Come Before All Else*, Haudenosaunee Environmental Task Force, Syracuse, New York, 1992.

22 (a) K. Bosselmann, J. Engel and P. Taylor, *Governance for Sustainability: Issues, Challenges, Cases,* Earthprint, Bonn, Germany, 2008; (b) C. Cullinan, *Wild Law: A Manifesto for Earth Justice,* Green Books, Totnes, Devon, 2003.

23 See, generally, (a) P. Hassan, 'Environmental Protection, Rule of Law and the Judicial Crisis in Pakistan', 35 *Pakistan Law Journal,* 2007, pp. 278–292; (b) 'Securing Environmental Rights through Public Interest Litigation in South Asia', *Virginia Environmental Law Journal,* 2004, pp. 216–236.

24 IUCN Commission on Environmental Law, *Draft International Covenant on Environment and Development,* 3rd edition, Environmental Policy and Law Paper, no. 31, World Conservation Union, Gland, Switzerland, 1995.

25 Y. Hassan, 'Rule of Law and the Judiciary in Pakistan', Remarks on occasion of receiving an award, for and on behalf of the judiciary and the legal fraternity in Pakistan, from the New York State Bar Association, January 30, 2008, available at http://www.nation.com.pk/daily/feb-2008/10/international.php.

26 See 'Statement: Restore the Rule of Law in Pakistan', available at http://www.earth-charterinaction.org/content/articles/99/1/Restore-the-Rule-of-Law-in-Pakistan/Page1.html.

27 'Chapter IX. A Short History of the Earth Charter Initiative'. *The Earth Charter Initiative Handbook,* Earth Charter International Secretariat, available at http://www.earthcharterinaction.org/invent/images/uploads/Handbook%20ENG.pdf.

Governance for Integrity?
A Distant but Necessary Goal

Laura Westra

Introduction to Ecological Integrity

Ecological or biological integrity originated as an ethical concept in the wake of Aldo Leopold (1949) and has been present in the law, both domestic and international, and part of public policy since its appearance in the 1972 U.S. Clean Water Act (CWA). Ecological integrity as also filtered into the language of a great number of mission and vision statements internationally, as well as being clearly present in the Great Lakes Water Quality Agreement between the United States and Canada, which was ratified in 1988.

The generic concept of integrity connotes a valuable whole, the state of being whole or undiminished, unimpaired, or in perfect condition. Integrity in common usage is thus an umbrella concept that encompasses a variety of other notions. Although integrity may be developed in other contexts, wild nature provides paradigmatic examples for applied reflection and research.

Because of the extent of human exploitation of the planet, examples are most often found in those places that, until recently, have been least hospitable to dense human occupancy and industrial development, such as deserts, the high Arctic, high-altitude mountain ranges, the ocean depths, and the less accessible reaches of forests. Wild nature is also found in locations such as national parks that have been deemed worthy of official protection.

Among the most important aspects of integrity are the autopoietic (self-creative) capacities of life to organise, regenerate, reproduce, sustain, adapt, develop, and evolve over time at a specific location. Thus integrity defines the evolutionary and biogeographical processes of a system as well as its parts or elements at a specific location.[1] Another aspect, discussed by James Karr in relation to water and Reed Noss (1992) regarding terrestrial systems, is the question of what spatial requirements are needed to maintain native ecosystems. Climatic conditions and other biophysical phenomena constitute further systems of interacting and interdependent components that can be analysed as an open hierarchy of systems. Every organism comprises a system of organic subsystems and interacts with other organisms and abiotic elements to constitute larger ecological systems of progressively wider scope up to the biosphere.

Ecological Integrity and Science

Ecological integrity is both 'valued and valuable as it bridges the concerns of science and public policy'.[2] For example, in response to the deteriorating condition of our fresh-waters, the CWA has the objective: 'to restore and maintain the chemical, physical, and biological integrity of the Nation's waters' (sec. 101[a]). Against this backdrop, Karr developed the multimetric Index of Biological Integrity (IBI) to give empirical meaning to the goal of the CWA.[3] Karr defines ecological integrity as 'the sum of physical, chemical, and biological integrity.' Biological integrity, in turn, is 'the capacity to support and maintain a balanced, integrated, adaptive biological system having full range of elements (genes, species, and assemblages) and processes (mutation, demography, biotic interactions, nutrient and energy dynamics, and metapopulation processes) expected in the natural habitat of a region'.[4] Scientists can measure the extent to which a biota deviates from integrity by employing an IBI that is calibrated from a baseline condition found 'at site with a biota that is the product of evolutionary biogeographic processes in the relative absence of the effects of modern human activity'[5] – in other words, wild nature. Degradation or loss of integrity is thus any human-induced positive or negative divergence from this baseline for a variety of biological attributes.[6] The Noss Wildlands Project, which aims to reconnect the wild in North America, from Mexico to Alaska,[7] utilises the ecosystem approach to argue the importance of conserving areas of integrity.

But the most salient aspect of ecosystem processes (including all their components) is their life-sustaining function, not only within wild nature or the corridor surrounding wild areas although these are the main concerns of conservation biologists. The significance of life-sustaining functions is that ultimately they support life everywhere. Gretchen Daily, for instance, specifies in some detail the functions provided by nature's services, and her work is crucial in the effort to connect respect for natural systems integrity and human rights.[8]

The interface between ecological integrity and human rights demonstrates the importance it should have in global governance. As it is vital for the right to life, to health, and other basic rights, ecological integrity should be foundational for all domestic and international law regimes, especially in order to protect Indigenous and local poor communities who live close to the land. These people are most often the victims of disintegrity fostered by the hazardous industrial operations that tend to represent so-called 'development' globally.

Some have argued that these activities amount to the latest developments in a long history of colonising mass violence, and raise the question of the relation between colonialism itself and its genocidal effects. It is paradoxical that, just as state sovereignty is fading in both importance and power; both the mindset and the effects of state colonialism surround us.[9]

Raphael Lemkin clearly saw that 'genocide' encompasses far more than the narrowly limited definition present in the UN convention[10], and, we will return to the topic at this time, as we compare genocide to 'crimes against humanity'.

The importance of addressing the *roots* of the ecological problems that confront us demands the re-examination of the interface between globalised colonialism and mass violence. Only then can we return to the subject matter of this chapter, in order to see (1) why is integrity the basis of any possible reversal of the current situations and trends; and (2) whether and how 'genocide' or other such concepts in international law, should and could be co-opted in order to redress some of the grievous present harms, and halt the ongoing mass violence.

Ecofootprints, Colonisation and Mass Violence

In Al Gore's movie 'An Inconvenient Truth', a great deal of time is spent demonstrating how, for example, the environment of polar bears has been devastated, while very little or no time at all, is given to what happened to the Native Americans reduced to living in reservations. After all, weren't the Natives of all conquered/colonised/occupied lands and accused to being to close to nature, and therefore in need of civilisation?[11]

Of course, in the case of Indigenous communities, the 'conquering/colonising/occupying' of their territories, is the clearest example of a Western ecofootprint that destroys a civilisation, and even a culture, as, with the disappearance of the animals and even the ecosystemic functions upon which traditional peoples depend, their 'right' to hunt or fish becomes moot, their lifestyle disappears as even aside from actual killings on the part of their occupiers – they are collectively put at grave risk or not surviving *as a people*.[12]

Genocide: The Question of Intent

In 1944 R. Lemkin provided a rich and complete definition of genocide:

Genocide is directed against national group as an entity, and the actions involved are directed against individuals, not in their individual capacity but as members of the national group. [13]

He also distinguished between different forms of genocide. Based upon his work, Pentassuglia says: 'taking examples from Nazi practice, "political genocide", "social genocide", "cultural genocide", "economic genocide", "biological genocide", "physical genocide", "religious genocide" and "moral genocide".'[14] Hence, Indigenous people do not only possess the right to life individually, but they also do so in an extended, richer sense, as a group. In fact, the biological and physical integrity of groups is clearly dependent on both their individual and communal dimensions.

Genocide is a primary example of an obligation *erga omnes*.[15,16] After considering the related ATCA jurisprudence[17], the argument to elevate attacks on Indigenous peoples, singly and collectively, from torts to international crime should not be too hard, as each of the cases discussed contains realistic, often first-person accounts of the material facts involved. In some cases, the non-state actors perpetrating the crimes attempt to evade responsibility by claiming that international law only applies to states, hence it cannot touch them.

But William Schabas reminds us that 'at the end of the Rome Conference in July 1998, the *Financial Times*, the prestigious British business daily, published an article warning 'commercial lawyers' that the treaty's accomplished liability provisions 'could create international criminal liability for employees, officers and directors of corporations'.[18]

In this section we will consider those aspects of the crime of genocide that are most relevant to Indigenous peoples, starting with the components of the crime itself.

> For a legal policy perspective, it is especially noteworthy that the magnitude of problems of traditional groups all over the world … still awaits a comprehensive treaty response, and many of them still struggle for recognition and appropriate protection at home.[19]

One of the possible causes of this anomaly lies in the economic/trade orientation of most legal instruments that presently deal with the issue, and with the powerful interests that militate against a serious consideration of Indigenous peoples' rights. Another possible answer, albeit a partial one, may be found in the 'intent' requirement that forms an integral part of the crime of genocide.

Mens rea is basic to all serious crimes, and it is certainly required for the act of genocide. The starting point now should be the definition of *mens rea* in Article 30 (2) and (3) of the Rome Statute of International Criminal Court.

2. For the purpose of this article, a person has intent here:

   (a) In relation to conduct, that person means to engage in the conduct.

   (b) In relation to a consequence, that person means to cause that consequence or is aware that it will occur in the ordinary course of events.

3. For the purpose of this Article, 'Knowledge' means awareness that a circumstance exists, or that a consequence will occur in the ordinary course of events. 'Know' and 'knowingly' shall be construed accordingly.

In addition, the Chapeau of Article II (Physical and Biological Genocide) of the Genocide Convention emphasises the required 'intent': the proscribed acts must be deliberate, and they have to be '… committed with the intent to destroy a national, racial, religious or political group, on grounds of the national or racial origin, religious belief, political opinion of its members'. That said, it remains to consider the meaning of 'intent', 'knowledge', 'awareness' and related concepts.

In fact, one of the most salient characteristics of genocide is that it cannot be committed by a single (natural) individual. It requires a large-scale operation and planning as Lemkin also argued.[20] Case law also speaks of 'widespread and systematic' crimes[21], or of involving a 'plan or policy'.[22] Hence, although the hatred or at least the deliberate intent to destroy or eliminate a group (in part or as a whole) may be hard to prove, the 'planning' aspect of *mens rea*, must be in place as it is for 'crimes against humanity'.[23]

At any rate, absent the *dolus specialis* or the specific aspect of *mens rea*, indicating the deliberate intent to eliminate a group, all or in part, the category of 'crimes against humanity' remains as a viable alternative.[24]

But the main difficulty inherent in our present understanding of genocide persists:

> Genocides are not just the product of deranged individuals; they occur in a context which cannot be ignored or bracketed as not pertinent.[25]

If there is more to mass violence and atrocities than 'deranged individuals', but we are faced instead with societal developmental goals, pursued through colonisation and globalisation, then we must recognise not only that genocide exceeds the purview of international war, but also the requirement for a *dolus specialis* component no longer fits the ongoing situation. Perhaps, '... the way of thinking economically and politically, that has ruled the world over the last few centuries – is coming to an end'.[26] Colonisers from the time of Columbus onward have not moved with the intent 'to destroy a group, whole or in part': their aim was simply to foster their own goals, and those of the powers that funded their enterprise.

Colonialism and Structural Ecoviolence

> [A]uthoritarian orientations and routinisation interact to make 'ordinary men into perpetrators of the most gruesome acts and crimes against humanity.[27]

Because of the restrictive definition of genocide present in the 1948 Convention, sometimes it is hard to apply the term correctly to various forms of colonial mass violence. I have used the expression 'ecoviolence' advisedly to characterise their occurrence as that concept incorporates both the required 'structural generality' and the 'space' aspects that are part of genocidal atrocities.[28] In fact, Jurgen Zimmerer examines the interface of 'colonialism and the holocaust'[29] and he isolates the two major components of genocide, when viewed in conjunction with the 'situation coloniale':

> If one considers Nazi policy in eastern Europe in its different dimensions – war of annihilation, occupation policy and genocide – two concepts bind them together. The first is racism ...; the second is the policy of space, mainly with regard to Eastern Europe and the economy of destruction planned for it.[30]

When this connection is fully understood, the question of intent re-emerges: seeking 'space', expansion, even economic advantage, are not illegal aims. But when these aspects of that quest are coupled with racism, those goals no longer appear benign and 'normal'. Yet, in order to pursue expansionist policies in areas where others live, is a practice that entails a total disregard for those persons as individuals and as communities. Whether the group under attack is an Indigenous community in North or South America, one in Africa, starting with the German extermination of the Herero in Namibia, or in the Middle East, with the destruction of Palestinians as their territories are under attack from Israel and the US, the racist disregard for their rights as 'peoples' is evident. Their very existence stands as an obstacle to colonial/imperialist goals, hence it is clear that some sort of 'intent' to subjugate, dispossess, or even eliminate the 'obstacle' is present.

Expansionist/imperialist goals are only sought where weak, impoverished or otherwise vulnerable people stand in the way. Hitler put it clearly:

> The struggle for hegemony in the world is decided for Europe by the possession of Russian territory; it make Europe the place in the world most secure from blockade ... The Slavic people on the other hand, are not destined for their own life ... The Russian territory is our India and, just as the English rule India with a handful of people, so we will govern this our colonial territory. We will supply the Ukrainians with headscarves, glass chains as jewelry, and whatever else colonial peoples like. My goals are not immoderate; basically these are all areas where Germans (Germanen) were previously settled. The German Volk is to grow into this territory.[31]

Nor was Hitler the first to articulate a German expansionist goal: the Herero holocaust dates back to the early nineteenth century.[32]

The presence of 'primitives', of Aboriginal peoples, is never deterrent enough for the imperial drive of states intent on expansion in fact, even the presence of politically organised societies does not eliminate the view that their presence does not alter the status of an area, considered as 'vacant land'. The very presence of local legal infrastructure is equally ignored;

> Primitive law and primitive groups as a political unit have to be destroyed and a colony has to become a state before a Western type of law can begin to rule.[33]

Joseph Pugliese adds: 'In its privileged meta-legality, Western law reserves the right to adjudicate all other legal (i.e. Western), proto-legal and a-legal (i.e. non-Western) systems'.[34]

Thus the 'role of colonial law'[35] becomes the racist standard from which other forms of governance are judged to be barbaric, primitive, and needing the transfer to and the control of Western law instead. Perhaps this approach to Western colonialism does not indicate, let alone prove, the intent to destroy. Yet the drive to control is racially motivated, and the intent to change or assimilate local populations is clearly intended to eliminate local laws, practices and policies. It is a form of cultural genocide, even if the 'change' proceeds without violence; otherwise, it is an example of the clear intent to eliminate a *people as such*, even if wholesale murder does not ensue.

The fact that 'reason of state' is primary, and it is not geared to respect the primacy of human rights, especially those of other states.[36]

Ralph Miliband analyses state power in detail, and he starts by citing Weber, as he remarks that the state '... in order to be, it must successfully claim the monopoly of the legitimate use of physical force within a given territory'.[37] I believe that there is an important distinction between 'force', which I take to be either (a) the expression of a legitimate regime or legal infrastructure (such as the police 'force' within a state); or (b) the popular expression of a protest against violence being brought against a peaceful people, as part of economic oppression or other form of colonial exploitation.

I have argued that resistance to 'ecoviolence' is justified as self-defense against unprovoked, violent attacks, not only by states, but also by corporate persons, so that activities such as those of Greenpeace and various forms of 'monkey-wrenching', could be termed the justified use of force in defense of all nature, including human beings.

The destructive ecofootprint of Western development and globalisation does not intend the physical annihilation of any population living in the area: they are far more valuable if they can be assimilated in some way, and in some cases turned into the consumers that will support further growth and expansion. Of course assimilation also eliminates a community as a distinct 'people'.[38] Essentially the quest for 'space', whether general or specific to certain desirable commodity in an area or region, *starts* by ignoring the ecology of the locality, in order to establish corporate economic activities or an expansion of a nation. The ecological integrity of the area, and the biological integrity of the inhabitants are not allowed to stand in the way of 'progress' or expansion. That is the basic argument of this chapter as we attempt to show what role integrity would play to bring about the changes that are necessary today.

An Unsustainable Society? Ecology Integrity as a Counter Proposition

> For capital in the abstract, space is simply stuff for profit-making or it is the distance that must be traversed faster and faster in order to decrease turnover time. Space (mainly the earth and its atmosphere), then is something to be altered at will in order to maximise profit and it is to be shrunk for the same reason.[39]

The last three sections have emphasised the grave harms imposed by our general mindset, the national ambitions of affluent countries, and capitalist globalisation. Increasingly, countries and institutions are becoming aware of the impact of ecological damage on both the resources on which the economy depends, and the human beings whose health and safety are under political areas that do not respect local ecosystem functions.

At best, as we argued, the concern is to retain an area's productivity for further exploitation forces those who advance economic activities, to research and support sustainability, that is, at best, ecosystem health. For the most part, even 'governance for sustainability',[40] does not necessarily imply respect for the foundational role of ecological integrity.

In essence, I have argued that an ecosystem can be said to possess integrity when it is wild, and free as much as possible today from human intervention, that is, when it is an 'unmanaged' ecosystem, although clearly not a necessarily pristine one. This aspect of integrity is the most significant one, in fact, it is the aspect that differentiates it from ecosystem health, which is compatible with support and manipulation instead.

I have discussed the meaning of 'health' in this context elsewhere.[41] Some have spoken of health as the capacity to resist adverse environmental impacts at the present time, and as 'the imputed capacity to perform tasks and roles adequately'.[42] But the capacity to perform certain 'roles' need not be dependent on specific or 'complete' structures. A carefully managed monoculture (such as a plantation forest), for instance, may fit the 'health' model quite well, yet it may have very little of the 'parts' or structure appropriate to that ecosystem, if not managed for a specific purpose.

Another point of divergence between health and integrity, is the time frame within which they are viewed. The health paradigm is concerned with the 'present time' and perhaps the immediate future. The integrity perspective poses no time limits, and envisions birth/maturity/death cycles that may also produce different paths and trajectories, according to the largely natural, evolutionary development of the system. James Karr acknowledges that even unmanaged ecosystems, in today's world, will not be pristine; further, the precise point where an unmanaged system can no longer be said to possess even diminished integrity, and has – at best – only health instead, is open for debate.[43] But whether or not 'integrity' (as a perfect paradigm case) exists, any meaningful debate must start with separate definitions for the two states. Notwithstanding the fact that many important concepts, such as 'justice' for instance, do not have a single, precise definition,[44] we need some working concept on which to base both laws and public policy, even as we continue our dialogue to clarify further the debated term.

I have also argued that ecosystem integrity, is defined through the aspects and characteristics it exhibits of health, the capacity to 'withstand' anthropogenic stress, but primarily the system's undiminished *optimum* capacity for the greatest possible ongoing developmental options within its time/location. The latter is fostered by the optimum possible biodiversity (dependent on contextual natural constraints), in its dual role as basis for genetic potential, and as locus of relational information and communication, both actual and potential. Thus, the system in a state of integrity will retain its ability to continue its ongoing change and development,[45] and will therefore retain its excellence *(ergon/function* in the Aristotelian sense), or capacity for an optimum number of options.

Practically, the requirements of integrity (a) will affect the protection of sustainability as such, in large, wild areas, thus also ensuring the protection of habitats and the goals of conservation biology; it will also (b) address the need for food production to eliminate world hunger; and, (c) specify the limits of all other non-essential human activities in 'culture'. Briefly, these activities must at least reflect integrity in a derivative way. No completely natural evolutionary processes may persist in an urban area (although reproduction of some species continues), but successful work has been done to show that such areas *can* be made compatible with integrity.[46]

In essence, there is a basic difference between landscapes *utilised* (however carefully) for the implementation of some human goals, hence *managed* rather than protected, and viewed as instruments and valued as such, and those that are not. The latter, in turn, can be said to be valued as life is, both in itself and as the basis of all other 'goods' that can be experienced. Their existence is valuable as it supports not only the life of all biota within them, but also because it supports the life of everything else in various ways. The concept of integrity is now also present in the American Fisheries Draft Position Statement on Biodiversity.[47]

To sum up, the ethics of integrity require the following: (1) principles of respect and preservation for wild (that is unexploited) areas; and (2) the protection of ecosystems, not only for their own intrinsic value, but also to permit them to maintain their life-support function. These first order principles in turn entail second order rules. They mandate: (3) restraints on all technological activities involving toxic, hazardous or genetically manipulated substances, as these cannot be permitted if there is any chance they might have either a negative or an unpredictable impact on core and buffer areas; and (4) the acceptance of additional limits to human 'culture' activities, that is, to all non-basic human wants and preferences, through (a) zoning regulations, and (b) qualitative and quantitative regulations regarding the use of so-called natural resources. This implies that limits must be imposed on, *how much* can be taken, as well as to *what* can be taken, from the standpoint of ecological sustainability.

Both first and second order principles appear to fly in the face of our accepted emphasis on individual/aggregate rights and freedoms and our current belief that *limits* are neither actually extant, nor acceptable. Natural limits, we tend to believe, are there to be circumvented or overcome, and 'expansion' and 'growth' are the only acceptable and 'right' goals a society should pursue. The ethics of integrity, in contrast, demands that these assumptions be recognised for what they are: a fiction with disastrous consequences. Given that we are dealing with a finite earth, entailing obviously finite quantities of available 'sources' and 'sinks', the only way we can even briefly continue to pursue expansionist economic goals, is at the expense of others on earth, human and non-human.

Today, 'sustainability' is indeed a 'growth industry';[48] that buzzword appears to earn a 'green star' for those who argue for it, as it confers an ethical and even politically correct aura its supporters. But long time-term sustainability depends on integrity itself: a 'sustainability society' is and can only be one that retains and protects enough areas of ecological integrity to ensure that systematic services and functions will persist. Ecological integrity, by definition, however, excludes human input and the manipulation of natural organisms and processes for one's interests, or even for the common economic or short term interests of a community and a nation.

Hence, governance for integrity, rather than sustainability, can only mean as I have proposed[49] that we commit ourselves to 'living in integrity', that is, to living as though all we do takes place as in a 'buffer zone'. Practically, the restoration of 'brownfields', for instance, ought to be a priority, as would be the promotion of legal instruments demanding the thorough testing by third (disinterested) parties of all substances being introduced into the environment, as well as the immediate evaluation of those that are already present – but have not been impartially and thoroughly tested prior to their approval: and there are dozens of substances in that category.

Then there is the question of eliminating the 'exception' in environmental legal instruments, such as those that follow the 'absolute prohibitions' found in such standard regulations as the Canadian Environmental Protection Act of 1999 (CEPA).[50]

Ecological sustainability, in the sense expressed for instance in the 'Sustainable Biosphere Initiative',[51] is the basis of both social and economic sustainability, and the emphasis on, and the primacy of ecological integrity support precisely that message. Sustainability as such instead may include the use of substances used to increase productivity in agriculture, for instance, or in forestry.

Sustainability implies some concern for the environmental conditions that form the background to our projected activity, but it does not either make integrity primary, or take a long-term view. The goal of sustainability implies, at most, the 'best', or ecologically safer way of achieving human goals; it does not suggest avoiding certain activities altogether because of the undesirable effects these activities may have on the environment itself, even if all precautions have been taken.

When ecological integrity and its maintenance (or restoration, as well as possible) form the starting point of public policy, then all activities and governance instruments would be judged by that standpoint whether these activities and instruments foster the respect for ecological processes, hence for the natural functions they support, should be the first consideration of any policy, not, at least an afterthought or additional consideration.

Conclusion

Much more could be said in support of taking 'ecological integrity' rather than 'sustainability', or any other related paradigm as foundational for just governance today. Such issues as the right to health, the right to food and the right to water, for instance, all rely on an ecologically sound assessment of the policies that would reflect justice and good governance today. Individuals and communities also need healthy and safe food and water for both the primacy of ecological integrity is absolutely basic. Their environment is not 'empty space' to be conquered and utilised for purposes that do not respect what is already there. The individual components of that space, soils, plants, grasses, insects, birds and other animals, together with the processes engendered by the ecosystems in each area are there and they are the basis of normal and healthy survival.

The dignity of humans and – at least – the function of non-human inhabitants of an area, are not temporarily there, open to removal or to forced displacement, as the natural systems upon which they depend are to put 'better' economic uses by various colonisers. The return to the primacy of ecological integrity, as advocated by such instruments as the Berlin Water Rules, or by a soft law instrument like the Earth Charter, would turn much of the present regulatory systems on their collective heads.

However, the thrust of recent decades of industrial growth and globalising capitalism, has reduced many to dire economic straits while few have flourished, and has placed the Earth's systems under grave stress, the results of which are increasingly felt, worldwide. A return to ecological integrity, unlikely though it might be in today's geopolitical climate, would represent, essentially, a return to supporting ecojustice[52] worldwide, and the life of present and future generations.

Notes

1   Angermeier and Karr 1994.

2   Westra et al. 2000, pp. 20–22.

3   Karr and Chu 1999.

4   Karr and Chu 1999, pp. 40–41.

5   Karr 1996, p. 97.

6   Westra et. al., 2000.

7   Noss 1992, Noss and Cooperrider 1994.

8   G. Daily & K Ellison, *Nature's Services: Societal Dependence On Natural Ecosystems*, Island Press, Washington, USA, 2002.

9   J. Depelchin, 'The History of Mass Violence Since Colonial Times – Trying to Understand the Roots of a Mindset', in *Development Dialogue*, Dag Hammarskjold Centre, Sweden, 2008, pp. 13–32

10 R. Kossler, 'Violence. Legitimacy and the Dynamics of Genocide – Notions of Mass Violence Examined', in *Development Dialogue*, Dag Hammarskjold Centre, Sweden, 2008, pp. 45–124; A. Court, 'Do We Need an Alternative to the Concept of Genocide?', in *Development Dialogue*, Dag Hammarskjold Centre, Sweden, 2008, pp. 125–154.

11 J. Depelchin, *History of Mass Violence*, p. 24.

12 L. Westra, Laura, *Environmental Justice & the Rights of Indigenous Peoples*, Earthscan, London, UK, 2007, ch. 8.

13 R. Lemkin, *Axis Rule in Occupied Europe*, Carnegie Endowment for International Peace, Washington, DC, 1944 p. 79.

14 G. Pentassuglia, *Minorities in International Law*, Council of Europe, Publisher: Strasbourg, 2002, p. 79.

15 *Barcelona Traction Light and Power Company Ltd* (Second Phase).

16 ICJ Reports 1970:32 see also W. Schabas, *On Genocide in International Law*, Kluwer Publishing, 2000, p. 82.

17 For instance Aguinda v. Texaco, Inc., 142 F.Supp. 2d 534 (S.D.N.Y. 2001); Bancoult v. McNamara, 217 FRD 280,2003; Doe/Roe v. Unocal Corp., 110 F. Supp. 2d 1294, 1306 (GD.Cal.).

18 W. Schabas, *On Genocide*; IRRC June 2001 Vol. 83, No 842, 439. See further Rome Statute of the International Criminal Court (ICC), of July 17, 1998, UN Doc. A/CONF.183/9.

19 Pentassuglia, *Minorities*, p. 247.

20 R. Lemkin, *Asis Rule,* p. 79.

21 Prosecutor v. Akayesu, (Case No. ICTR-96–4-T) Judgment, 2 September 1998, para. 477.

22 ibid. para. 651.

23 Draft Elements of Crimes', UN Doc. PCNICC/1999/DP.4, p. 7; see discussion in Schabas, 2000:20009.

24 W. Schabas, *On Genocide*, p. 253.

25 J. Depelchin, *History of Mass Violience*, p. 25.

26 ibid., p. 28.

27 R. Kossler, *Violence*, p. 43.

28 L. Westra, *Ecoviolence and the Law*, Transnational Publishers, Inc., Ardsley, New York, 2004.

29 J. Zimmerer, 'Colonialism and the Holocaust – Towards an Archaeology of Genocide', in *Development Dialogue*, Dag Hammarskjold Centre, Sweden, 2008, pp. 95–117.

30 Ibid., p. 99; see also G. Ali, *Final Solution: Nazi Population Policy and the Murder of European Jews*, Arnold, London and New York, 1999.

31 Adolf Hitler, 18 September, 1941, in W. Jochmann, *Adolf Hitler: Monologue im Fuhrenhauptquartier*, Orbis, Hamburg, 1980, pp. 60–64.

32 J. Silvester, W.H. Illebrecht, and C. Erichsen, 'The Herero Holocaust? The Disputed History of the 1904 Genocide', *The Namibian Weekender*, 20 August 2001; see also R. Anderson, 'Redressing Colonial Genocide Under International Law: The Herero' Cause of action Against Germany', California Law Review, vol. 93 2005, p. 1155.

33 P.G. Sak, *Land Between Two Laws*, Australian National University Press, Canberra, 1972, p. 18.

34 J. Pugliese, 'Rationalized Violence and Legal Colonialism: Nietzsche contra Nietzsche', *Cordozo Studies in Law and Literature* vol. 8, 1996, p. 277–280.

35 ibid.

36 R. Kossler, 'Violence, Legitimacy and the Dynamics of Genocide – Notions of mass Violence in *Development Dialogue*, Dag Hammarskjold Centre, Sweden, 2008, pp. 33–52 37.

37 R. Miliband, *The State in Capitalist Society*, The Camelot Press Limited, London and Spothampton, 1970, p. 49.

38 Westra, *Environmental Justice*.

39 R. Albritton, 'Eating the Future', in *Political Economy and Global Capitalism*, eds R. Albritton, R. Jessop and R. Westra Anthem Press, London, UK. 2007, p. 47.

40 K. Bosselmann, *The Principle of Sustinability*, Ashgate Publishing Company, Burlington, 2008. Although Bosselmann, as we noted, qualifies the term with 'ecological', thus escaping this critique.

41 L. Westra, *An Environmental Proposal For Ethics: The Principle of Integrity*. Rowman, Littlefield, Lanham, MD, 1994.

42 T. Parson, 'Definitions of Health and Illness in the Light of American Values and Social Structure', in *Social Structure and Personality*, ed E.G. Jaco, Free Press of Glencoe, New York, 1964.

43 J.R. Karr, 'Landscapes and Management for Ecological Integrity', in *Biodiversity and Landscape: a Paradox of Humanity*, Cambridge University Press, New York, 1994, pp. 227–249.

44 J. Rawls, *A Theory of Justice*. Harvard University Press, Cambridge, MA, 1971; R.C. Solomon, & M.C. Murphy, *What is Justice?* Oxford University Press, New York, NY, 1990.

45 L. Westra, *An Environmental Proposal*.

46 J. Kay, & E. Schneider, 'The Challenge of the Ecosystem Approach', *Alternatives*, vol. 20 no. 3, 1995, p. 1–4; H.A. Regier, 'The Notion of Natural and Cultural Integrity', in *Ecological Integrity and the Management of Ecosystems*, eds S. Woodley, J. Francis & J. Kay, St. Lucie Press, Delray Beach, 1993, pp. 3–18.

47 D.B. Winter & R.M. Hughes, 'AFS Draft Position Statement on Biodiversity', *Fisheries*, vol. 20 no. 4, 1995 pp. 20–26.

48 D.B. Winter & R.M. Hughes, 'AFS Draft Position Statement on Biodiversity', *Fisheries*, vol. 20 no. 4, 1995 pp. 20–26.

49 L. Westra, *Living in Integrity*, Rowman Littlefield, Lanham, MD, 1998.

50 See for instance the definition of 'toxic' in s. 64, 64.a, 64.b and 64.c ... but it gives no definition of inherently toxic'; see also s.93(1)(a) to (r); *The Ontario Environmental Protection Act*, RSO 1990,C.E.19,s1(1), for definitions of contaminants. In all cases, both additives and wastes are 'excepted' from consideration if they are part of 'normal farming practices', although there is nothing in these acts to distinguish regular from industrial farming, the latter using practices and having effects that are different *in kind* from those of normal farming. See discussion on in Westra, *Ecoviolence*, Chapter 4.

51 J. Lubchenko, 'The Sustainable Biosphere Initiative – An Ecological Research agenda', *Ecology*, vol. 72 pp. 371 –412.

52 L. Westra, *Environmental Justice and the Rights of Unborn and Future Generations*, Earthscan, London, UK, 2006, ch. 6.

Epilogue

Island Civilisation:
A Vision for Human Occupancy of
Earth in the Fourth Millennium

Roderick Frazier Nash

> What we call wildness is a civilisation other than our own.
> – Henry David Thoreau

The new, third millennium we are just entering affords an excellent opportunity to think big about the history and future of wilderness and civilisation on planet Earth. Of course a millennium is an entirely synthetic (as opposed to astronomical) concept. Measuring time in thousand-year units only began in 1582 when Christian officials arbitrarily fixed a date for the birth of Christ. So there was nothing special about December 31, 999; it wasn't even recognised as the end of the first millennium. But we made a big deal about the end of the second one a thousand years later on December 31, 1999. Here was an opportunity to transcend our species' characteristic myopia. Rarely do humans make plans more than a couple of years in advance. And we don't do history very well either. Similarly, we don't often think in the wider angles that encompass our species as a whole, but now is an excellent time to begin. One way to look at the opportunity and the responsibility we have with regard to the environment is in terms of legacy. As an historian I am concerned about how the future will regard what happened to the planet on our watch. What will my great grandchildren (and their's) think when they learn the truth about passenger pigeons, salmon, whales and coral reefs?

My mission in this essay is to review the history of human – nature relations and to extend the discussion into a quite distant future. I want to stretch our minds a bit. What could the human tenure on Earth be like a thousand years from now – at the start of the fourth millennium? My proposal involves some really major changes. I expect it to be controversial. At first glance you may think Island Civilisation is crazy and impossible, but don't stop with criticism. The whole purpose of this essay is to advance for discussion a strategy for occupation of this planet that will work in the very long run, and for the whole ecosystem. This is simply the greatest challenge facing our species, and, in a sense, facing evolution on Earth. If you disagree with some or all of my vision of an Island Civilisation, create your own. Particularly, if you think staying the present course is the way to go, put forward your evidence and reasoning. The essential thing is that we occasionally

lift our eyes from everyday details and five-year plans to the far horizons of planetary possibility. Having such a goal is a vital first step to solving problems. Without
it we lack direction and the means to evaluate options as they come into focus.

As a starting point let's consider wilderness. It's a state of mind, a perception,
rather than a geographical reality, and prior to the advent of herding and agriculture about 10,000 years before the present, it didn't exist. But after we began
to draw mental lines between ourselves and nature, and to place walls and fences
on the land, the idea of controlled versus uncontrolled environments acquired
meaning. The root of the word 'wilderness' in Old English was something that had
its own will. The adjective that came to be used was 'wild'. For example, wildfire,
wild (undammed) rivers, and wildcats that you can't herd. The other important
part of the word, 'ness,' indicates a condition or place. So 'wilderness' literally
means self-willed land, a place where wild (undomesticated) animals roam and
where natural processes proceed unencumbered by human interference.

After humans created farms, and literally bet our survival on them instead of
on hunting and gathering, uncontrolled nature became the enemy of the new civilisation. Pastoral societies, like those that produced the Old and New Testaments,
became obsessed with making the crooked straight and the rough places plain.
For thousands of years the success of civilisation seemed to mandate the destruction of wild places, wild animals and wild peoples. The game plan was to break
their 'wills'. In the Bible 'wilderness' was the land God cursed. Its antipode was
called 'paradise'. Adam and Eve lost paradise when they angered God and found
themselves banished into the wild. The first European colonists of the New World
carried in their intellectual baggage a full load of bias against wilderness. The last
thing settlers of the eastern seaboard had in mind was protecting wild nature or
establishing national parks! Indians were savages who needed to be 'civilised' or
eliminated. After a rocky start, these pioneers became very good at breaking the
'will' of uncontrolled land and peoples. Axes, rifles and barbed wire – and more
recently railroads, dams and freeways – were the celebrated tools of an environmental transformation that left the wilderness in scattered remnants.

Lost in the celebration of westward expansion, however, was the possible irony
in the process. When does success in too great a dose produce failure? We always
thought of growth as synonymous with progress, but maybe bigger is not better
if it creates a civilisation that is unsustainable. Maybe what really needs to be
conquered is not wilderness but rather our capitalist-driven culture with its cancer-
like tendency to self-destruct.

Americans began to explore these revolutionary ideas as the second millennium drew to a close in the nineteenth and twentieth centuries. As early as 1851
Henry David Thoreau thought that wildness held the key to the preservation of
the world. George Perkins Marsh, a well-travelled diplomat who spoke twenty-one
languages, understood in 1864 in his remarkable book *Man and Nature* that with
their improved technology, untempered by ethics, humans had become a new and

destructive force of nature. He suspected that what humans assumed to be victory against the forest primeval could result in floods, droughts and desertification that would defeat their dreams of progress and prosperity. Beginning in the 1870s, John Muir reversed thousands of years of Judeo-Christian attitude by publicising mountain forests as temples and cathedrals. What shocked Americans of this generation the most was the United States Census' pronouncement in 1890 that there was no more frontier. With the Indians crushed, the buffalo almost gone and big, industrial cities losing their lustre, it was possible to think that the cherished civilising process could go too far. The appearance in the early twentieth century of best-selling books with a primitivistic slant like Jack London's *The Call of the Wild* (1903) and Edgar Rice Burroughs' *Tarzan* (1913) indicated that the relative valuations of wilderness and civilisation were changing.

As the twentieth Century began a scarcity theory of value began to reshape the relative importance of wilderness and civilisation in the United States. It explains the national angst over the ending of the frontier. Attitude toward wilderness was passing over a tipping point from liability to asset. Of course the pioneers did not go camping for fun! Wilderness appreciation, and later preservation, began in the cities where wild country was perceived as a relative novelty and substantially less threatening.

The rationale of the early movement for wilderness was almost entirely anthropocentric. Scenery, recreation and the economics of a new nature-based tourism underlay the growing popularity of wild places. More sophisticated, but no less utilitarian, were ideas of wilderness as a church, a museum of national history, a stimulant to a unique art and literature and a psychological aid. These were good arguments for their time and they underlay the establishment of the first national parks and wilderness. The Wilderness Act of 1964 was revolutionary but, make no mistake, its point was the benefit of people.

A new, biocentric rationale for wilderness emerged in the last fifty years of the Second Millennium. At its core was the idea that wilderness had intrinsic value, that its protection was not about us at all! Rather, it was a place where our species took a badly needed 'time out' from our ten thousand year old obsession with the control and modification of the planet. In honouring wilderness we manifested a capacity for restraint. Preserved wilderness was a gesture of planetary modesty, a way to share the spaceship on which all life travels together.

The roots of this valuation of wilderness run back in the United States to Henry David Thoreau's belief that 'wildness is a civilisation other than our own'. John Muir wrote about 'the rights of all the rest of creation' that civilised humans had consistently ignored. The case for the rights of certain animals had been vigorously made in England the United States in the nineteenth Century, and in 1915 Albert Schweitzer extended the ideal to 'reverence for life'. The implication here and in Cornell University botanist Liberty Hyde Bailey's book *The Holy Earth*, also 1915, was not just being a good manager or 'steward' of nature but respecting it as an ethical equal because it had been created by God. As Bailey put it, humans should 'put our dominion into the realm of morals. It is now in the realm of trade'. This theological holism, which has a long history in Western thought and, even longer, in Asian cultures, received major support from the new science of ecology. The phrase 'food chains' first appeared in 1927 and 'ecosystem' in 1935. Focusing on interdependence, ecologists gave scientific reason to believe that nature was a community to which mankind belonged not a commodity it possessed.

In essays written in the 1920s and 1930s, and particularly in his book *A Sand County Almanac* (1949) wildlife ecologist Aldo Leopold became the major American articulator of what he called 'the land ethic'. It is significant that wilderness preservation was one of Leopold's highest priorities. It constituted, Leopold argued, 'an act of national contrition' on the part of a species notorious for 'biotic arrogance'. In the 1960s the emergence of Leopold's book as a best-seller, along with the popularity of ecologist Rachel Carson, particularly her *Silent Spring* (1962), evidenced a changing American attitude toward nature. 'Conservation,' around as a term since 1907, had been strictly utilitarian in its emphasis on national strength and prosperity. 'Preservation,' which John Muir favoured, implied human benefit from uncontrolled and unutilised environments. A new 1960s word, 'environmentalism,' took a broader view of utility, gave rise to the term 'pollution' (which impacts many species), and added momentum to the idea of the rights of nature. Theologians and philosophers joined environmentalists in arguing that the nation's natural rights tradition, which had extended the moral community in the past to include black people, natives and women, should now turn to the task of liberating another oppressed minority: nature. The phrase 'deep ecology' appeared in 1973 to describe a belief in the right of every life form to function normally in a shared ecosystem. Some philosophers extended their application of natural rights to land forms like rivers and mountains and to ecosystems.

This line of ethical thinking suggested that just as John Locke's 'social contract' mandated restrictions on individual freedom in the interest of creating a sustainable society, so an 'ecological contract' might restrain the human species in its relations to the ecosystem.. The passage of the Marine Mammals Protection Act (1972) and the Endangered Species Act (1973) were remarkable in that they endowed non-human species with rights to life, liberty and the pursuit of happiness (in appropriate terms of course). Significantly, many of the species protected were not considered cute or useful to humans in any way; their value was intrinsic and their membership in the biotic community indisputable.

The appearance of biocentrism and environmental ethics were encouraging, but an avalanche of evidence suggested that civilisation continued to wreck havoc with natural rhythms and balances as the Third Millennium began. Awareness of the problems has penetrated deeply into contemporary thought and discussion. Accelerated human-caused decline in biodiversity amounts in the opinion of many biologists to a Sixth Great Extinction. More humans than existed since the start of the species occupied the planet in 1950 and population surged upward at a billion every fifteen years. Sprawling into open space at the rate in the United States alone of 6,000 acres each day, people dominated most of the preferred locales in the temperate latitudes. Climate change now seems to be at least partially human-induced. Fresh water, soil, forest and food issues make headlines daily. Lurking just over the horizon are concerns over massive epidemics and the dark, cold specter of a nuclear war that would take down most life on the planet. Civilisation, in a word, appears vulnerable. Making the point explicit, Jared Diamond's book *Collapse* (2005) underscores the lack of sustainability in many human cultures over the last 10,000 years, and suggests strongly that we are not exempt. There will be a resolution of environmental problems, he argues, if not by intelligent choice then by ecological disaster and social disintegration. My proposal for Island Civilisation, below, responds to the concerns Diamond raises.

As for wilderness, where most of the thirty-odd million species sharing Earth reside, it's now an endangered geographical species. Only about two per cent of the contiguous forty-eight states are legally wild, and the same amount is paved! Much of the American landscape has been modified to some degree. And the United States is a leader in national parks and wilderness preservation and is only a little more than a century beyond its frontier era. In other, older regions (France and Japan come to mind) environmental control is near total. At least in the temperate latitudes we are dealing with remnants of a once-wild world, and we face irreversible decisions about their future on a planet that suddenly seems small and vulnerable. In a century wilderness could disappear or become so fragmented as to be ecologically meaningless. Some now view this not just as a violation of the rights of humans to enjoy wild nature but of the rights of other species and self-willed environments themselves.

Looking toward the Fourth Millennium, a thousand years ahead, there seem to be several ways that the natural world we evolved in could end. The wasteland scenario anticipates a trashed, poisoned and used-up planet that can support only a pathetic remnant of its once-miraculous biodiversity and civilisation. Humans have proved to be terrible neighbours to most of the rest of life on the planet. We did not share well. Growth was confused with progress. Centuries of deficit environmental financing of too large and sprawling civilisation has brought the ecosystem, ourselves included of course, to its knees. Maybe, in the height of ingratitude and irresponsibility, we have abandoned and discarded this planet. A vanguard of humans, no wiser for their history, treks through the stars seeking new frontiers to plunder. Perhaps wilderness conditions eventually return to what Alan Weisman calls a world without humans, but the setback to evolution would be profound and slow in healing.

The second possible future is the garden scenario. Imagine by the Fourth Millennium human control of nature is total, but this time it's beneficent. Our species has occupied and modified every square mile and every planetary process from the oceans to weather to the creation and evolution of life. It is finally, as some feared, all about us. We're no longer part of nature; we've stepped off, or more exactly, over the biotic team. Scores or even hundreds of billions of people occupy this planetary garden. Dammed rivers flow clean and cold (but without much diversity of life) and waving fields of grain stretch to the horizon. The only big animals around are those we eat! Maybe such a world could be made sustainable for a few species, but the wilderness, and the diversity of life that depends on it, is long gone. So may be environmental health long thought linked to the normal and natural functioning of ecosystems. The gardeners of Eden may not be quite as sapient apes as they imagined and become victims of homogenisation, biotic impoverishment and their own excessive appetites.

There is a third scenario that has captured the imagination of some thoughtful environmental philosophers. It might be called the future primitive. It involves writing off technological civilisation as a 10,000 year bad experiment. Either by choice or necessity small numbers of humans resume the kind of hunter and gatherer existence that indeed worked quite well for our species for millions of years. But the downside is that the extraordinary achievements and breath-taking potential of civilisation are lost. A better goal, I feel, is Henry David Thoreau's who wished 'to secure all the advantages' of civilisation 'without suffering any of the disadvantages'. Don't humans have as much right to fulfil their evolutionary potential as other species? The vital proviso is that in so doing we don't compromise or eliminate the opportunity of other members of the biotic community to fulfil theirs. This means not discarding technology but using it responsibly.

The fourth scenario for the Fourth Millennium I call Island Civilisation. It's a vision, a dream if you prefer Martin Luther King's rhetoric, and it means clustering on a planetary scale. Boundaries are drawn around the human presence not around wilderness. Advanced technology permits humans to reduce their environmental impact. For the first time in human history, better tools mean peace rather than war with nature. Of course Island Civilisation means the end of the idea of integrating our civilisation into nature. The divorce that began with herding and agriculture is final! Since we proved clever enough to create our environment, rather than adapt to what nature provided, we've taken that option to the logical extreme. We impact only a tiny part of the planet. The rest is self-willed. The matrix is wild not civilised.

Of course a change like this one involves compromises with human freedom. On a finite planet, shared with millions of other species, only limited numbers of humans can enjoy unlimited opportunities. The first step toward Island Civilisation is to check population growth and turn it back to a total of about 1.5 billion or a quarter of the present level. Of course this can be done! Here's one problem for which we know the cause and the solution. It's the motivation that is thus far lacking. A new, expanded earth ethic and plain fear about the crash of a bloated species might change things around. The essential first step is to put nature above people: 'Earth First!' was the name Dave Foreman, Mike Roselle and their colleagues gave to their program in 1980. As it is, humans increase and multiply at the rate of 10,000 per hour, a rate that wipes out any gains friends of wildlife and wilderness try to make today.

The other need for restraint is in the realm of living space. We've historically demanded too much of a planet we supposedly share with other species. We've pushed the wild things into the least desirable corners of the environment. It's time our species took some of the 'marginal' lands which we can modify with our intelligence. The fact is that we've been horrible roommates in the earth household. What species would support an endangered species act for us? One version of Island Civilisation might mandate that the 1.5 billion people live in five hundred concentrated habitats scattered widely over Earth. Food production, energy generation, waste treatment and cultural activities take place in 100-mile closed-circle units supporting three million humans. 'Cities' cannot begin to describe the new living arrangements that the architects and engineers of the Fourth Millennium could create. They might be on the poles, around mountains, in the air, underground and undersea. Rivers might run through some of them. Some of the islands might float in water or in the air. There would be cultural exchange, of course, but no need for global trade in food, energy or materials among the islands. Economies would be re-localised; the concept of 'hundred mile meals' would be a reality. We would get back to an arrangement that worked well on a small scale for Greek city-states, medieval monasteries and pueblos of the Southwest. Sure, wild nature will be severely altered on the islands we occupy, but isn't that fairer and better than a planet-wide sacrifice to a single species? The concept of an island means that human impact is completely contained. The kind of sprawl from which the planet suffers today would be gone. And, I am hoping, no more war. At least border tensions and territorial expansion would not be factors!

Exciting as the possibilities are for this new way for humans to live, it is what's outside the islands (or more clearly what is not outside them!) that is especially compelling. The human presence has imploded. Fences are down. Dams are gone. Roads, railroads, pipelines, telephone lines, ocean-going ships indeed all terrestrial forms of transportation could be unnecessary in a millennium. I'm counting on amazing new technology to make all this possible. Nuclear fusion may be just the tip of the new technological iceberg. Utopian science fiction? Well, consider what was said about television and computers a century ago. And the pace of technological change is accelerating dramatically. Of course I can't prove marvels such as transportation by teleportation will exist in a thousand years, but by the same token you can't deny they won't. Turn our best minds loose on the technological challenges of Island Civilisation (rather than repairing the old, dead-end paths) and miracles will happen. It is not necessary to go back to the Pleistocene to live with a low ecological impact. Technology is essentially neutral; it's what we do with it that is the problem. So why not expand our ethics, and mind pollution and take the high tech road to minimal impact? The result could be the conservation biology dream. The frontier reappears, and this time it is permanent. Rivers are full of salmon and the deer and antelope play on the plains. The big predators are back too and, without human interference, perhaps evolving into some of the Pleistocene mega fauna we never got to know. As we were before herding and agriculture, say 10,000 BC, humans in the year 4000 are once again good neighbours in the ecological community. Homo sapiens is healthy and enjoying its version of liberty and the pursuit of happiness and so are all the other components of the natural world.

But what, the question frequently arises, are your options if you don't want to live on densely-populated islands in a matrix of wilderness? The short response is that if you wanted to live a technological lifestyle in 4000 you wouldn't have a choice. According to the terms of a new, ecological contract, we'd surrender some freedoms like herding cows on the open range or living in a sprawling ski resort. (If you wanted to ski you'd chose to live on the island built into, say, part of the Alps.) But you could leave the islands to enjoy minimum-impact vacations in high-quality wilderness. You could even live out there for a while or forever. The condition is that you'd have to do it as part of the wilderness. That means a resumption of the old pre-pastoral ways. No herds or settling down, no towns and walls, not even cottages in the woods. Those who opted off the island would take only what they needed from nature; profits and growth would not figure into the equation. We would have finally learned what the 1964 Wilderness Act meant about people being ' visitors' who do not remain in someone else's home. Perhaps humans of the distant future could choose on a seasonal basis between ways of life centred on computers or campfires. And young people of that society might be encouraged to take a two-year mission into the wild. Completely out of contact with the civilised islands, they would learn the old hunting/gathering ways and the old land ethics. Here is where some humans might go back to the Pleistocene and live in the 'future primitive' way I described a bit earlier. But is it possible people could support themselves out there for that long, living off the land? The answer is of course they could, considering that the healthy land and sea on which our ancestors built a very sustainable culture for hundreds of thousands of years was back again.

Island Civilisation is a response to the history of Homo sapiens on Earth. For some five million years the planet was self-willed. Humans were just another hunter and gatherer and population remained small and stable. It was a successful lifestyle that weathered just as severe climate changes as the one that scares us now. About 10,000 years before the present our species began to experiment with controlling nature and reshaping our habitat. More precisely, humans stopped adapting to their environment and began to create it. Parts of this experiment resulted in impressive pinnacles of evolutionary achievement. But over time irony kicked in. Human success, especially the idea that bigger was better, carried the seeds of its own destruction as well as that of many other life forms. From the standpoint of the rest of life, the growth of our civilisation amounts to a cancer in the ecosystem. We no longer belong to the ecological team; we've checked off the biotic ark! Isn't this exactly what biologist Edward O. Wilson meant in saying 'Darwin's dice have rolled badly for Earth'? Island Civilisation makes the needed the correction. It permits human beings to realise their cultural and technological potential while safeguarding the same right of self-realisation for all the other beings.

I have long been a supporter of the wilderness preservation movement and, more recently, of conservation biology and the rewilding idea. But it seems increasingly evident that the admirable scientists, philosophers and public servants involved in these efforts shy away from the full implications of their own ideas. Worrying about fragmentation of wildlife habitat, they neglect the option of frag-menting us! Trying to create connections between wild islands, they pass up the possibility of making civilisation an island on a wild Earth. It is hard for me to see the important goals of conservation biologists for the self-willed components of this planet being realised without a major restructuring of human lifestyles and ambi-tions. Island Civilisation may not be the only answer to the big questions hanging over our species, but you can't deny it is an answer.

Biologists warn us that evolution has discarded thousands of promising starts such as ours, and that we should be worried about the future of our present life-style. The upward-trending curves cannot be sustained. There will be major changes. The rub is whether they will be made deliberately or desperately. In this context it is well to remember Winston Churchill's observation that if you play for more than your afford to lose, you will learn the game. Well, the stakes have gotten pretty high; nothing less than the future of life on Earth and that, of course, includes ours too.

So we stand at a crossroads not merely of human history but of the entire evolutionary process. Life evolved from stardust, water and fire over billions of years until one clever species developed the capacity to bring down the whole biological miracle. But amidst the fear associated with this reality of a sinking ark, there is one comfort. Earth is not threatened as in the age of the dinosaurs by an errant asteroid, a death star. Now we are the death star, but we are also capable of changing its course. And it may be appropriate at this point in the paper to observe that the environmental movement has been mostly negative. It's against things. With Island Civilisation I am trying to look at the half-full part of the glass. I am not just talking about problems. Island Civilisation is something to be for.

Imagine, in conclusion, this planet, in the desperate frame of mind contem-porary conditions warrant, putting a 'personals' advertisement on a hypothetical, intergalactic cyber dating service, a kind of E-Harmony.Com for the universe.

TEMPERATE BUT ENDANGERED PLANET

ENJOYS WEATHER, PHOTOSYNTHESIS, EVOLUTION, CONTINENTAL DRIFT

SEEKS CARING LONG-TERM RELATIONSHIP WITH COMPASSIONATE LIFEFORM

Well, maybe it could still be us! Maybe biocentric ethics and respect for self-willed nature (along with a healthy does of fear for our future!) could turn us from cancerous to caring. Maybe we should answer that personals ad. Earth might just be ready to receive a proposal for Island Civilisation.

Appendix

Additional Materials

Books

Nicholas Agar, *Life's Intrinsic Value: Science, Ethics and Nature*, Columbia University Press, New York, 2001.

Janine M. Benyus, *Biomimicry: Innovation Inspired by Nature*, Harper Perennial, New York, 1997.

Thomas Berry, *The Dream of the Earth*, Sierra Club Books, San Francisco, 1988.

Thomas Berry & Brian Swimme, *The Universe Story: From the Primordial Flaring Forth to the Ecozoic Era*, Harper SanFrancisco, San Francisco, 1992.

Thomas Berry, *The Great Work*, Bell Tower, New York, 1999.

Thomas Berry, *Evening Thoughts: Reflecting on Earth as Sacred Community*, Sierra Club Books, San Francisco, 2006.

Fritjof Capra, *The Web of Life: A New Scientific Understanding of Living Systems*, Anchor Books, New York, 1996,

Cormac Cullinan, *Wild Law: A Manifesto for Earth Justice*, Green Books, Devon, 2002.

Stephan Harding, *Animate Earth: Science, Intuition and Gaia*, Devon, Green Books, 2006.

Lawrence E. Johnson, *A Morally Deep World: An Essay on Moral Significance and Environmental Ethics*, Oxford University Press, Oxford, 1991.

Aldo Leopold, *A Sand Country Almanac*, Bellantine Books, New York, 1966.

Thomas Linzey & Anneke Campbell, *Be The Change: How To Get What You Want In Your Community*, Gibbs Smith, Utah, 2009.

Anne Lonergan & Caroline Richards (eds), *Thomas Berry and the New Cosmology*, Twenty-Third Publications, Connecticut, 1991.

James Lovelock, *Gaia: A New Look at Life on Earth*, Oxford University Press, Oxford, 1979.

Lynn Margulis & Dorion Sagan, *What is Life?*, Simon & Schuster, New York, 1995.

Freya Mathews, *The Ecological Self*, Barnes & Noble Books, Maryland, 1991.

Roderick Nash, *The Rights of Nature*, The University of Wisconsin Press, Wisconsin, 1986.

Kilpatrick Sale, *Dwellers in the Land: The Bioregional Vision*, Sierra Club Books, San Francisco, 1985.

George Sessions (ed), *Deep Ecology for the 21st Century*, Shambhalah, Boston, 1995.

Vandana Shiva, *Earth Democracy: Justice Sustainability and Peace,* South End Press: New York, 2005.

Christopher D Stone, *Should Trees Have Standing? Law Morality & The Environment,* Oxford University Press, Oxford, 2010.

Brian Swimme, *The Hidden Heart of the Cosmos: Humanity and the New Story*, Orbis Books, New York, 1988.

Brian Swimme, *The Universe is a Green Dragon: A Cosmic Creation Story*, Bear & Company, Vermont, 2001.

Gary Snyder, *Turtle Island*, New Direction Books, New York, 1974.
Edward O. Wilson, *Biodiversity*, National Academy Press, Washington, 1988.
Edward O. Wilson, *The Diversity of Life,* W.W. Norton & Company, New York, 1992.

Websites
African Biodiversity Network: http://www.africanbiodiversity.org
Community Environment Legal Defence Fund: http://www.celdf.org
Earth Jurisprudence Resource Center: www.earthjurisprudence.org
Enact International: http://www.enactinternational.org
Friends of the Earth Adelaide: http://www.adelaide.foe.org.au
Gaia Foundation: http://www.gaiafoundation.org
Global Alliance for the Rights of Nature: http://therightsofnature.org
Earth Laws Inc: http://keepingthefire.org
Navdanya & Earth Democracy: http://www.navdanya.org
Schumacher College: http://www.schumachercollege.org.uk
St Thomas and Berry University, Center for Earth Jurisprudence: http://www.earthjuris.org
Trees have Rights Too: http://www.treeshaverightstoo.com
United Kingdom, Environmental Law Association: http://www.ukela.org
Wild Frontiers (Ning): http://wildfrontiers.ning.com
Yale University, Forum of Religion and Ecology, Thomas Berry: http://www.thomasberry.org

Notes on the Authors

JULES CASHFORD is co-author, with Anne Baring, of *The Myth of the Goddess: Evolution of an Image*, 1993. She is author of *The Moon: Myth and Image*, 2003, and translated *The Homeric Hymns for Penguin Classics*, 2003. She has written two books for children, *The Myth of Isis*, 1992 and *Osiris*, 1994. She has contributed chapters and articles to various books, newspapers and journals including 'Joseph Campbell and the Grail Myth,' in John Matthews, ed., *The Household of the Grail*; 'Homo Duplex: An Epilogue to Joseph Conrad's "The Secret Sharer"', 2005; 'Imagining Eternity: Weaving "the heaven's embroidered cloths"', in *Cosmos and Psyche*, ed., Nicholas Campion, 2006. With Kingfisher Art Productions she made a DVD exploring the symbolism of the early Renaissance painter Jan van Eyck, called *The Mystery of Jan van Eyck*. She now writes and lectures on Myth and Literature.

CORMAC CULLINAN is an environmental lawyer based in Cape Town, South Africa, and author of *Wild Law: A Manifesto for Earth Justice*, 2002. He is a director of the leading South African environmental law firm, Winstanley & Cullinan Inc, and CEO of EnAct International, an environmental governance consultancy. In 2008 he was listed among the world's most extraordinary environmental champions in *Planet Savers: 301 Extraordinary Environmentalists*, which lists 301 people in history to be commended for their important role in saving and conserving the environment and promoting sustainable governance

LIZ HOSKEN is the director of the Gaia Foundation UK, an international organisation working in partnership to restore cultural and biological diversity and ecological governance. Born in South Africa, Liz was active from a young age in both ecological issues and the antiapartheid movement. She was exiled to the UK in her early twenties, and co-founded the Gaia Foundation in 1984, now at the hub of a global network of grassroots actors committed to rebuilding cultural and ecological integrity as the foundation for resilience. Gaia's commitment over the last two decades has been to identify and nurture initiatives and associates that recognise our need to shift from a human-centred to an Earth-centred mode of governance. In 2009 she was an instructor at the 'Earth Jurisprudence and Community Resilience: Learning from Africa' Schumacher Course.

IAN MASON is a practising barrister whose interest in Earth jurisprudence stems from life-long studies in practical philosophy and economics-with-justice with the School of Economic Science in London. Ian is the Director of the Gaia Foundation's Earth Jurisprudence Resource Centre. For the last three years he has given annual workshops on Earth jurisprudence for the Gaia Learning Centre. He has also worked with the UK Environmental Law Association and co-authored with Begonia Filguera, the 2009 report, *Wild Law: Is There Any Evidence of Earth Jurisprudence in Existing Law and Practice?* In 2008 he was one of the teachers of the first Earth jurisprudence course at Schumacher

College in Devon, UK and has been a regular participant in and contributor to the annual UKELA Wild Law weekend conferences.

JUDITH E. KOONS came to Barry University School of Law, Florida USA, after receiving a Master of Theological Studies degree from Harvard Divinity School. While at Harvard, Professor Koons studied philosophy, ethics and theology, and explored the intersection of faith and social justice. At Harvard, Professor Koons also served as a teaching fellow in the Sociology Department, focusing on social stratification by gender, race, and class. Professor Koons serves as the chair of the governing committee of the Center for Earth Jurisprudence, a collaborative initiative of Barry and St Thomas Universities. The Center for Earth Jurisprudence invites an interdisciplinary reevaluation of the premises of law and governance to foster mutually-enhancing relationships among all members of the Earth community.

PETER BURDON is a Lecturer at the Adelaide Law School and a member of the management committee of the University of Adelaide Research Unit for the Study of Society, Law and Religion (RUSSLR). He is currently in the final stages of a PhD dissertation at the Adelaide Law School on Earth Jurisprudence. This research focuses on legal theory, critical theory, property theory and environmental philosophy.

RODERICK FRAZIER NASH is Professor Emeritus of History and Environmental Studies, University of California Santa Barbara. He is considered America's foremost wilderness historian. He is regarded as a national leader in the field of environmental history and management and environmental education. Among his numerous books and over 150 essays, Professor Nash is best known for *Wilderness and the American Mind,* 1967 and *The Rights of Nature,* 1989. A past Lindbergh Fellow, he has served on the board of directors of the Yosemite Institute and as a member of the advisory committee to the U.S. National Park Service.

STEPHAN HARDING holds a doctorate in ecology from the University of Oxford. He is the Co-ordinator of the MSc in Holistic Science at Schumacher College in Devon, UK where is also Resident Ecologist and teacher on the short course programme. He is the author of *Animate Earth: Science, Intuition and Gaia,* 2006.

JOEL CATCHLOVE is an educator and co-ordinator of 'Reclaim the Food Chain' with Friends of the Earth Adelaide.

GLORIA L. SCHAAB is an Assistant Professor of Systematic Theology and Associate Dean for General Education at Barry University Florida USA. She serves as a consultant for the Center of Earth Jurisprudence, a joint venture of Saint Thomas University and Barry University, Miami, Florida, and on the Board of Directors for the Center for Earth Jurisprudence, Inc. She also convened the Theology and Natural Science Topic Area of the Catholic Theological Society of America and co-convenes the Theology, Ecology and Natural Science section of the College Theology Society.

JASON JOHN is an ordained minister in the Uniting Church in Australia, working as an ecominister, and as a congregational minister, in Bellingen, New South Wales.

PAMELA LYON is an Australian Postdoctoral Fellow in the Discipline of Philosophy at the University of Adelaide. She earned her PhD at the Australian National University, Canberra, in 2006 with a thesis that argued cognition should be approached like other biological functions, that is, by first understanding in detail the simplest possible examples, such as potentially cognitive behaviour in bacteria. The thesis won the Crawford Prize, the university medal, for that year and now forms the basis of a multidisciplinary collaboration involving philosophers and scientists at a dozen universities worldwide.

LAWRENCE E. JOHNSON is the author of *A Morally Deep World: An Essay on Moral Significance and Environmental Ethics*, 1991 and *A Life-Centered Approach to Bioethics: Biocentric Ethics*, 2010. Johnson is an Affiliate Research Fellow in the humanities at the University of Adelaide in South Australia. He has also taught at West Virginia University and Flinders University.

HERMAN GREENE is a corporate, tax and securities attorney with Greene & Franklin, North Carlonia, USA. He began his legal career with Shearman & Sterling, New York and later joined Mayer, Brown, & Platt where he became a partner. In 1990–91 he served as director of public responsibility, American Express Company, also in New York. He returned to North Carolina in 1992. In addition to practising law, he serves as president, Center for Ecozoic Studies (N.C.), program consultant, Center for Earth Jurisprudence (Fla.), and governing board member, International Process Network (Cal.).

MELLESE DAMTIE is an Ethiopian lawyer and biologist who has been pioneering the research and implementation of Earth Jurisprudence from his unique position as Dean of the Ethiopian Civil Service College's Department of Law, Addis Ababa. Mellese initiated a Masters research programme at the college, supervising university students to study the customary laws that traditional people use to govern their relationship with nature as part of a course on Earth Jurisprudence, which he introduced into the syllabus and has been teaching on for several years. He is also Chief Legal Advisor for Ethiopian NGO, Movement for Ecological Learning and Community Action (MELCA), and plays a key role in paralegal training for local communities on their rights and legal opportunities to gain recognition for governing and protecting their territories. Mellese helped to establish the Sheka Forest Alliance (SheFA), which brings together lawyers, civil society organisations and NGOs to advocate for the strengthening of environmental legislation in Ethiopia. In 2008 he was an instructor at 'Earth Jurisprudence: Making the Law Work for Nature' at Schumacher College.

NG'ANG'A THIONG'O was co-founder and Legal Policy and Advocacy Officer for Porini Trust, Kenya. A respected barefoot lawyer and activist, he was Chair of Kenya's Release of Political Prisoners campaign and a member of Kenya's High Court. As a former Legal Advisor to the Green Belt Movement he worked alongside Nobel Peace Prize winner Prof.

Wangari Maathai during her most difficult times. He was also a leading light on Earth Jurisprudence thinking and practice in Africa, and a popular participant at international Earth Jurisprudence events. Sadly, with many activities on the go and a young family to support, Ng'ang'a Thiong'o died in February 2010.

NICOLE ROGERS is a Senior Lecturer in the School of Law and Justice, Southern Cross University, NSW. She is the editor of *Green Paradigms and the Law*, 1998.

BEGONIA FILGUEIRA is co-director and founder of Eric Group, UK. She is a solicitor of the Supreme Court of England and Wales and a Spanish abogada. Begonia is a council member of the United Kingdom Environmental Lawyers Association and sits on the Institute of Environmental Management & Assessment experts' panel. She advises Northern Ireland's DOE on environmental law, including the implementation of European Directives, and edits LexisNexis' EF&P Environment Volume. She also teaches banking and project finance at City University, London.

KLAUS BOSSELMANN is a professor of Law at the University of Auckland, New Zealand. He has been teaching in the areas of public international law, European law, constitutional law, jurisprudence and comparative and international environmental law. His research focus is on the conceptual and international dimensions of environmental law and governance. He is particularly interested in sustainability ethics with respect to climate change, biodiversity, justice, human rights, legislation, democracy and international law. Klaus has been the Director of the New Zealand Centre for Environmental Law since its establishment in 1999. As Chair of the Ethics Specialist Group of the IUCN (The World Conservation Union) Commission on Environmental Law, he currently coordinates a number of international research collaborations in the area of sustainability law and governance. Klaus also has an active role in projects and annual conferences of the Global Ecological Integrity Group, a network of 250 environmental scholars. He has authored or edited 24 books on environmental law, political ecology, and sustainability law and governance; his numerous articles appeared in many of the world's leading law journals. In 2009, he was the Inaugural Winner of the Senior Scholarship Prize of the IUCN Academy of Environmental Law, the global body of environmental law scholarship.

ELIZABETH RIVERS is a mediator and educator based in London. She has worked as a facilitator and educator for the UKELA Wild Law Workshop and Schumacher College. In 2009 she was the keynote speaker at 'Wild Law: Australia's First Conference on Earth Jurisprudence'.

THOMAS BERRY received his Ph.D. from the Catholic University of America in European intellectual history with a thesis on Giambattista Vico. Widely read in Western history and theology, he also spent many years studying and teaching the cultures and religions of Asia. Thomas authored two books on Asian religions, *Buddhism* and *Religions of India*, both of which are distributed by Columbia University Press. For more than twenty years, Thomas directed the Riverdale Center of Religious Research along the Hudson River.

From 1975 to 1987 he was President of the American Teilhard Association, and it was from Teilhard de Chardin that he was inspired to develop his idea of a universe story. With Brian Swimme he wrote *The Universe Story*, 1992. His major contributions to the discussions on the environment are in his books *The Dream of the Earth* 1998, *The Great Work: Our Way Into the Future*, 1999 and *Evening Thoughts: Reflecting on Earth as Sacred Community* 2006.

MARI MARGIL is the Associate Director of the Community Environmental Legal Defense Fund, Pennsylvania USA. She conducts campaign and organisational strategy, media and public outreach, and leads the organisation's fundraising efforts. In 2008 she travelled to Ecuador where she advised the country's Constitutional Assembly on the re-writing of their Constitution. She is also a Lecturer for the Legal Defense Fund's Democracy Schools. She is the author of a chapter titled 'A New Democracy in Action' in the book *The Public Health or the Bottom Line*, to be published by Oxford University Press in February 2010.

NICOLE GRAHAM is a lecturer in law at the University of Technology Sydney. Her research explores the relationship between property and the natural environment in legal and cultural discourses and land use practices. In particular, Nicole is interested in the use of property rights and environmental markets in natural resource management, drought, and climate change policy. She is the author of *Lawscape: Property, Environment, Law*, 2011.

ERIC T. FREYFOGLE, is the Max L. Rowe Professor of Law at the University of Illinois College of Law. Professor Freyfogle is the author or editor of various books dealing with issues of humans and nature, some focused on legal aspects, others reaching to larger cultural and social issues. His works on conservation thought includes *Why Conservation is Failing and How It Can Regain Ground*, 2006; *The New Agrarianism*, 2001, *Bounded People, Boundless Lands*, 1998; and *Justice and the Earth* (1993). Writings focused particularly on private property rights in nature include *On Private Property: Finding Common Ground on the Ownership of Land* 2007; and *The Land We Share: Private Property and the Common Good*, 2003.

PAUL BABIE is Senior Lecturer (tenured) of the Adelaide Law School and Founder and Director of the University of Adelaide Research Unit for the Study of Society, Law and Religion (RUSSLR). RUSSLR is the first research centre or institution of its kind in Australia, devoted to the sustained and focused study of the relationship between society, law and religion/theology. Paul's teaching and research are in the areas of legal theory, property law and theory, religious understandings of property and possessions, and Eastern Catholic canon law. He is an internationally recognised expert on the relationship between the concept of private property and anthropogenic climate change and has spoken on these topics in the United States, Canada, Norway, Malaysia, and Australia. Paul is currently writing a book for University of British Columbia Press (Canada) entitled *Private Property, Climate Change and the Children of Abraham*.

SAMUEL ALEXANDER is a sessional lecturer and doctoral student at the Melbourne Law School. He is also the editor of *Voluntary Simplicity: The Poetic Alternative to Consumer Culture*, 2009. He is also the founder of the Life Poet's Simplicity Collective, a grassroots network of imaginations dedicated to the organisation and advancement of the Voluntary Simplicity Movement.

POLLY HIGGINS is a London-based international environmental lawyer, barrister and author of *Eradicating Ecocide: Laws and Governance to Stop the Destruction of the Planet*, 2010. In November 2008 Polly was invited to address the United Nations on the call for a Universal Declaration of Planetary Rights. Polly founded the campaign Trees Have Rights Too. In April 2008 Polly also founded WISE Women in Sustainability and the Environment – an international network set up to provide a voice for more women in the environmental field. In 2009 Polly was voted by *The Ecologist* magazine as 'One of the Top Ten Visionaries to Save the Planet' for her work on Planetary Rights, Polly is a renowned international speaker and expert on the integration of Earth Jurisprudential principles into global governance systems.

BRENDAN MACKEY is a professor of Environmental biogeography, environmental conservation, cross-disciplinary studies in sustainability at Australian National University. He is also the director of the ANU Wild Country Research and Policy Hub, a member of the IUCN council and a member of the Earth Charter International Council. He is co-author of *Green Carbon: The Role of Natural Forests in Carbon Storage*, 2008; *The Nature of Northern Australia: Its Natural Values, Ecology and Future Prospects*, 2007; 'The Earth Charter and Ecological Integrity: A Commitment to Life on Earth' in Peter Corcoran (ed), *The Earth Charter in Action*, 2005; and *Wildlife, Fire and Future Climate: A Forest Ecosystem Analysis*, 2002.

LINDA SHEEHAN brings 20 years of environmental law and policy experience to her work as Executive Director of the California Coastkeeper Alliance. Over the past 15 years, Linda has focused on protecting and enhancing the health of California's waterways and its world-renowned coast and ocean, implementing programs to ensure clean coastal waters with healthy flows and to safeguard marine ecosystems. She has achieved notable success in protecting the health of coastal waters by advancing legislation and policies to reduce polluted runoff, curtail sewage spills, increase coastal water quality monitoring, heighten enforcement of water laws, and make state water data readily available to all. For her efforts in 'fight[ing] pollution of the Pacific and the streams and rivers that flow into it', Linda was recognised as a 2009 'California Coastal Hero' by *Sunset* magazine and the California Coastal Commission.

RON ENGEL is Professor Emeritus at Meadville Lombard and Senior Research Consultant, The Center for Humans and Nature, with offices in New York and Chicago. Ron became active in international work on behalf of global ethics in the course of research with UNESCO. He was a core member of the international drafting committee for the Earth Charter, and is currently co-chair of the Ethics Specialist Group of the Commission on Environmental Law for the World Conservation Union. Ron is the author of *Sacred Sands: The Struggle for Community in the Indiana Dunes*, 1986 which won several book awards, including the Meltzer National Book Award; editor of *Voluntary Associations: Socio-cultural Analyses* and *Theological Interpretation*; co-editor of *Ethics of Environment and Development: Global Challenge, International Response*; and co-author of *Justice, Ecology, and Christian Faith: A Critical Guide to the Literature*.

LAURA WESTRA is a professor of law the University of Windsor, Ontario, Canada. She is the founder of the Global Ecological Integrity Group (GEIG), and has organised several conferences of that group in conjunction with the IUCN Commission on Law and Environment (CEL), Special Ethics Group (ESG), to which she has belonged for the last five years. She is also the co-chair of the IUCN-CEL Specialist Indigenous Peoples Group with John Scott. She has held offices/served in the International Society for Environmental Ethics, the science for Peace Group, the Occupational Ethics Group, the Society for the History and Philosophy of Science and Technology (Ontario), and the York centre for Applied Sustainability. Westra has been the Principal Investigator for SSHRC (1992–1999) and for NATO's Advanced Scientific Research Workshop (1999) and has organised numerous meetings and sessions in that capacity, in Europe, Australia, the U.S. and Canada. Most of Westra's work is on environmental ethics, policy and law, with special emphasis on human rights and global justice. Westra has published more than 80 articles and chapters in books, and 20 books/monographs.

Wakefield Press is an independent publishing and
distribution company based in Adelaide, South Australia.
We love good stories and publish beautiful books.
To see our full range of books, please visit our website at
www.wakefieldpress.com.au
where all titles are available for purchase.

Find us!

Twitter: www.twitter.com/wakefieldpress
Facebook: www.facebook.com/wakefield.press
Instagram: instagram.com/wakefieldpress